Mikrochimica Acta

Supplementum 10

Proceedings of the 11th Colloquium on
Metallurgical Analysis, Institute for Analytical Chemistry,
Technical University in Vienna, November 3–5, 1982

Springer Science+Business Media, LLC

Progress

in Materials Analysis

Vol. 1

**Edited by M. Grasserbauer
and M. K. Zacherl**

Springer-Verlag Wien GmbH

Scientific and Organisation Committee

Prof. Dr. M. Grasserbauer, Technical University of Vienna; Chairman
Dr. S. Baumgartl, Thyssen AG, Duisburg
Prof. Dr. Th. Hehenkamp, University of Göttingen
Dr. K. H. Koch, Hoesch AG, Dortmund
Prof. Dr. H. Malissa, Technical University of Vienna

© 1983 by Springer-Verlag Wien
Originally published by Springer-Verlag/Wien in 1983

With 201 Figures

ISBN 978-3-211-81759-9   ISBN 978-3-7091-3943-1 (eBook)
DOI 10.1007/978-3-7091-3943-1

Preface

The 11th Colloquium on Metallurgical Analysis — a joint venture of the Institute of Analytical Chemistry of the Technical University in Vienna, the Austrian Society for Analytical Chemistry and Microchemistry, the German Metals Society (DGM), and the Society of German Iron and Steel Engineers (VDEh) — was attended by 120 scientists from 12 nations.

The major topics covered were surface, micro and trace analysis of materials with a heavy emphasis on metals. According to the strategy of the meeting attention was focussed on an interdisciplinary approach to materials science — combining analytical chemistry, solid state physics and technology.

Therefore progress reports on analytical techniques (like SIMS, SNMS, Positron Annihilation Spectroscopy, AES, XPS) were given as well as presentations on the development of materials (like for the fusion reactor). The majority of the discussion papers centered on the treatment of important technical problems in materials science and technology by a (mostly sophisticated) combination of physical and chemical analytical techniques.

The intensive exchange of ideas and results between the scientists oriented towards basic research and the industrial materials technologists was very fruitful and resulted in the establishment of several scientific cooperations. Major trends in materials analysis were also dealt with in a plenary discussion of which a short summary is contained in this volume.

In order to facilitate international communication in the field of materials analysis and in view of the important questions treated in the various contributions this proceedings volume was edited in English.

On behalf of the scientific committee I would like to express my gratitude to all authors for their contributions and support in publishing the proceedings. Special thanks go to Doz. Dr. W. Wegscheider, Technical University Graz, for linguistic editorial work.

M. Grasserbauer
Technical University Vienna
Chairman of Scientific Committee

Contents

Summary of Plenary Discussion on "Materials Development — a Challenge to Analytical Chemistry"

The 11th Colloquium on Metallurgical Analysis was concluded by a plenary discussion on "Materials Development — a Challenge to Analytical Chemistry" in which experts from industry and research institutions treated important areas of materials research and the resulting demands for materials characterization. The discussion did not cover all important areas of materials development but rather a selection of topics which seemed important to the participants. The summary reported here is based mainly on contributions from K. H. Koch, Hoesch Hüttenwerke, Dortmund, G. Kraft, Deutsche Metallgesellschaft, Frankfurt, K. H. Schmitz, Mannesmann-Röhren-Werke, Duisburg, V. Thien, Kraftwerks-Union, Mülheim, H. Beske, Kernforschungs-anlage Jülich, W. Wintsch, Sulzer, Winterthur, E. Kny, Metallwerk Plansee, A. P. von Rosenstiel, TNO Apeldoorn, L. Rinderer, Technical University, Lausanne, G. Tölg, Institute for Applied Spectroscopy and Spectrochemistry, Dortmund, H. Straube, H. Malissa, and M. Grasserbauer, Technical University, Vienna.
Development of the following analytical areas and techniques were found to be particularly important:

1. **Development of High Speed Routine Methods** — especially for iron and steel production in order to cope with the increasing speed of the process.

2. **Automated Routine Measurements for Gases** — especially oxygen — in metal melts. Although substantial progress has been achieved the accuracy of the results is still usually unsatisfactory.

3. **Surface Analysis** by combination of all major techniques for characterization of surface treated materials — e.g. wear resistant coatings on hard metals or steel. High accuracy, high lateral and depth resolution are usually required. For routine applications large area techniques (like GDOES) seem useful. The transfer of surface analysis from research institutions to industrial laboratories is just under way but — despite exceptions — usually not sufficient to cope with industrial needs. Main hindrance are the high costs of establishing surface analytical laboratories.

In-situ-analysis of liquid-solid interfaces for study of reaction phenomena (electrochemical processes in corrosion) by FTIR-spectrometry seems very promising.

4. Multi-Element Trace Analysis especially by contamination free techniques (like SIMS, SNMS, NAA) for low impurity levels (ng/g – μg/g range). The rapid increase in purity requirements in the course of refinement of materials demands not only the development of more sensitive methods but also higher accuracy of trace analytical results. Combination of chemical enrichment steps with physical detection methods seems particularly promising. Due to the development of new materials (e.g. for nuclear power plants) extreme trace analysis has to be developed even for rare elements (e.g. Gd in Zr).
Reference materials and recommended methods are still missing for many problems in multielement trace analysis.
Recycling of materials may introduce a variety of impurities into the process. The influence of many of these impurities is not well understood. On the other hand controlled doping of materials with trace elements can be used to generate specific microstructures and properties. Systematic study of the behaviour of such trace elements for sinter materials, cast alloys etc. is necessary – starting with accurate multielement characterization, including the use of mathematical procedures to correlate properties and trace elements (e.g. by clustering techniques).

5. Distribution Analysis of Trace Elements. Since not only the type and bulk concentration of trace elements but also their distribution influence material properties techniques which allow distribution characterization with maximum spatial resolution have to be applied or developed. Scanning AES and SIMS (Ion Probes, Ion Microprobes) seem to be particularly useful. Major progress has already been achieved in semiconductor analysis and grain boundary segregation studies of some elements (e.g. P in iron). Major limitations however are still the limited sensitivity of AES and the often unsufficient lateral resolution of SIMS. Major developments in this direction, e.g. by use of metal ion sources (In, Ga) may open up the area of sub-micrometer surface distribution analysis of trace elements.
This is particularly useful for characterization of semiconductor devices (VLSI-circuits), grain boundaries in metals, and compound structures.

Development of quantification methods is necessary – especially production of trace element standards for distribution analysis in metals. Ion implantation may offer large possibilities.

6. Ultra-Microanalysis. Microstructures of dimensions well below 1 μm are important for the properties of many materials – e.g. precipitations at grain boundaries in metals, interface domains in semiconductors. Means to characterize these structures by geometrical or chemical parameters – especially through electron microscopy – have to be further developed and incorporated more into industrial laboratories. High voltage TEM as well as use of X-ray microanalysis in STEM are the major developments.

7. **Stereometric Analysis.** For accurate description of the microstructure of solids and for establishing precise relationships between structure and properties stereometric analysis techniques have to be further developed. Automatic electron probe systems for (lateral) distribution analysis yielding chemical as well as geometric parameters are becoming more and more important in industrial laboratories. Three-dimensional stereometric analysis (based on SIMS, AES etc.) is in its initial stage, but rapidly progressing. The incorporation of holography into stereometric analysis could be of great interest.

8. **High Precision Stoichiometric Analysis.** There is still a need for increasing accuracy of many techniques in order to determine precisely the stoichiometry of compounds. Especially small deviations from stoichiometry are often very important but cannot be characterized adequately. Also for precious metals (Cu, Ag, etc.) an accuracy in the order of 0.01% is often required. Refinement of quantitative chemical techniques offers the best chances.

9. **Mathematical Methods for Determining Homogeneity of Materials, Representativity of Analytical Results and Sampling Errors.** Suitable algorithms or mathematical procedures have to be developed which allow to establish objective and quantitative relations between homogeneity, sampling and analytical results.

10. **Increase of Cooperation Between Research Institutions and Industry.** Since industry often finds it difficult to incorporate expensive analytical methods when they are still in the stage of development, even if these techniques do already yield important information (especially for research) a close cooperation with research institutions is necessary. Such cooperations should be increased for mutual benefit. For basic methodological research and treatment of complex frontier area analytical problems which demand the cooperative use of a variety of sophisticated techniques, high standard analytical laboratories (like at NBS) should be established. Such laboratories should also offer high quality service to industry on special problems.

M. Grasserbauer

Mikrochimica Acta [Wien], Suppl. 10, 1–8 (1983)

Projekt Kernfusion der Kernforschungsanlage Jülich *, Jülich, Federal Republic of Germany

Materials for the First Wall of Fusion Reactors and Their Analytical Characterization

By

Wolfgang O. Hofer

(Received January 19, 1983)

The critical path towards controlled thermonuclear fusion as an energy source has often been sub-divided into three phases which end in demonstration of the physical, technical and economic feasibility, respectively. It is expected that the first step will be reached within this century. The proving of technical feasibility, however, has already started since many problems have been clearly recognized. They are, to a large extent, concerned with materials, in particular materials of the *first wall*. By first wall is meant generally not only the vacuum vessel but all components therein which are − or might come − in contact with the plasma. These components are therefore subjected to quantum, electron, ion and neutral particle irradiation. This plasma radiation is absorbed in a surface layer of $10-100$ μm thickness. Moreover, it is not uniformly distributed over the surface of the wall but imposes locally a particularly high power load on specific components. The effect of plasma radiation is thus of an entirely different nature than that of the 14.1 MeV fusion neutrons; owing to their large mean free path, fusion neutrons essentially constitute a uniform volume load.

Two methods of plasma confinement are considered: magnetic and inertial confinement. This short review restricts itself to *magnetic confinement in toroidally symmetric devices*, since both linear magnetic (i.e. mirror) and inertial confinement presently appear to be powerful breeder or hybrid systems, but are hardly capable of direct power output. With toroidal machines most of the effort by far goes into *Tokamaks*. The ring-shaped plasma is used here as the secondary winding of a transformer which induces a plasma current in order both to stabilize the plasma and to heat it "ohmically".

---

* Assoziation EURATOM-KFA

The shape of the plasma column can be defined either by magnetic means or by solid plasma scrapers, so-called limiters. In the first case the magnetic configuration is designed in such a way that particles from the plasma edge are "diverted" into a separate chamber, the *divertor*, where the main contact with the solid wall occurs. Tokamaks equipped with divertors presently yield the best plasma conditions as regards purity, discharge length and reproducibility. Nevertheless, divertors seem to be structures too complex to be feasible in reactors. It might turn out that they are indispensable, but present day hopes are with *limiters*: of large area, arranged toroidally along the torus and shaped to match the plasma boundary, protected with refractory coatings to withstand chemical erosion and high heat fluxes, and equipped with pumping facilities — in the form of pump limiters as they are called — in order to control plasma and impurity density as well as to remove the helium "ash".

Both the limiters and the collector plates in divertors will be subjected to considerable surface erosion. The main erosion mechanisms for solids exposed to high-temperature plasmas are: physical as well as chemical sputtering, evaporation/sublimation, and unipolar arcing. Both physical sputtering and arcing depend strongly on the plasma temperature. Hence any means of lowering the temperature in the plasma boundary will alleviate the problems with these erosion components. Proper conditioning of the surface, on the other hand, can reduce chemical erosion to some extent and arcing by orders of magnitude.

Materials in Forthcoming Fusion Machines

The strategy concerning the materials to be used for in-vessel components as well as for the vacuum wall has changed repeatedly during the last decade: while *refractory metals* such as Nb and Mo were the prime choice in the late sixties, its much higher tolerable concentrations of low-Z impurities called for *carbon* as the main wall material. When the high rate of chemical erosion of carbon at elevated temperatures was found, thin surface coatings of TiC or TiB_2 on graphite were designed. TiC layers especially, of $10-20$ μm thickness, are frequently in use and will be the main coating in the Tokamak Fusion Test Reactor (TFTR) in Princeton, USA. Such thin coatings, however, will not withstand the erosion in Tokamaks with discharge times of 20 seconds or longer (JET, INTOR). Thicker layers on actively cooled structures will be required then. This largely excludes carbon, which cannot be used as coolable substrate without protection against erosion by the coolant. The development of (coated) graphite-metal devices or of metal-metal structures (eg. V or Ni on Cu) is the response to this requirement. While graphite-metal combinations require a compliant layer in between in order to relieve the stress brought about by differing thermal expansion, metal-

metal devices can be fabricated directly by explosive cladding or diffusion bonding. The first set of limiters for the Joint European Torus (JET), for instance, will be made of (Cr-precipitation hardened) copper explosion-cladded with a 1 mm thick nickel layer.

Meanwhile, the quest for ever-lower low-Z materials has somewhat abated: it was recognized recently that a limited amount of low- to middle-Z impurities in the boundary layer will have a positive effect on the power load of high heat flux materials, in that the enhanced quantum radiation from these highly ionized atoms results in a more uniform distribution of the thermal flux on the wall. Hence a reduction of the edge-plasma temperature is achieved, though at the expense of enhanced electromagnetic radiation. But a step towards equipartition of the power load is achieved. Moreover, reduction of the kinetic energy of the particles impinging on the wall decreases sputtering and thus erosion. It is still an open question, however, whether such an impurity-seeded plasma boundary or, more generally, a cool plasma mantle, can be kept stable during the discharge.

Distributing the particle and heat load as uniformly as possible over the largest possible area is the method of coping with the high power levels of fusion plasmas presently pursued. Up to several 10^2 W/cm^2 are generally considered to be manageable. On the other hand, there are components and conditions where even higher loads must be handled. As regards *components*, limiters, divertor plates, and possibly armour panels (which protect against the beam shine-through from neutral injectors) will have to be *cooled actively* for machines with pulse lengths exceeding 20 s. At shorter pulse duration the thermal inertia of these components prevents excessive surface heating during exposure ("inertial cooling").Cooling of the support structure is then mostly effective during pulse intervals.

There are, however, also conditions under which components not designed as high heat flux structures are subjected to heat loads exceeding 10 kW/cm^2. These *off-normal conditions* are met in run-away discharges and hard disruptions. In the former case electrons ("run-away electrons") are accelerated without imparting energy to ions in collisions; with increasing energy the collision cross-section drops so that eventually high-energy electrons ($\geqslant$ 10 MeV) impinge on the torus wall with electron beam melting effects. Low-density discharges are particularly prone to run-away disruptions. Careful cleaning of the vacuum vessel after exposure to air is a prerequisite to – at least partly – override this regime.

Hard disruptions, on the other hand, are instabilities, where the plasma current drops within some milliseconds, resulting in complete loss of the stored energy (in JET, for instance, of the order of 50 MJ). This energy is deposited in an unpredictable way on a fraction of the wall. Such events, when occurring in full power discharges of reactors, are beyond what solids can withstand, and the only way to cope with hard disruptions appears to be to prevent them. It should also be mentioned that disruptions cause strong electromagnetic forces in electrically conducting components. As a consequence these components may fail either by fatique or, in extreme cases, after a single hard disruption.

It is interesting to note that energy dissipation from solids under pulsed high heat fluxes is mostly by evaporation (in the case of carbon: sublimation). Thus, while radiation cooling of surfaces predominates under normal conditions, *evaporation cooling* is the far more important mechanism in disruptions. In this case the temperature of the surface is therefore determined by material loss. This is in conflict, of course, with plasma purity as well as with the integrity of the wall. As a precaution against such events the thickness of actively cooled components cannot be made as thin as would be optimal for heat removal under normal discharge situations.

The possibility of shock-loads restricts, on the other hand, the applicability of *ceramics* – often a recourse in high temperature material technology. In addition, the thermal expansion coefficient of ceramics is so different from that of metals, that direct contact with the actively cooled metal substrate is not a viable solution. *Compliant layers* in between are again required to absorb the thermal stresses; arrays of copper pins, felts, or carbon layers are presently considered.

A Comparison with the Thermal Protection System of Re-Entry Vehicles

There is an apparently striking similarity between high-heat-flux components of the first wall and the thermal protection system of re-entry vehicles. Not only is the visual impression – determined by the tile structure – suggestive of such a comparison, but the solutions to some of the problems are also similar, sometimes even the same. These similarities should not conceal the fact, however, that the problem itself is rather different: while for re-entry vehicles heat is to be prevented from penetrating the shield, the first wall of a fusion reactor will have to let the heat flux penetrate the wall structure so that cooling media can either dissipate it away from high-load components or, preferably, take advantage of this thermal power output (20% of the total fusion power!). Re-radiating the heat – the method used with re-entry vehicles – would be no solution at all for the first wall of fusion devices.

Only for special components (limiters, divertor plates) can radiation cooling be an acceptable means. The same is true of ablative cooling – the main cooling method applied for the Gemini- and Apollo missions. It was abandoned, however, when multiusable vehicles such as the space shuttles were designed. Erosion of the protection system would require too frequent replacements. In fusion devices erosion would, furthermore, lead to plasma quenching owing to impurity radiation. Ablative cooling – in the form of evaporation cooling – occurs here only during off-normal conditions, as in hard disruptions.

Another common feature is the use of *ceramics*. The space shuttle Columbia is covered with more than 30.000 silica tiles, coated with borosilicate glass (approx. 300 μm thickness). The main function of the glass coating is to revert the heat flux by radiation while the bulk of the tile serves as heat insulator. It consists of high purity, amorphous silica fibers and is more than 90% void.

It appears that all the favourable properties of ceramics, in particular the low weight and the excellent stationary heat load capacity, can be utilized in space applications. In fusion applications the poor thermal shock resistance is of considerable concern, but what is worse at present is the lack of knowledge of whether the poor electrical conductivity can be tolerated at components in contact with the plasma.

The means of overcoming thermal stresses at the tile-to-substrate-interface are about the same (compliant layers, strain isolation pads), and in no case will ceramic components be heavy mechanical load-bearing elements.

While ceramic tiles are used on space shuttles up to temperatures of 1200 °C, *carbon* elements are used at parts where the heat load during ascent or entry will raise the surface temperature beyond this value. Here again is a parallel with fusion applications. These reinforced carbon/carbon components have excellent thermal characteristics, including thermal shock resistance but must be protected against chemical degradation at higher temperatures. This is achieved by a silicon carbide coating – similar to the TiC coating presently used on graphite components in Tokamaks.

Chemical and Structural Characterization

Chemical analysis of first wall materials is generally carried out for two purposes: material testing and developing, and plasma edge diagnostics. With present-day low-temperature, non-nuclear machines the plasma contact results predominantly in surface effects. The analysis techniques will therefore be mostly surface sensitive techniques such as Secondary Ion Mass Spectrometry (SIMS), Auger Electron Spectroscopy (AES), and Rutherford Backscattering (RBS). When changes in the bulk composition or structure are observed, they are generally due to thermal effects, often combined with mechanical stresses.

When plasma edge diagnostics is to be performed with the aid of surface analytical techniques, specially designed samples are introduced into the plasma region to be probed. The areal density of plasma particles (hydrogen isotopes) and impurities collected on these probes is subsequently determined. Since erosion on these collection probes must be avoided, measurements of this kind can only be performed in the limiter shadow, i.e. in the scrape-off region behind the limiter.

The quantities measured with collection probes are fluxes. In order to get information on plasma densities, the velocity of the particles – the plasma temperature – must be known. In some cases, especially with hydrogen, this quantity can be extracted from the depth distribution of the collected/implanted particles. Hydrogen analysis is obviously of particular interest in this field. This excludes many of the standard techniques as AES, RBS and X-ray analysis. SIMS, on the other hand, is capable not only of detecting hydrogen isotopes but also allows depth-resolutions reached by no other hydrogen detection technique. SIMS therefore plays a unique role in materials analysis in fusion research.

A controversial point for all surface-analysis groups at plasma laboratories is whether the analysis should be performed in-situ, i.e. directly attached to the torus. This allows – in principle at least – much faster analysis of the exposed samples in vacuum-suitcases is required less often than at first expected, which again facilitates the procedure.

than an offsite evaluation of the probes, and the environment of a large plasma device generally prevents analysis near the detection limits. Hence, the cleaner the plasmas achieved, the more restricted will be the use of in-situ systems. For these reasons the majority of scientists and engineers in the field of plasma surface interaction tend towards analysis at dedicated remote instruments. Fortunately, it was found recently that transport of exposed samples in vacuum-suitcases is required less often than at first expected, which again facilitates the procedure.

Outlook

The materials problems of fusion devices up to those of the next generation (JET, TFTR, JT-60) are imposed by the direct plasma-wall contact. The main problem here is that the power load is essentially a surface load. By contrast, the total neutron fluence for the tritium-burning JET will be some four orders of magnitude smaller than in fission reactors. The neutron load will thus be more of a radiological consideration, whereas the plasma power load on the limiter surface will be decisive for the performance of the Tokamak.

This energy deposition in near-surface layers will also be the crux of the choice of materials in future fusion reactors, even though the power load by the plasma comprises only 1/5 of the total power output. The then predominant neutron load, however, constitutes a volume load and thus appears to be easier to cope with; helium embrittlement and, to a smaller degree, displacement damage are anticipated to be the key issues for structure materials. The first wall, however, will have to be capable of withstanding both the neutron radiation and the plasma contact.

For the sake of demonstrating the physical feasibility of controlled thermonuclear fusion, materials which are anticipated to be ultimately unacceptable in DEMO-reactors are currently considered for high load first wall components. It is important to be aware of the temporary nature of such choices of materials; anisotropic graphite, for instance, is a promising candidate for high heat flux components, but will be out of the question when neutron fluences exceed 10^{21} cm^{-2}: degradation of the characteristically high thermal conductivity due to radiation damage presumably renders oriented graphite a short term solution. It is conceivable that beryllium will then be considered — despite its high toxical hazard. Being a low-Z metal with high thermal and electrical conductivity as well as good thermal shock resistance, beryllium comes nearer to providing an ideal solution. It is expected that most of the safety measures against its toxicity will have been covered by those precautions imposed by neutron activation and tritium.

Even the best high load materials would not be able to withstand the heat and particle fluxes in next-generation machines, if this load were not distributed over larger areas. All efforts in shaping the plasma boundary layer aim finally towards spatial equipartition of the wall load and reduction of the particle energies. Means of homogenizing the power load are to break up the magnetic surfaces near the wall ("ergodization" of the magnetic flux), to adapt the macroscopic shape of limiters to the particle and energy flux distribution in the plasma boundary, and to transform kinetic particle energy into quantum radiation (impurity seeded boundary layer, "photosphere"). The "cold plasma mantle" would be a major goal for achieving both spatial equipartition and reduction of the level of kinetic particle energy. It remains to be seen, however, whether such a cold plasma boundary layer can be kept stable: the range of parameters where neither the hot centre plasma is extinguished by an expanding boundary, nor the cold mantle consumed by the hot core appears to be rather small. No experimental studies have been performed on this problem as yet.

Of crucial importance, however, will be the experience of how to avoid hard disruptions. This experience must be acquired with plasma machines presently in operation, since major damage will be incurred in machines of the next generation if under off-normal conditions a major fraction of the stored energy is deposited locally on parts of the first wall; these machines will withstand only a limited number of disruptions.

Acknowledgements

I am grateful for valuable discussions with, and comments by M. M. Ferguson, A.W. Mullendore, K.-G. Tschersich, and G.H. Wolf.

Summary

Materials for the First Wall of Fusion-Reactors and Their Analytical Characterization

A short overview of the present materials strategy for magnetically confined fusion plasma is given. Emphasis is laid on the materials problems of the first wall and in-vessel components subjected to high particle and heat loads, while the effect of impurity load on the plasma, as well as neutron radiation effects in structural materials are not discussed here. Some apparent similarities between the thermal protection system of re-entry vehicles and high heat flux components in fusion devices are compared.

References

Instead of a detailed list of references we refer to the proceedings of two series of conferences, all published in the Journal of Nuclear Materials by the North-Holland Publishing Company, Amsterdam:
Plasma Surface Interaction in Controlled Fusion Devices, J. Nucl. Mater. 63 (1976), 76/77 (1978), 93/94 (1980), 111/112 (1982), and
Fusion Reactor Materials, J. Nucl. Mater. 85/86 (1979), 103/104 (1981).

As a general introduction the following recent monographies can be recommended
J. Raeder et al., Kontrollierte Kernfusion. Stuttgart: Teubner. 1981, and
Th. J. Dolan, Fusion Research. New York: Pergamon Press, 1982.

Correspondence and reprints: Wolfgang O. Hofer, Kernforschungsanlage Jülich, D-5170 Jülich, Federal Republic of Germany.

Mikrochimica Acta [Wien], Suppl. 10, 9–20 (1983)

Institut für Metallphysik der Universität Göttingen and SFB 126, Göttingen/Clausthal, Federal Republic of Germany

Application of Positron Annihilation to Metallic Alloys

By

Th. Hehenkamp, W. Lühr-Tanck, and A. Sager

With 12 Figures

(Received April 25, 1983)

Segregation of impurity atoms to or depletion from surfaces or grain boundaries plays a significant role in the mechanical and corrosion resistance of metallic materials. A great deal of effort has been spent on detecting the extremely thin layers of impurity enrichment or depletion at those boundaries by suitable analytical techniques on open or freshly broken intergranular surfaces, this volume containing many references. It is the purpose of this paper to show that not only such more open and extended structures like surfaces or grain boundaries but also holes on an atomic scale like single vacancies attract or reject impurity atoms and that their behavior at these local sites is very similar. Since the number of vacancies which are generated in solids in thermal equilibrium is relatively small, the study of the phenomena of impurity-vacancy interaction requires a technique, which is extremely sensitive to and somewhat specific for vacancies, the positron annihilation. After introducing two possible techniques to utilize this effect, the positron lifetime measurements and the Doppler-shift of the annihilation radiation, data will be presented for the study of thermal vacancy formation in pure metals and particularly some noble metal alloys.

Positron Annihilation

It is well known that positrons are trapped very effectively by vacancies. The positrons supplied usually by a radioactive source annihilate in metals not containing vacancies with a lifetime of 100 to 400 ps. Positrons being trapped by vacancies exhibit a lifetime longer by 30 to 50%. This effect may be utilized to detect positron capture by vacancies employing a lifetime spectrometer.

Positron Lifetime Spectrometer

Fig. 1 indicates the situation schematically. Na 22 mechanically encapsulated but not dissolved in the specimen emits a γ-ray of 1.27 MeV and a positron almost simultaneously. A fraction of 5 to 50% of the positrons is backscat-

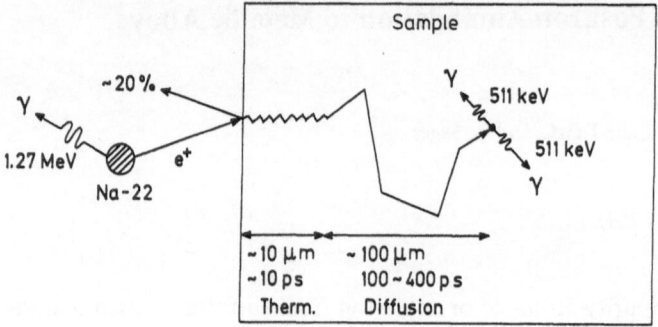

Fig. 1. Arrangement of sample and positron source for the determination of positron lifetimes

tered from the surface of the specimen, depending on its atomic number. Many of them penetrate into the sample with high energies. They become thermalized within a range of 10 μm and 10 ps. The positrons then diffuse for some 100 μm with thermal energies before they annihilate with electrons after 100 to 400 ps in metals. Then the total mass of the positron-electron

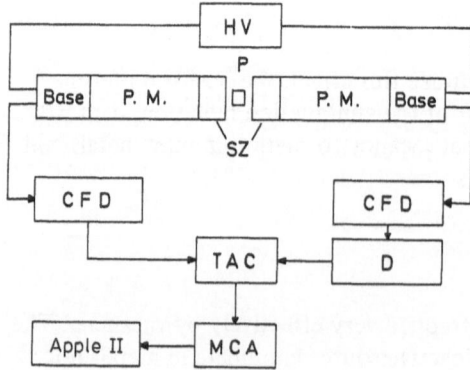

Fig. 2. Components and configuration of a positron lifetime spectrometer

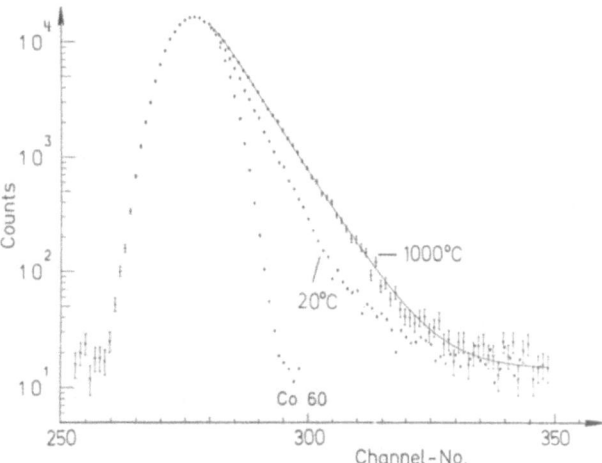

Fig. 3. Positron lifetime spectrum for prompt curve (Co 60) indicating the time resolution of the spectrometer, for pure copper at 20 °C containing no thermal vacancies and at 1000 °C with positron trapping almost exclusively at thermally generated vacancies

pair is transformed in most cases into two γ-quanta of 511 keV if both particles were at rest before the annihilation. The configuration of the lifetime spectrometer is shown schematically in Fig. 2. The sample P is positioned between two fast plastic scintillators SZ which are optically coupled to photo-multiplier tubes P. M. with only few dynodes providing high gain at low walk and drift. These tubes are operated from the same high voltage power supply HV. The dynode impedance has to be matched to the pulse cable impedance of 50 ohms. The output pulses are fed to constant fraction differential discriminators CFD, one set to the γ-energy of 1.27 MeV, the other to the annihilation radiation of 511 keV. A delay line D permits an adjustment of the relative time scale. The outputs of the CFD's are connected to a time to amplitude converter TAC. This unit accepts a start pulse from the 1.27 MeV channel indicating the birth of a positron and starts charging a capacitor at a constant rate. The end of the life of the positron is indicated by the 511 keV annihilation radiation, the pulse of which stops the charging. The charge accumulated between start and stop is represented by the voltage across the capacitor, which is fed to a multichannel analyzer MCA for display, coupled to a microcomputer Apple II. This analyzer integrates all singly measured times for many thousands of these events at the appropriate pulse height level (channel #) and displays a spectrum of lifetimes. Examples are given in Fig. 3, one channel being equivalent to a time interval of 23.8 ps. The lowest symmetrical curve is the so called "prompt" curve obtained by a normal γ-source with two energy levels which are emitted simultaneously. (Co 60, 1.17 and 1.33 MeV.) In this way the full width at half maximum

(FWHM) may be determined, which characterizes the time resolution of the lifetime spectrometer. The time resolution in Fig. 3 is 220 ps. The second curve with closed circles represents measurements in pure copper at 20 °C, the one with squares and error bars data for 1000 °C. The increase in positron lifetime as compared to the room temperature result is clearly visible. The positron lifetime then is obtained from the slope of the respective curves. If more than one lifetime component is present, a computer fit for the superimposed slopes is necessary to determine the different lifetimes and their relative contribution to the spectrum. Fig. 4 is shown as an example. The lifetime component τ_1 represents the positron lifetime in the bulk material, whereas τ_2 indicates the lifetime of positrons trapped at thermally

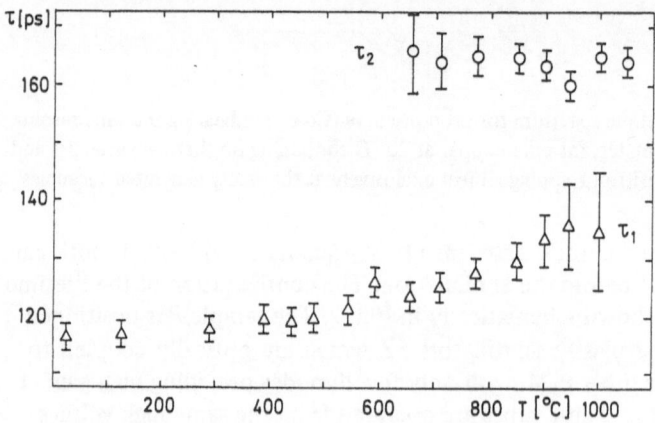

Fig. 4. Positron lifetimes in pure copper, τ_1 lifetime in the bulk, τ_2 lifetime for positrons being trapped at vacancies as function of temperature

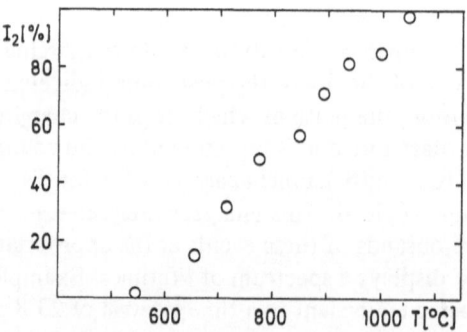

Fig. 5. Intensity of the longer lifetime component for positrons being trapped at vacancies as function of temperature

generated vacancies. Since the number of vacancies is very small at low temperatures, increasing, however, at an exponential rate upon increasing the temperature both components may be identified simultaneously over a certain range of temperatures. This increasing number of vacancies leads to a rising intensity of the second lifetime component as function of temperature as plotted in Fig. 5.

Doppler-Shift Measurements

Another means of gaining information about vacancies in metals from positron annihilation is by the Doppler-shift originating from the conservation of momentum of the annihilated electron-positron pair. If both particles were at rest the γ-energy would be 511 keV. The momentum of the pair is provided essentially by the electron, since, as it is indicated already in Fig. 1, the positron is supposed to be completely thermalized in a time very short compared to its lifetime. The momentum of the electron depends on the type being a core or conduction electron. The electron momentum shifts the 511 keV-line by some keV for the core electrons, whereas the shift by conduction electrons is much smaller. Employing solid state Ge-detectors it is possible to measure γ-energies effectively with a resolution of about 1.2 keV and hence the Doppler broadening. In Fig. 6 the experimental configuration

Sample / e^+- Source

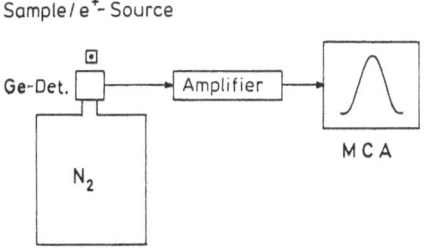

Fig. 6. Configuration of a Doppler-shift apparatus

of a Doppler-shift apparatus is shown schematically. The annihilation radiation generates pulses in a Ge-detector the heights of which are proportional to the γ-energy. The pulses are amplified linearly and fed into a multichannel analyzer for display and evaluation. The interesting difference in the annihilation of a positron in the bulk of the lattice as compared to the one bound to a vacancy lies in the fact, that the probability of annihilation with a core electron is higher in the bulk than it is at the vacancy, where the core electrons of the empty lattice site are missing. In Fig. 7 the Doppler-broadened

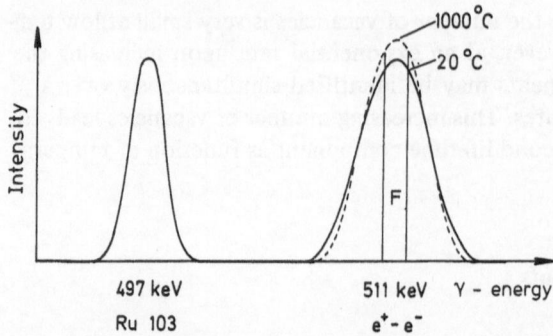

Fig. 7. Comparison of a normal γ-line (Ru 103 at 497 keV) with the Doppler-shifted annihilation line at 511 keV

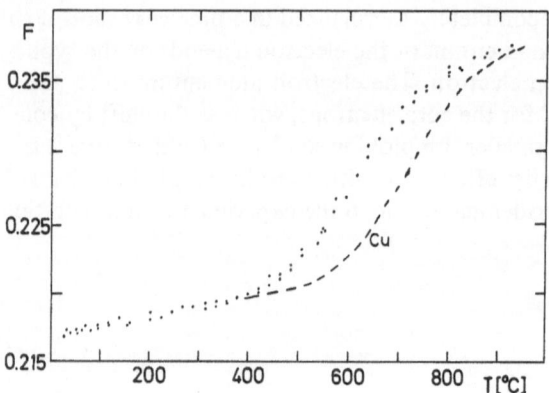

Fig. 8. Comparison of F-parameter for pure copper (dotted line) with a measurement for a Cu-6.6 at.% Ge alloy (points)

511 keV line is compared to a line emitted by a normal γ-source Ru 103 at 497 keV, indicating the resolution of the detector system. The broadening of the 511 keV annihilation line by the Doppler-shift is clearly visible. The broadening, however, is different for a temperature of the copper sample of 20 °C and 1000 °C. At the higher temperature the line becomes narrower due to the preferential annihilation of the positrons at vacancies. This narrowing may be expressed in terms of a relative intensity of the central line channels F. In Fig. 8 this parameter F is plotted versus temperature for pure copper (dotted line). At low temperatures F increases only slightly with temperature due to various effects[1-3], at least in part by a decrease in electron density upon thermal expansion of the lattice. Above 600 °C the thermal

formation of vacancies sets in substantially and F increases faster with T. That F does not continue to rise exponentially as the vacancy concentration is the result of a saturation of positrons with vacancies. All positrons provided by the source are being trapped by vacancies and annihilate there. The experimental points in Fig. 8 show the same kind of experiment but for a Cu-Ge alloy containing 6.6 at.% Ge. The behavior is very similar to the pure copper except that vacancy formation starts already at a lower temperature. This indicates a gain in free binding energy between the vacancy and the impurity Ge atom. By such a binding the total free energy required to form a vacancy becomes reduced and less thermal energy is necessary. If a repulsive force acts between vacancy and impurity atom a higher thermal energy has to be provided. In the noble and many transition metals impurity atoms which are positioned to the right of the host metal in the periodic chart of the chemical elements and hence have a positive excess charge generally reduce the formation energy whereas transition metal impurities in the noble metals increase it, as was explained many years ago by Lazarus[4] and Le Claire[5].

Now it is a question of the range in which these binding or repulsive forces are operative in metallic lattices, how complicated the formal description of vacancy formation becomes in alloys. If the mutual range of interaction between the vacancy and impurity atoms is very small (order of magnitude of one atomic distance) the formal treatment may be simplified and restricted to nearest neighbor positions of a given vacancy. Then one forms vacancies in an alloy in many different local environments, which do not overlap as long as the vacancy concentration remains small which usually is the case. These environments may be characterized by the number of impurity atoms being present in the first coordination shell of a vacancy. These aggregates are called complexes. A complex i means that i impurity atoms are present in that shell i ranging from 0 to 12 in a fcc lattice. In thermal equilibrium there exists an equilibrium population of the different complexes. The molar fraction x_i of such aggregates and their degree of association α_i may be expressed formally by a model developed by Hehenkamp, Schlett, and Sander[6,7] on the basis of a statistical thermodynamic treatment by Dorn and Mitchell[8].

$$x_i = \binom{Z}{i} x_A^{Z-i} \; x_B^{i} \; \exp(-g_i/kT) \tag{1}$$

$$x_v = \sum_{i=0}^{Z} x_i \qquad \qquad \alpha_i = \frac{x_i}{x_v} \tag{2}$$

where x_A and x_B are the molar fractions of matrix A and impurity B-atoms respectively and x_v the total of vacancies in all complexes, and g_i is the free enthalpy of formation of a vacancy in a complex with i impurity atoms. Since this quantity is different in the various possible complexes one should expect a strong temperature dependence of the effective formation enthalpy in alloys, which is given by

$$H_{eff} = \frac{d \ln x_v}{d \, 1/T} = H_0 - \sum_{i=1}^{z} \alpha_i \, h_{bi} \tag{3}$$

where H_0 denotes the formation enthalpy of vacancies in the pure base metal and h_{bi} the binding enthalpy in the *i-th* complex.
Whereas the slope of an evaluation of the positron data according to a simple trapping model[9, 10] yields a straight line giving a formation enthalpy of 1.29 eV, Fig. 9 for pure copper, the corresponding Arrhenius plots for Cu-Sb

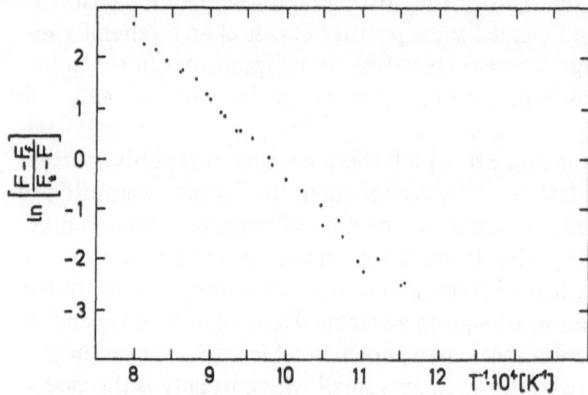

Fig. 9. Arrhenius plot for pure copper

alloys are more or less curved depending on the concentration of antimony and the temperature in best accord with the complex model. The curvature is plotted in Fig. 10 versus temperature. The physical reason for the curvature is the drastic change in the population of the various complexes as function of temperature as indicated in Fig. 11 which shows the degrees of association. These may be calculated from the effective formation enthalpies as either function of temperature (formation enthalpy) or as function of composition (formation free enthalpy) at a given temperature by minimizing the error in fitting h_{bi} or g_{bi} values to Eq. (3) respectively. In this way experimental values of the vacancy-impurity binding may be obtained, listed in Table 1 for various impurities in copper and silver. At higher temperatures there is a tendency for the complexes to dissociate, at lower ones to associate, the number i of impurity atoms involved in the binding at a given vacancy being strongly dependent on the available impurity concentration.
A very striking effect may be identified in Cu-Sb alloys an example being

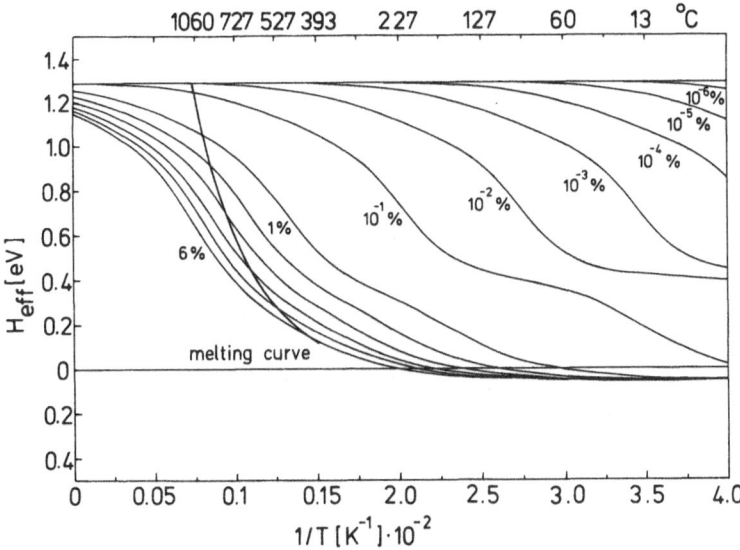

Fig. 10. Effective formation enthalpy of vacancies in various Cu-Sb alloys as function of (reciprocal) temperature with Sb concentration in at.% as a parameter

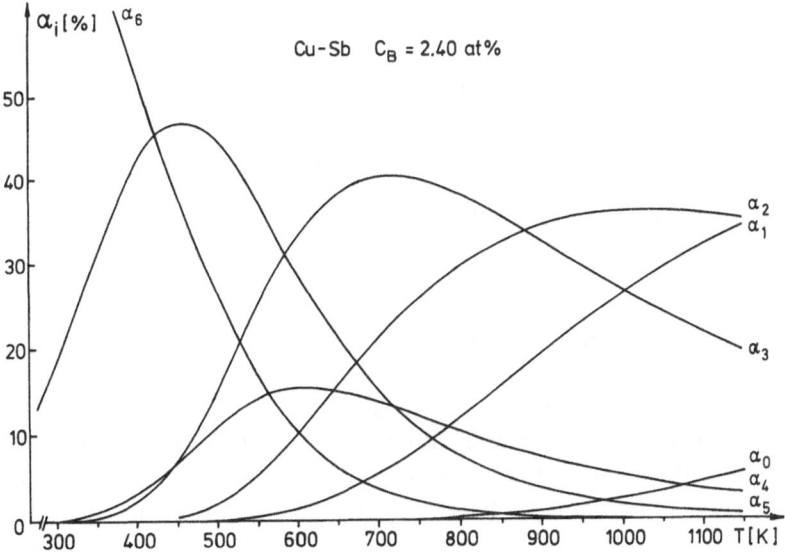

Fig. 11. Degrees of association of various impurity – vacancy complexes as function of temperature. In this example the composition is 2.4 at.% Sb in copper

Table 1. Binding-Enthalpies [eV] (Experimental)

	h_{B1}	h_{B2}	h_{B3}	h_{B4}
Cu-Ge	0.19	0.32	0.43	0.48
Cu-Sn	0.18	0.34	0.46	0.54
Cu-Sb	0.30	0.50	0.68	1.02
Ag-Ge	0.19	0.34	0.45	0.51
Ag-Sn	0.18	0.35	0.43	0.47
Ag-Sb	0.20	0.36	0.50	0.60
error	±0.02	±0.03	±0.04	±0.04

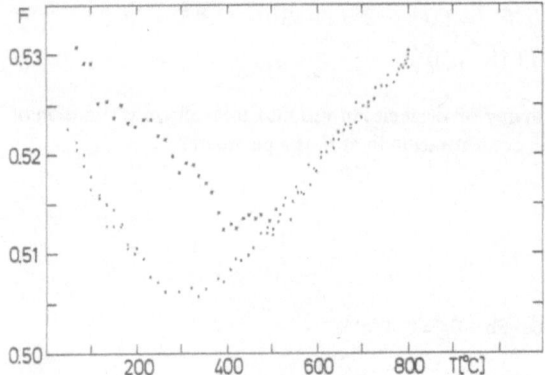

Fig. 12. F-parameter for a Cu-2.6 at.% Sb alloy as function of temperature. The lower curve is completely reversible if temperature stays above 55 °C. Below there appears segregation, when the sample is held for some days. Upon isochronal annealing the upper curve is obtained

given in Fig. 12. Here the F-parameter shows a very unusual behavior. At the higher temperatures part of the ordinary S-shape of the Doppler-shift is visible, at lower ones, however, the F-parameter starts to increase again. This phenomenon may be interpreted in different ways so far. It is possible that the binding enthalpy of higher impurity vacancy complexes becomes larger than the formation enthalpy for vacancies in the pure base metal. This would lead immediately to the observed rise in the F-parameter at lower tempera-tures, where according to Fig. 11 the association of vacancies with 5 or 6 im-purity atoms becomes very significant. In this context an ordering effect of the impurities at the vacancies may play an important role additionally, reducing the configurational entropy $\binom{Z}{i}$ for the higher complexes drastically. $\binom{Z}{i}$ gives this entropy for a statistical distribution of the impurity atoms around the vacancy and should be applicable at high temperatures. The interpretation of the experimental data with a reduced configurational entropy due to ordering would lead automatically to a higher value for the binding enthalpy in the higher complexes.

Another possible interpretation would be the segregation of an intermetallic phase containing structural vacancies which also may act as traps for positrons. The type of curve observed experimentally, however, is not in favor of such an interpretation.

Also segregated clusters of impurity atoms could act as shallow traps for positrons, shallow, because only little thermal energy is required to liberate the positrons from these traps, whereas the binding to vacancies may involve binding energies as high as $3-4$ eV.

Further experimental work is certainly necessary to clarify the situation. It should be pointed out, however, that close links to segregation theories like spinodal decomposition exist, if the increase of the F-parameter at lower temperatures could be verified as being due to vacancies.

In this way the excellent capability of different positron annihilation techniques to investigate vacancy formation in alloys is clearly demonstrated. There exist, however, still some problems in using these techniques for quantitative evaluations, which have to be further investigated namely the possible change in trapping rates and lifetimes in the different impurity vacancy complexes and the bulk lifetime behavior as function of temperature. At present, nevertheless, the possible influence of such uncertainties on the data obtained seems to be small.

Summary

Application of Positron Annihilation to Metallic Alloys

Two techniques to apply measurements of positron annihilation to metallic alloys are presented, the determination of positron lifetimes and the Doppler-shift. Study of positron annihilation provides very sensitive means to check for the presence of vacancies generated either thermally, by cold work, radiation damage or being present for reasons of structural stability in some intermetallic phases (structural vacancies). Some examples are given for thermal vacancies in pure metals and Cu-Sb-alloys.

References

1. M.J. Stott and R.N. West, J. Phys. F 8, 635 (1978).
2. P. Hautöjärvi (ed.), Positrons in Solids. Berlin-Heidelberg-New York: Springer-Verlag. 1979.

3. A. Seeger, J. Phys. F **3**, 248 (1973).
4. D. Lazarus, Phys. Rev. **93**, 973 (1954).
5. A.D. Le Claire, Phil. Mag. **7**, 141 (1962).
6. Th. Hehenkamp and L. Sander, Z. Metallkde. **70**, 202 (1979).
7. Th. Hehenkamp, Mikrochim. Acta [Wien], Suppl. IX, **1981**, 15.
8. J.E. Dorn and J.B. Mitchell, Acta Metall. **14**, 71 (1966).
9. W. Brandt, in Positron Annihilation, A.T. Stewart (ed.). New York: Academic Press. 1967.
10. B. Bergersen and M.J. Stott, Solid State Comm. **7**, 1023 (1969).

Correspondence and reprints: Prof. Dr. Th. Hehenkamp, Institut für Metallphysik, Hospitalstrasse 12, D-3400 Göttingen, Federal Republic of Germany.

Mikrochimica Acta [Wien], Suppl. 10, 21–49 (1983)
© by Springer-Verlag 1983

Physics Department and SIMS-Laboratory, Chalmers University of Technology, Gothenburg, Sweden

Applications of SIMS in Interdisciplinary Materials Characterization

By

Alexander Lodding and **Hans Odelius**

With 13 Figures

(Received January 19, 1983)

Although pioneering work on secondary ion mass spectrometry was performed already in the 1930-s, one may say that SIMS as a microstructural-microanalytical technique for materials research saw the light of day about 21 years ago[1,2]. As a commercially available tool, SIMS is only some 14 years old[3-6]. Since about half that time, its typical assets, mechanisms and artifacts have been illuminated by extensive and thorough international discussion[7,8], and today SIMS is widely accepted as an off-age technique for very sensitive three-dimensional characterization of materials. Current trends[9-11] are directed mainly towards perfected quantitation, towards efficient routines in daily applications, and towards apparative development for further improved detection sensitivity and spacial resolution.

SIMS is multi-disciplinary in application. Globally about 25 % of the users are in metal, steel and ceramic industry, about 40 % are associated to semiconductor and electronic component industry, approx. 15 % are engaged in interdisciplinary materials research in major industries, and about 20 % in interdisciplinary and basic research at academic institutions.

Existing SIMS equipment may be divided into three main classes. *1) Non-imaging ion probes* are used either for specialized surface studies ("static SIMS", see e.g. Ref.[12]) or for in-depth profiling on systems with undifferentiated surface topography. Such equipment, both commercial and of laboratory design, is of relatively common occurence, i.a. as accessory to other surface analytical techniques (such as AES or ESCA). *2) Ion microprobes* (IMMA) employ a small (ca $0.5 - 10~\mu m$ diam.) beam of primary ions[3]. The beam size is essentially determining for lateral resolution. Ion images are obtained by TV-type beam scan over sample surface. Presently about 40 microprobe-type SIMS instruments are effectively in operation. *3) Ion microscopes-microanalyzers,* employing a relatively wide primary ion beam (ca 5 – 300 μm), are capable of stigmatic ion-optic imaging[1]. The obtainable lateral resolution in microanalysis is given by available microscopic magnification, by gating of the imaged area, or by image processing. About 70 such instruments are in operation 1982.

The relatively sparse global occurence of three-dimensional SIMS-equipment (ion microprobes, ion microscopes) is perhaps not as much due to the high price of such equipment as to the common view that SIMS is seldom an easy routine. In fact, successful use of the technique does usually require considerable experience on the part of the operator. Each individual system to be investigated involves particular stratagems of experimental approach and evaluation.

The present paper is intended to illustrate different aspects of practical SIMS by example of application on different materials. Most of the examples are from work performed with commercial ion microscope equipment of the first generation (Cameca IMS 300 at the Gothenburg SIMS laboratory); however, recent and potential improvements in procedures and apparatus are to be discussed where relevant. The different facets of analytical requirements to be discussed also motivate a brief but referenced initial orientation about the present capabilities of SIMS and about the main factors, including artifacts, affecting the assets (sensitivity, quantitation, spacial resolution) of the technique.

Sensitivity and Quantitation Parameters in SIMS

a) Ion Current Versus Concentration

The functioning principles of SIMS may be summarized as *a)* sputtering, *b)* mass spectrometric ion separation, and *c)* collection (as total current or as image) and counting of separated ions. Nearly monoenergetic primary ions, total current I_p, current density i_p, strike the sample surface, ejecting secondary species with a sputtering yield S. An isotope, the concentration c of which is to be measured, emerges in part ionized from the sputtering cascade. The positively or negatively charged ions are extracted into the mass spectrometer, and counted as an ion current I_s, of the isotope (or of a molecule containing the element). If the ionizability (ions per emerging atom of the isotope) is γ, and the transmission of the instrument is η, then the ion current of the isotope, arriving at the collector from a surface area A is

$$I_s = i_p A \cdot S \cdot \gamma \cdot \eta \cdot c . \tag{1}$$

An expression for the most striking feature of SIMS, viz. the low limit of isotopic detection, may be arbitrarily formulated[11] i.a. by putting the limit at a factor of 3 above the electronic noise level of the detector, ΔI, so that, from Eq. (1),

$$c_{min} = 3 \, \Delta I \, (\eta \, S \, \gamma \, \alpha)^{-1} (i_p A)^{-1} \, , \tag{2}$$

where the factor α accounts for the fact that a mass spectral peak may contain other contamination than electronic noise. The primary current density is by simple proportionality[11] determining for the sputtering speed v_{sp}, so that one may write

$$c_{min} = K_1 / i_p A = K_2 / v_{sp} A \, , \tag{3}$$

with $K_1 = 3 \, \Delta I \cdot S^{-1} \, (\eta \, \gamma \, \alpha)^{-1}$ and $K_2 = 3 \times 10^{-5} \cdot \Delta I \cdot M \, (z \, \rho)^{-1} (\eta \, \gamma \, \alpha)^{-1}$, where M and ρ are, respectively, the atom mass and density of the matrix, and $z.q_e$ is the charge of the primary ion. With α near to unity and with typical operational values for the ingoing factors, one may expect K_1 and K_2

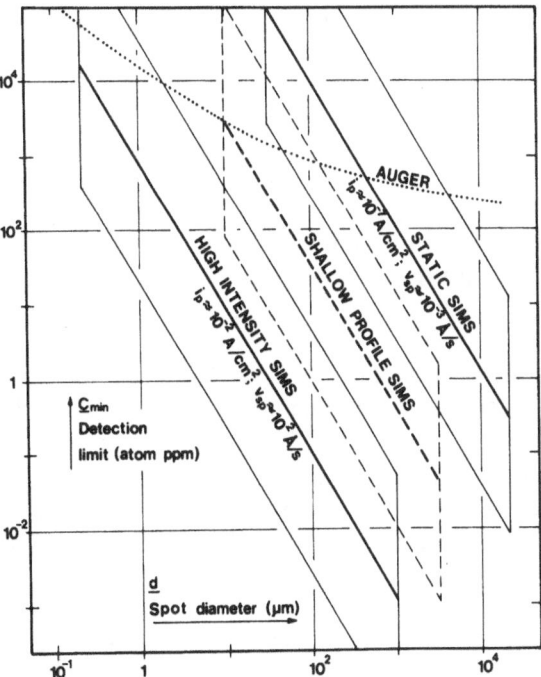

Fig. 1. Isotope detection limits in SIMS, in dependence of diameter of analyzed area

for different systems to range within about 4 orders of ten about 10^{-13} and 10^{-17}, respectively (using Amps and cm as units).

The proportionality of the sensitivity in SIMS to the total substance sputtered from the sputtered area[8,11] is illustrated in Fig. 1 for three cases of i_p or v_{sp}. Clearly if high sensitivity is to be obtained when very small areas are analysed, high sputtering rates are necessary. Similarly, when only the outermost atomic layer is studied, relative large surface areas are required. The three chief assets of SIMS, i.e. sensitivity, depth resolution and lateral resolution, are mutually competitive.

b) Ionizabilities

The thin-drawn lines in Fig. 1 approximate the practical variations in K_1 and K_2. The dominating parameters here are the ionizabilities γ and the background factor α, both dependent on the element to be detected as well as on the matrix.

Empirically it is found (see e.g. Refs.[11,13]) that the positive and negative ion yields of sputtering often reasonably well obey the so called LTE formalism[14]:

$$\gamma^+ = \text{const.} \cdot n_e^{-1} \cdot (B/B_0)\exp(-E_i/kT_i) \ , \tag{4a}$$

$$\gamma^- = \text{const.} \cdot n_e \cdot (g/g_0)\exp(+E_a/kT_i) \ , \tag{4b}$$

where n_e denotes the electron concentration in the collision cascade of sputtering, B/B_0 and g/g_0, respectively, the ratios of partition functions and of statistical weights in the excited and ground states, E_i is the first ionization potential of the sputtered atom (or molecule), and E_a is the electron affinity. Molecular binding energies may also enter the Boltzmann terms. T_i, "ionization temperature", is an entity typical of each matrix. Table 1 gives approximate values of T_i as measured for components and impurities in several one-phase matrix systems. The values of γ^+ and γ^- for different (metallic) elements in these systems may range from near-unity down through 3 to 5 orders of magnitude.

Although the theoretical premises of the original "local thermal equilibrium" model[14] are tentative, several more rigorous theories have later arrived at exponential expressions reminiscent of Eq. (4). The formalism in principle offers possibilities of automatic calibration (ion current versus concentration), using measured T_i values, alt. one or two internal standards. At least semi-quantitative results have often been obtained in this way (see e.g., Ref.[13]). However, essential caution is in place: The T_i entity varies not only from one matrix to another, but may in a given matrix be different for positive and negative ions, singly and doubly charged ions, atomic and molecular

Table 1. LTE-Formalism: Approximate "Ionization Temperatures" in Different Matrices. Obtained with an Energy Pass-band of 0 – ca 100 eV for the Secondary Ions

Matrix	Polarity of ions	Ioniz. temp. (K)	Remark	Ref.
Borosilicate glasses	+	10000 – 13000		11
Calcium apatites	+	8500	F^+, Cl^+ *not* in systematics	11
Noble metal alloys	+	3000 – 4000	Oxygen leak	11
Al	+	11000	" "	15
Fe	+	7000	" "	15
Si	+	8000	" "	15
GaAs	+	6500	" "	15
Ni base alloys	–	1600	Caesium leak	16
Sn, GaAs, Ge, Si	+	7500	Oxygen primary ions; γ^+ extrapolated to $v_{sp} = 0$	17
Sn, GaAs, Ge, Si	–	2700	Caesium prim. ions; γ^- extrapolated to $v_{sp} = 0$	17

ions, etc.; effective ionization temperature also appears influenced by crystallographic effects and by the probability of molecule formation[11, 18-21].

Of particular importance is the "chemical enhancement" effect on ionization. The formalism of Eq. (4a) is found reasonably valid only in ionic matrices, or if the sample surface is maintained ionic by introduction of a reactive species (by primary ion implantation or by a gas leak). E.g., it is well known that the presence of oxygen greatly enhances the positive ion yields from metals. Similarly in negative ion emission, cesium stimulates the yields, probably by a lowering of the electron exit work function[16]. As an example of chemical enhancement, Fig. 2 shows the oxygen dependence of γ^+ for different elements in a range of Au-Ag-Pd-Pt alloys[11].

If the surface is saturated with the reactive species, considerable reproducibility in ion yields may be achieved. Most of the values in Table 1 show T_i values for maximal saturation, using O_2 gas or Cs vapour leak for the metals and semiconductors. If, instead, the reactive species is introduced only by implantation of the primary ions, the degree of saturation is dependent

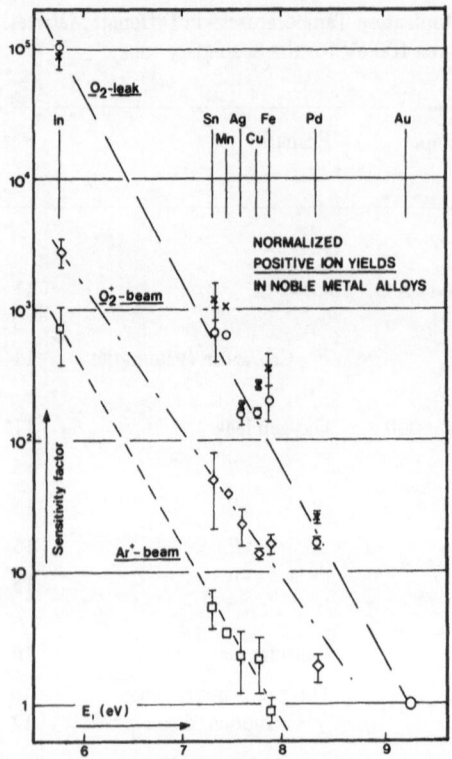

Fig. 2. Positive ion yields of elements in noble metal alloys; chemical enhancement effect. Crosses: O_2^+ primary ions, 10^{-2} A/cm^2; O_2 leak, 5×10^{-5} Torr. Rings: Ar$^+$ ions, 5×10^{-3} A/ cm^2; O_2 leak. Diamonds: O_2^+ ions, no gas leak. Squares: Ar$^+$ ions, no gas leak

on the rate of sputtering, and $\gamma^{\pm}$ is found to decrease steeply with increasing ν_{sp}[17].

Attempts at formulating a theoretical background of the chemical enhancement effect have hitherto allowed only qualitative interpretations (see e.g. Refs.[12, 16, 18, 21]).

c) Spectral Background

The intrinsic secondary ion mass spectrum of each matrix contains, in addition to peaks of atomic ions, also contributions from atomic dimers, trimers, etc. Molecular combinations may form large ion mass peaks when the resp. E_i or E_i values, together with the respective dissociation energies, are favourable. In positive spectra doubly or multiply charged ions also give rise to mass peaks. Moreover, even under conditions of good vacuum, pure primary

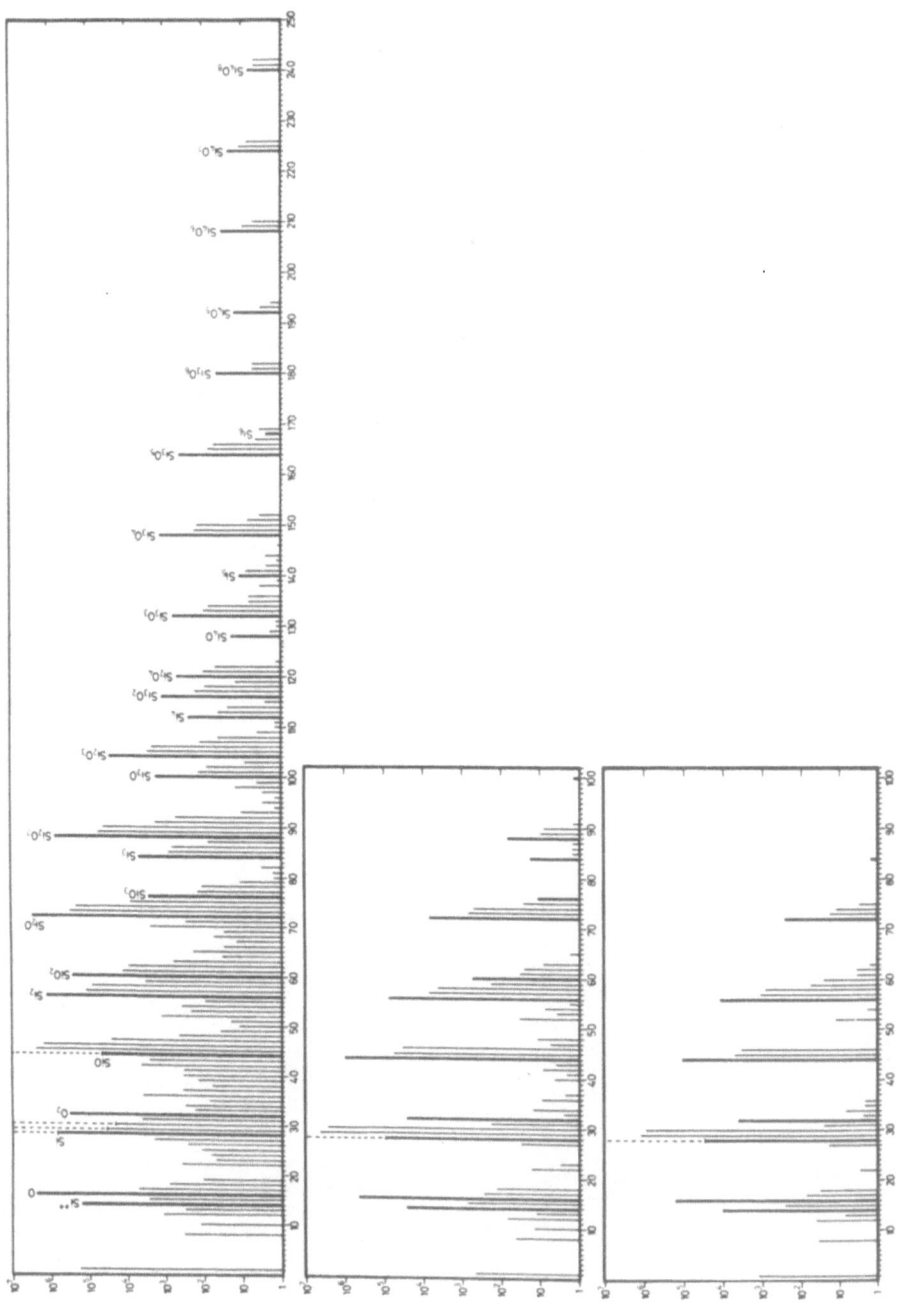

Fig. 3. Positive secondary ion spectra (counts/sec versus mass number) of silicon. Primary ion beam: O_2^+, 10 μA, 150 μm diam., rastered 150 x 150 μm, accelerating voltage 15 kV. O_2 gas-leak 10^{-5} Torr. Analyzed area: 150 μm diam. First row: no offset in secondary ion energy. Second row: offset 25 eV. Third row: offset 50 eV. Cut-off in bar, continued by dashed line: attenuation by a factor 10^3

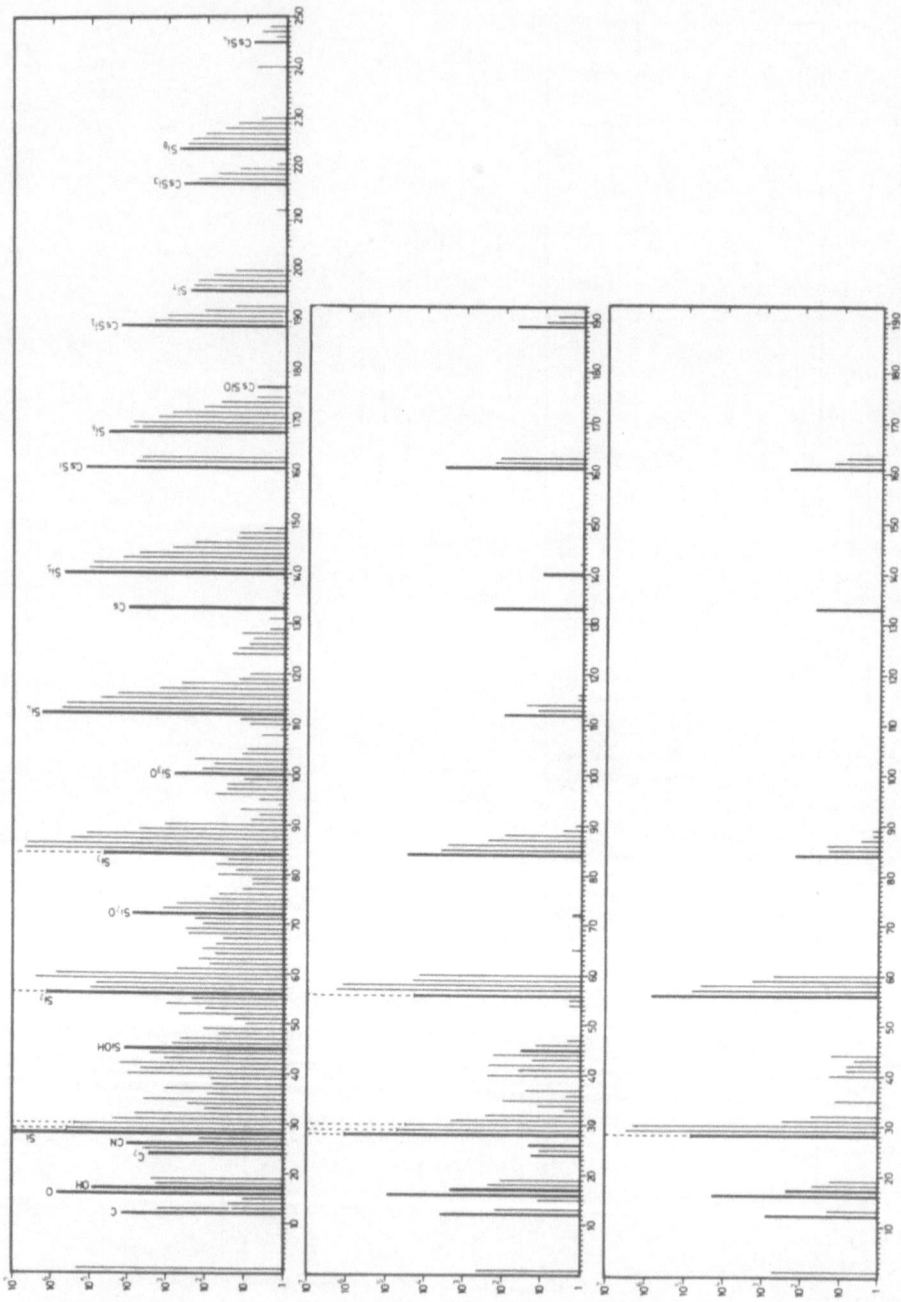

Fig. 4. Negative secondary ion spectra (counts/sec vs. mass number) of silicon. Primary ion beam: Cs$^+$, 0.2 μA, 150 μm diam., rastered 150 x 150 μm, acc. voltage 10 kV. Analyzed area: 150 μm diam. First row: no offset in secondary ion energy. Second row: offset 25 eV. Third row: offset 50 eV. Cut-off in bar, continued by dashed line: attenuation by a factor 10^3

beam and clean sample environment, contamination peaks arise from e.g. pump oil, ion optic lenses, and "memory" effects. Typical positive and negative mass spectra from practically pure silicon are seen in Figs. 3 and 4. The detectability of a particular isotope requires that its ion signal significantly exceeds the background level at its mass number. The factor α ($\leqslant 1$) in Eqs. (2) and (3) expresses the ratio of electronic noise to spectral background. To maximize α for the detection of species in doublet, triplet or multiplet peaks, isotope "peak stripping" (i.e. subtraction of background contributions as calculated from neighbouring isotope peaks) may sometimes be necessary. Modern SIMS equipment, however, offers two major facilities to eliminate background level, viz. high mass resolution and discrimination in secondary ion energy.

The mass resolution in most equipment available up to a few years ago was limited to about $M/\Delta M = 250$ in normal operation. Today many instruments are capable of $M/\Delta M$ of the order of 4000 without substantial loss of signal. Above this, however, the loss in transmission (η in Eqs. (1)–(3)) becomes rapidly more serious. Moreover, many significant doublets may require resolving powers in excess of 10000.

Although some elements may under particular circumstances be most sensitively detected as molecular ions (e.g. CN^-, BO_2^-, $CsZn^+$), normally one wishes to measure the atomic ion peak and to eliminate its molecular background. Here one may exploit the fact that the energy distribution of molecular species usually is considerably more narrow than that of monoatomic ions (see e.g. Refs.[11, 22, 23]). By imposing an offset in sample voltage, low energy ions may be prevented from reaching the mass spectrometer, so that practically only monoatomic ions are admitted to analysis. This valuable procedure may be conveniently effected in recent SIMS design. In Figs. 3, 4, the second and third rows of the Si spectra show the effect of two different offset voltages. E.g. in Fig. 4 one may see that the offset of 25 V reduces the signal of Si^- (i.e. impairs the effective transmission η) by approx. a factor of 8; that however e.g. for the detection of $^{56}Fe^-$ the suppression of Si_2 increases α by a factor 65; similarly, the suppression of the 3-atom Si_2O^- background results in a relative gain of as much as ca 5000 for $^{72}Ge^-$ or $^{75}As^-$.

d) Some Aspects of Quantitation

Basic features of quantitation in SIMS have received detailed treatment in recent reviews (see Refs.[13, 21]). In the present context, a few particular points in regard of practice and reproducibility may be of relevance. Via Eq. (1) one may define a "sensitivity factor" for isotope a, as

$$SF = (dI_s/dc)_a = i_p A \cdot (S_a \eta_a \gamma_a) \ . \tag{5a}$$

In practical analysis one most often utilises relative sensitivity factors, ρ_{ar}, i.e. normalised to a reference isotope or element r, whose concentration c_r in the matrix is known:

$$\rho_{ar} = (I_{sa}/I_{ra})(c_r/c_a) = S_a \eta_a \gamma_a / S_r \eta_r \gamma_r \ . \tag{5b}$$

As long as the S, η and γ terms may be assumed to be nearly constant within a range of varying c_a, one may obtain reliable concentration *profiles* in each sample; absolute calibration may moreover be obtained by means of a single external standard. Such circumstances may also allow more or less quantitative treatment of several impurities at a time via phenomenological models like the LTE formalism. Constant c_{ar} values have indeed been found to apply satisfactorily in wide concentration ranges for matrix systems such as binary alloys[16], steels or apatites[11]. However, in the general case γ_a/γ_r changes significantly with the concentration of each component; consequently many standards, calibration curves and complicated iterative procedures may be necessary in quantitation[21].

The deviations of S_a/S_r from unity are usually small in sputter equilibrium, but crystallographic effects may sometimes complicate the comparison of two samples. The transmission ratio η_a/η_r may be influenced i.a. by a) element dependence of exit angle of secondary ions; b) mass fractionization and counting deadtime of detector; c) surface charge build-up on insulating samples, and d) narrowness of acceptance window for energy of secondary ions. The latter factor is of importance in several types of SIMS equipment (e.g. several quadrupole instruments; also early design of ion microscope) where the energy passband may be some 15 eV or less. As different elements have different ion energy distributions, relative ion currents detected will depend on the position of the window[16]. A narrow passband not only entails low η values and irreproducible η_a/η_r ratios, but may, for insulating samples, totally prevent ion detection if the whole energy "drifts past" the acceptance window[25].

The most important irreproducibility in ρ_{ar} however follows, under most conditions, from the variations in γ_a/γ_r. As pointed out before, the degree of saturation with reactive species (such as oxygen or cesium) is a crucial parameter. Factors such as sputtering speed and residual atmosphere, for crystalline samples also the angle of beam incidence, must therefore be under control.

Although the point is obvious, it should be emphasised that an unknown sample and a reference sample should always be analyzed under near-identical experimental conditions. Particular care must be taken to distinguish separate phases (such as inclusions and precipitations), as calibration is likely to be different for each. This accentuates the usefulness of the image mode of SIMS even in quantitative studies.

In-Depth Resolution and Dynamic Range in Profiling

SIMS has an information depth of the order of one atomic layer, which to-gether with extreme detection sensitivity offers a unique potential in profil-ing. Aspects of depth resolution, dynamic range and sensitivity have been thoroughly discussed in recent reviews[26−28]. To precede the application examples of the following sections, the major artifacts should here be briefly mentioned.

a) Atomic mixing provides the ultimate limitation of depth resolution. Atoms involved in each collision cascade will "homogenise" its effective depth by "cascade mixing" (typi-cally ca 5 − 50 ÅU), in a smaller extent also by "recoil mixing" (up to some 200 ÅU)[26]. The mixing range is given by the energy and direction of the primary ions, and is not depth dependent.

b) Roughening, differential sputtering. Even on monocrystalline or amorphous samples, small irregularities of the original surface (ledges, facets, dislocations) entail local differ-ences in sputtering rate, and aggravate with milling depth. The effects (expresses e.g. as Δz, the broadening of a step-function) typically amount to 1 to 5% of the crater depth z in well prepared samples[29].

In single-phase polycrystalline samples, the local erosion rate is dependent on the angle between primary beam and the crystal axes. If the analysed area is of the same order as the crystal size, depth resolution may be as bad as ca 50% of z. If the area is much greater, the roughening is of the order of crystallite size.

Ion etching around local impurities and defects destroys depth resolution by inducing spectacular cones or ripples on crater bottom, even in single crystals[30]. When samples containing several phases with different hardness etc. are sputtered, all meaning of depth resolution may be lost, unless the equipment is able of positioning very small beams, or of selecting and gating very small areas.

c) Chemical effects; surfaces, interfaces. The factors v_{sp}, S and γ, presented above, are sensitive both to chemical composition and topographic discontinuities in the sample. Im-purities at surfaces, boundaries and other defects may cause considerable chemical en-hancement in ion yield. If reactive primary ions are implanted, equilibrium is not attained at depths less than the projected range. When an interface is being penetrated, earlier de-veloped roughening may affect not only resolution but also effective ionizability. In multi-layer or near-surface profiles the calibration of sputtering rates may be a relatively complicated procedure.

d) Crater-edge effects, neutral beam effects, redeposition, memory. Both resolution and dynamic range are impaired if ions are detected, originating from localities other than the bottom of the sputter crater; e.g. from 1) crater walls, 2) surface next to the crater, 3) residual atmosphere, 4) primary beam, or 5) ion-optic lenses close to the sample. To counter these effects, depth-profiling SIMS today tends to exploit the following facilities: 1) scanning of primary beam (providing flat crater bottom); extraction of ion for analysis only from central portion of crater (ion-optic or electronic gating); 2) suppression of neu-tral atoms entrained in the primary beam; surface coating; 3) optimal vacuum; 4) primary beam mass filtering; and 5) improved routines in cleaning and "baking" of ion-optic details.

e) Sensitivity of detection; dynamic range versus depth resolution. Slow sputtering furthers depth resolution by allowing numerous measuring cycles within a given depth interval. Likewise, as mentioned above, wide beam scan is often useful. However, this is at the expense of the signal-to-background ratio, which is the crucial entity for the attainable dynamic range. A maximisation of γ and α is therefore normally a primary concern in profiling. One very often employs a reactive gas leak and/or a reactive sputtering species, as this both enhances and stabilises ion emission, and is generally beneficial in limiting crystallographic and other adverse effects mentioned under *b)* and *c)* above.

Also at the upper concentration limits, high dynamic range profiling may encounter practical problems, due to the receptivity range of the collector and amplifier. In most applications of quantitative SIMS the effective counting deadtime must be taken into account.

Aspects of Surface Mapping, Imaging

Element distributions on sample surfaces may be mapped either by step or line scan of a small (or gated) ion beam, or by two-dimensional imaging. The latter may be effected by TV-type scan (ion microprobes[3, 31-33]), or by stigmatic ion micrographs (ion microscopes[1, 4, 34, 35]). The quality of such characterization is mainly given by *a)* lateral resolution of imaged detail or of smallest spot for practical quantitation; *b)* intensity of mass resolved ion signal; and *c)* contrast artifacts.

In ion microprobe development, the trend has always been to focus high ion currents onto very small sample areas. A theoretical limit has been assessed to lie at spot sizes around 200 ÅU, i.e. close to the lateral action range of a collision cascade[31]. By the development of electrohydrodynamic ion sources, beam diameters of a few hundred ÅU have in fact been achieved[36], while EHD beams recently used for SIMS purposes have been of the order of 0.5 μm[37]. In commercial microprobe design the lowest size limit appears to lie just under 1 μm. In ion microscopes the smallest beam diameters are about 3 μm, but since a few years areas as small as 0.5 μm may often be selected for convenient analysis, due to the introduction of an ion optical zoom system, "transfer optics"[34].

As seen above (Fig. 1), detection limits in SIMS are inversely proportional to analyzed area. Because of the relatively low η, γ, and α resources instrumentally available in the laterally resolving SIMS mode, its sensitivity has in recent past generally been by orders of magnitude lower than in bulk-analyzing or profiling SIMS. For ion micrographs, long exposures were often necessary, at the expense of topographic sharpness. In more recent equipment, lateral resolution has been improved considerably without essentially impairing η; at the same time effective transmission was much improved by image-amplifying multichannel-plate detectors. Further, the assets in regard of α, i.e. the facilities of energy offset and high mass resolution, are now also standard in the imaging mode.

Microscope contrast effects in SIMS, as well as faceting, preferential sputtering, chemical effects, etc., usually entail the need of evaluating ion micrographs in terms of relative rather than absolute local signal. This, together with abovementioned advances in sensitivity and lateral resolution, has stimulated recent efforts towards computer-assisted image processing in both ion microprobe[38] and ion microscope[35] applications.

Examples of Qualitative and Semi-Quantitative Applications

1) Phases of Al-Based Alloy

System: Al-alloy containing 7 % (wt) Si, 0.3 % Mg, 0.2 % Fe and 0.2 % Cr. In earlier SEM-examination, the following phases were identified: *A)* an Al-rich dendrite; *B)* a brittle β-phase, $Al_5Si(Fe_x Cr_y)$ with x and y less than unity; and *C)* a eutectic mixture of Al and Si.
Task for SIMS: Search for further phases and topographic details. Semi-quantitative determination of alloying elements in the resp. phases.

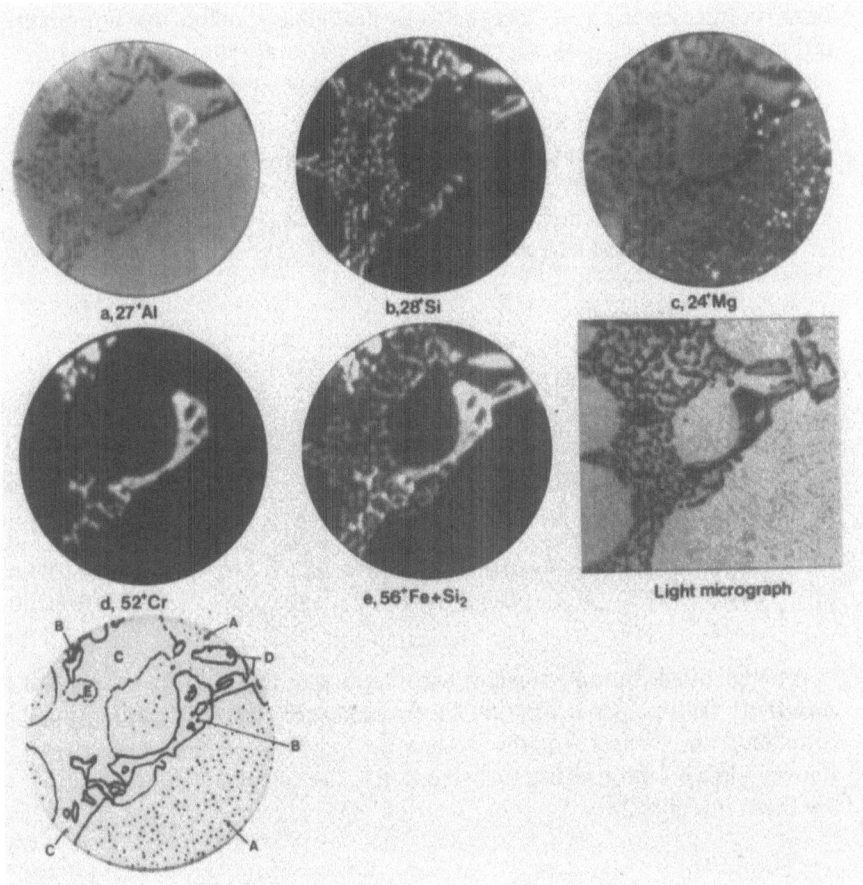

Fig. 5. Element distribution in an Al-base alloy. The ion mass micrographs depict an area 150 μm in diameter. Zone *A*: Al-rich dendrite. *B*: β-phase. *C*: Al-Si eutectic. *D*: Si-phase. *E*: porous region

Procedure

a) Ion micrographs were recorded of a representative surface spot, 150 μm in diameter. Fig. 5 shows the micrographs for the following ions: $^{27}Al^+$, $^{28}Si^+$, $^{26}Mg^+$, $^{52}Cr^+$, and $^{56}(Fe + Si_2)^+$. The primary ion beam (O_2^+, 4.5 kV, ca 3 μA) had a diameter of ca 200 μm and was rastered 300 x 300 μm. After SIMS, a light optical micrograph of the sputtered area was taken (Fig. 5f).

b) Secondary positive ion currents were recorded from areas 10 μm in diameter in each separate phase. The following mass numbers were of interest: ^{27}Al; ^{28}Si ; ^{24}Mg; $^{52}(Cr + AlMg)$; $^{51}AlMg$; $^{56}(Fe + Si_2)$; and $^{57}(Fe + Si_2)$. The molecular contributions on the 52 and 56 peaks were subtracted by using the 51 and 57 peaks for "isotope stripping". Mg_2 and MgSi contributions on the 52 peak were checked to be negligible in all phases. For quantitation reasons an O_2-leak was used (ca 10^{-5} Torr at sample surface). In Table 2, all counts are background- and isotope corrected.

Table 2. Aluminium-Base-Alloy. Recorded secondary ion counts (from 10 μm diameter spot) and semi-quantitatively calculated concentrations of Al, Si, Mg, Fe, and Cr in different phases (A_1: inner dendrite; A_2: outer dendrite; B: Fe-Cr-phase; C: eutectic Al-Si; D: Si-phase; WN: nominal total concentrations in alloy)

| | Counts/s ($x\ 10^5$) | | | | | Concentration, at% | | | | |
	A_1	A_2	B	C	D	A_1	A_2	B	D	WN
Al	230	230	280	140	9	98.9	98.4	70.0	≤4.5	92.7
Si	0.43	0.44	5.2	10.0	15.3	0.65	0.88	14.0	95.4	6.8
Mg	1.30	2.05	0.4	0.94	0.06	0.40	0.70	0.15	0.04	0.33
Fe	0.015	0.005	12.0	0.01	0.0015	0.01	0.01	10.5±2	≤0.002	0.10
Cr	0.048	0.07	12.3	0.045	0.002	0.03	0.05	5.5±2	≤0.002	0.10

c) A rough quantitation procedure was employed, based partly on available sensitivity factors, partly on the LTE formalism (T_i for nearly pure Al and Si taken from Table 1; for the β-phase T_i = 8650 K, obtained with the known Al/Si = 5 ratio as internal standard). The calculated concentrations are listed in Table 2.

Main Results

a) In addition to the A,B,C phases, a well defined phase, rich in Si, was identified (D in Fig. 5).

b) Fairly extensive porosities were noted in the eutectic phase (see *E* in Fig. 5).

c) In the Al-rich dendrite phase *A*, small dispersed particles containing Mg and Si were found, see Fig. 5c and b; these could also be found by SEM as crystallites some 50 ÅU in diameter, a few 100 Åu in length. As seen in Fig. 5c, they are abundant in the interior of the major dendrites, but much less common in the outskirts or in dendrites less than ca 20 μm across.

d) Semi-quantitative analysis described the *D*-phase as nearly pure Si.

e) In the β-phase, the concentration of Fe was found to be about twice that of Cr, corresponding to the approximate stoichiometry $x = 0.75$, $y = 0.4$ for Fe and Cr, respectively.

Remarks on Technique

The work is an example of routine application of imaging SIMS as a metallurgical tool. The recording of the micrographs took about half an hour, that of the ion counts in Table 2 less than 2 hours.

The high intensities of Al^+ necessitated (as ca 10^7 cts/s is the upper limit of detection by electron multiplier) that this be recorded with a defocused beam.

The isotope stripping procedure was required because the mass resolution of the instrument was only about 250 and no routine facility for offset in sample potential (see earlier section) was available. A SIMS equipment of latest design would have provided better convenience in this respect, and also permitted separate analysis at much smaller spot sizes than the 10 μm diam. in this example, with great practical use i.a. in line-scan across several phases.

II) Inclusions and Segregations in Steel

System: Steel with the following composition (by weight). 26.5 % Cr, 5.5 % Ni, 1.45 % Mo, 0.47 % Mn, 0.40 % Si, 0.15 % Ti, 900 ppm C, 550 ppm N, 200 ppm Al, 200 ppm B and 50 ppm S. Microscope examination reveals numerous inclusions of different sizes. Locally also two distinct zones are seen along grain boundaries, an inner zone 0 to ca 15 μm thick, an outer one stretching on each side up to ca 30 μm into the grains.

Task for SIMS: Identification of inclusions. Assessment of main element concentrations in the two phases along grain boundaries.

Procedure

Ion micrographs (150 μm diam. spot) and selective mass spectra (10 μm spot) were recorded using O_2^+ primary ions as in previous example. Both negative and positive secondary ions were utilized.

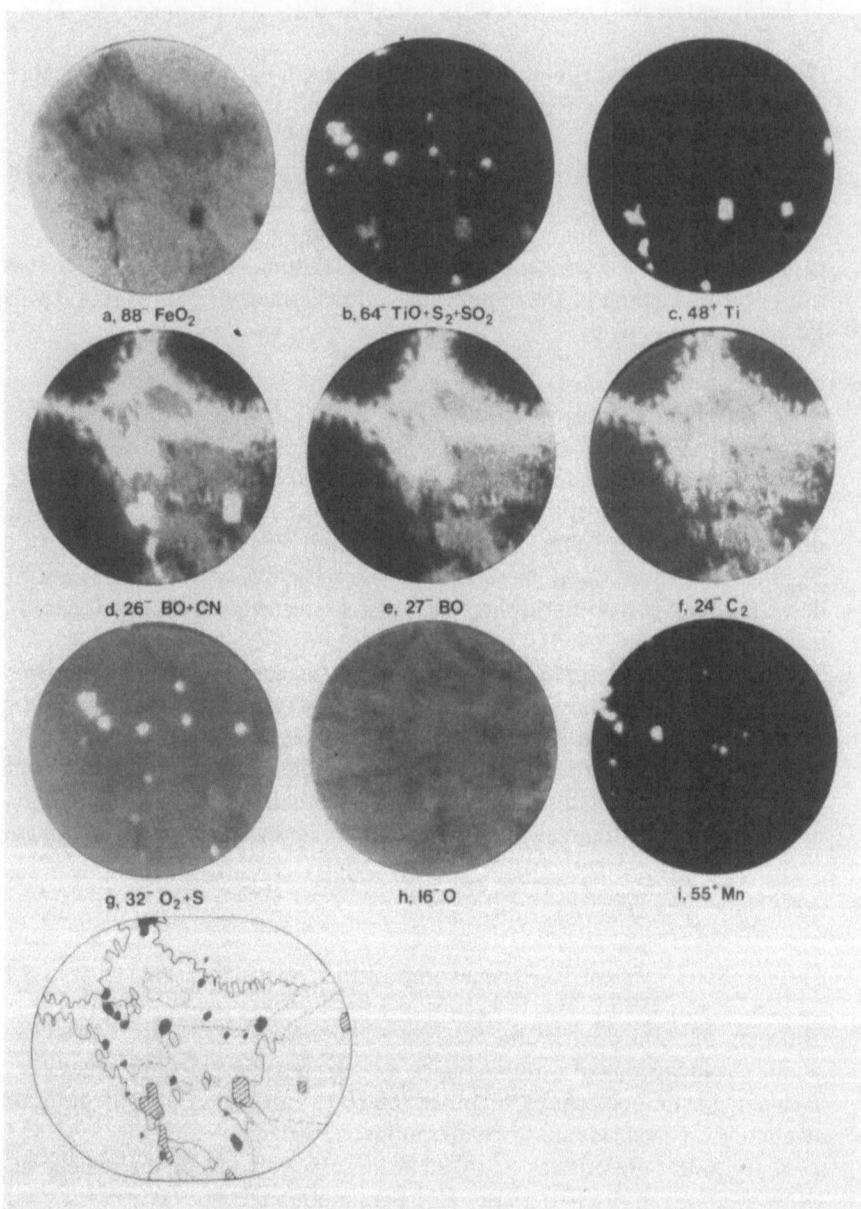

Fig. 6. Inclusions and boride phases in a stainless steel. The ion mass micrographs show an area 150 μm in diameter

Inclusions

In Fig. 6a (matrix element Fe), i.a., the typical inclusions are seen as dark contrast. Fig. 6b shows them to be of two classes with different ion yields. Fig. 6c indicates Ti to be represented in the relatively large (ca 3 – 30 μm) crystals. In addition to masses containing Ti, only $^{14}N^+$ shows the same topographic distribution. The nitrogen is particularly well detected by a comparison of 6d and 6e; while both show the identical topography of ^{10}B and ^{11}B, the 26 mass distribution also shows CN^-(by far the most intense N containing ion) on the sites of the large inclusions, which thus may be identified as TiN (subsequently checked by comparison of spectra with a standard sample). The other class of inclusions may be seen, by comparison of Figs. 6g and 6h, to contain sulphur. Nearly the same topography (the slight differences being due to continued sputtering) is revealed in the ion image of Mn^+, and the small particles may be identified as MnS.

Grain boundary phases: Figs. 6e and 6f show the morphological patterns of boron and carbon to be nearly the same. The quantitation of carbon from the C^+ signal is difficult, as the 1st ionization potential is much higher than that of the metallic elements and boron, and the uncertainty regarding T_i in the LTE formalism introduces large errors. In the case of C^-, C_2^-, and CN^-, quantitation, may lie between ca 1.3 and 13 atom percent in one case, between 0.9 and 9 in the other.

the differences in E_i are relatively small, and semi-quantitative analysis via the positive spectrum is possible (as checked by the good agreement with the given nominal concentrations in the steel). In Table 3, the calculated atomic concentrations are listed. The two grain boundary zones may be seen to be boride phases. The figures given for carbon may only be regarded as order-of-magnitude; actually the C concentration, according to uncertainties in the quantitation, may lie between ca 1.3 and 13 atom percent in one case, between 0.9 and 9 in the other.

Table 3. Concentrations (Atom %) of Principal Elements in Two Boride Zones and in Bulk of a Stainless Steel

Element	Inner zone at grain boundary	Outer zone at grain boundary	Bulk analysis by SIMS	Bulk composition, nominal
Fe	31.5 ± 4	44.1 ± 3.5	65.0	63.5
Ni	7.0 ± 1.5	11.9 ± 1.5	5.1	5.0
Cr	32.5 ± 10	28.5 ± 7.5	25.5	27.9
B	22.7 ± 4	9.1 ± 2	0.15	0.10
C	(4.1)	(2.9)	0.2	0.45

Remarks on Technique

The example shows the convenience of ion micrographs for rapid qualitative identification of phases and inclusions. In this study totally about 50 micrographs were recorded, at three different locations on the sample, utilizing both positive and negative spectra, molecular as well as atomic ion species. The entire imaging procedure took some 90 minutes. Owing to the sensitivity of the channel-plate collector, an exposure time of 1 second was nearly alsways sufficient, and surface topography was not essentially changed by sputtering. This safeguarded image sharpness and facilitated topographic comparison.

Non-metallic phases in metals are usually easily seen in ion micrographs, owing to superior ion emissivity. Even particles much smaller than the minimum area selected by gating (here 10 μm; in some recent equipment less than 1 μm) may therefore often be separately analysed.

As regards quantitative use of SIMS for steels, however, some limitations are set by the variety of phases, segregations and inclusions, by the complicated intrinsic spectrum, and by the great sensitivity of the main elements to chemical enhancement effect in ion emission. Among recent advances in SIMS instrumentation, beyond the facilities of the above example, the improvement in ion-optic magnification and the availability of cesium ion source appears of particular importance for steel studies.

Examples of In-Depth Profiling

I) Tracer Diffusion of Ge Isotopes in Silicon

System: Polished, (111)-oriented single crystal Si wafers (n-type, 500 cm), coated with vapor-deposited Ge film (8 to 200 ÅU), annealed at temperatures between ca 1150 and 1660 K.

Task for SIMS: Determination of the bulk diffusivity of Ge in Si, as well as its isotope effect. This is a part of a systematic investigation by SIMS of tracer diffusion in semiconductors[39-41]. The material of the present example has in part been published[42, 43].

Procedure

The depth profiles of Ge were measured in two ways. With the Cameca IMS-300 instrument, without offset facility, the $^{70}Ge^+$ and $^{30}Si^+$ currents were recorded alternatively, using O_2^+ primary ions and an O_2 gas leak. With a Cameca IMS-3f ion analyser all five isotopes of Ge^- were cyclically recorded, using also the $^{59}Si_2^-$ signal as a matrix reference; Cs^+ primary ions were used, and the secondary ions had an energy offset of 50 eV.

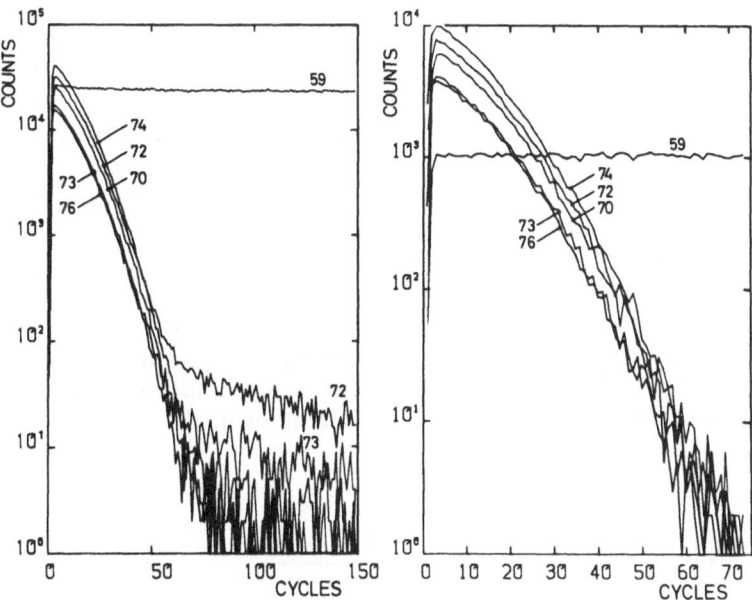

Fig. 7. Diffusion of the Ge isotopes in silicon. Raw data: in-depth profiles of Ge⁻. Primary ions: Cs⁺. Left: 25 eV offset in secondary ion energy. Right: 50 eV offset

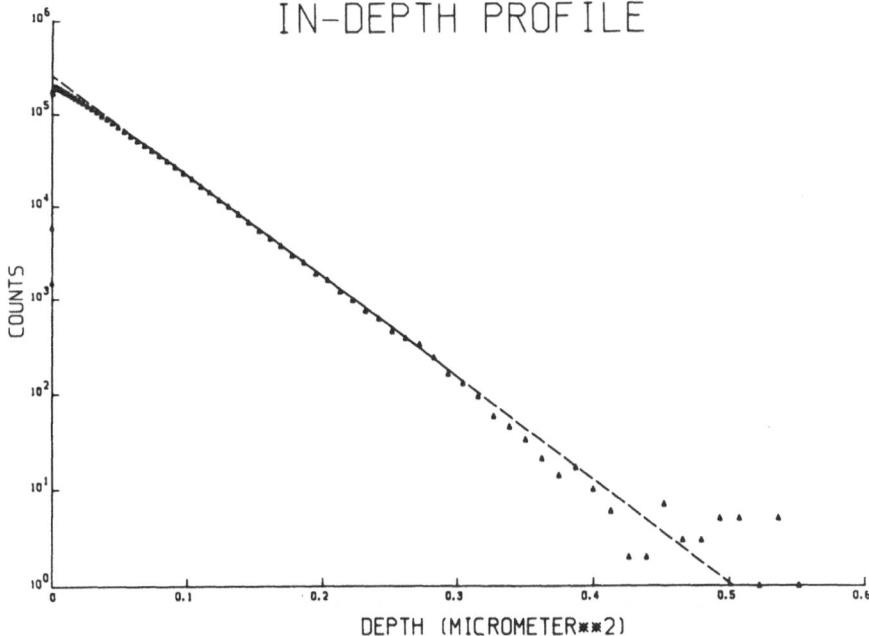

Fig. 8. Diffusion of ^{70}Ge in silicon at 1323 K. Ion counts, corrected for background and for counting deadtime, are plotted against square of depth. From slope: $D = (5.25 \pm 0.35)$ x 10^{-16}cm^2/s

The crater depths necessary to study the profile down to the background noise level in Ge ranged between ca 0.15 and 7.5 μm. The instrumental parameters (beam current, diameter and scan; aperture of gating) were always chosen so as to yield a dynamic range of at least two (for the isotope measurements four) orders of magnitude, at the same time permitting only ions from the flat central area of the crater (diameter less than half that of the crater) to be detected. Fig. 7 shows typical raw data for [70, 72, 73, 74, 76]Ge. The total surface concentration of Ge is here of the order of 1000 atom ppm. Only profiles with a nearly constant matrix signal were accepted.

Depth calibration was obtained (with an accuracy of the order of 2%) from mechanical measurement (Talysurf) of crater depth. Computer programs were designed to connect the running times of the cyclically repeated isotope peak heights with the corresponding penetration depths z.

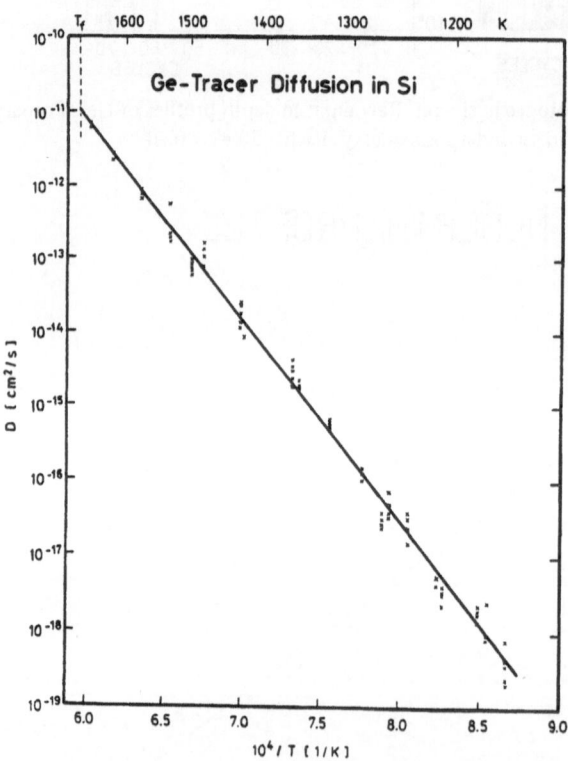

Fig. 9. Arrhenius plot of Ge diffusion in silicon, determined by SIMS

Main Results

Each diffusion coefficient was obtained, according to standard solution of
Fick's 2nd law for thin-film geometry, from the slope of the logarithmic plot
of c_{Ge} versus z^2. Fig. 8 illustrates the linearity and the dynamic range. In
Fig. 9 the diffusion coefficients are seen as measured by SIMS within
8 orders of magnitude. The activation energy, 5.335 eV, could be determined with an accuracy better than 0.9 %. Fig. 10 shows the isotope effect
of diffusion. The heaviest Ge isotope is seen to diffuse about 2 % slower than
the fastest. The relative accuracy of each individual D-measurement is here
better than ±0.4 %. This accuracy required relatively advanced electronics as
well as a computer program to account for the effects of amplifier deadtime,
cycling time algorithm, and background noise.

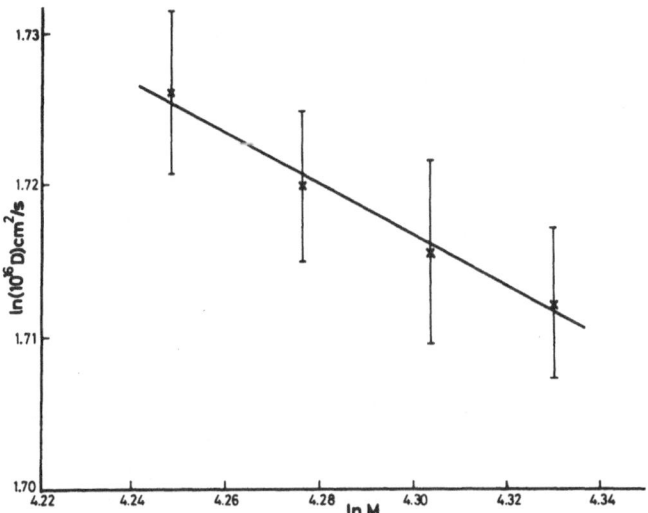

Fig. 10. Mass dependence of Ge diffusion in Si at 1323 K. The specific isotope effect is
$E = dlnD/dlnM = 0.33 \pm 0.07$

Remarks on Technique

It is clear from Fig. 6 that the relatively high offset, 50 eV, is necessary if all
five Ge isotope peaks are to be measured satisfactorily free from molecular
contamination (Si_2O, SiO_3). As may be seen by a glance at Figs. 4 and 5, in
the positive spectrum even higher offset would be needed, and available ion
counts would be much lower.

The example shows how the combination of tracer sensitivity and in-depth resolution, typical of SIMS, makes the technique particularly suited for quantitative measurements of diffusion. The lowest limit of D measureable in single crystals or amorphous specimens today, ca 10^{-20} cm^2/s, is essentially given by cascade mixing and by elementary surface discontinuities (kinks, ledges, etc.). Further, as regards measurements of isotope effects, the use of SIMS has expanded the range of possibilities considerably beyond that given by radioactive techniques.

II) Impurity Diffusion in Bulk and Along Grain Boundaries of Copper

System: a) Cu single crystal discs with vapor-deposited indium layers 100–250 ÅU thick; diffusion annealed at different temperatures. b) Cu bicrystals with symmetric tilt grain boundaries; In-layer ca 150 ÅU thick, deposited on the surface perpendicular to grain boundary; diffusion annealed at different temperatures.

Task for SIMS: a) Determination of the volume diffusion coefficients. b) Measurement of the penetration profiles of In along grain boundaries in order to determine the triple product of intergranular diffusion. Most of the material discussed in the example has been published[44, 45].

Procedure

The work was conducted with a Cameca IMS-300 ion probe, using O_2^+ as primary ions, recording the peaks of In$^+$ and Cu$^+$. For bulk diffusion at temperatures $\leqslant 873$ K, the procedure was essentially similar to that in the preceding example, involving cyclical profiling with beam scan and gating, and only one crater per profile. At higher temperatures, and for all grain-boundary profiles, the penetration depth of In was such as to necessitate, if studied by "head-on" profiling, craters $\geqslant 30$ μm in depth. For this reason, the surface was ground to a bevel angle, ca $1°$ to $5°$ with the original $z=0$ surface. The In penetration could then be obtained by step-scan along the beveled surface. As each step corresponded to a particular value of diffusion depth z, there was no restriction on sampling time, which safeguarded very good counting statistics and eliminated most of the common artifacts of shallow profiling. The diameter of each sampling spot was 60 or 100 μm and the step distance was of the order of 0.1 mm.

In the study of intergranular diffusion, the bicrystals had to be positioned symmetrically in the sample chamber, so that, despite two different crystal orientations, the erosion rate was the same on both sides of the boundary. This was effected by using the microscope mode of SIMS, i.e. visually comparing the brightness of the image on both sides of the monocrystal interface.

Main Results

The upper part of Fig. 11 shows the profiles of the background-corrected In^+
signal (Cu^+-normalised). In accordance with diffusion theory, straight lines
are obtained when $\ln c_{In}$ is plotted versus z^2 in bulk diffusion, versus $z^{6/5}$ in
grain boundary diffusion.
The Arrhenius representation of volume diffusion (bottom left in Fig. 11)
has a "mean" activation energy of 1.87 eV, but the expectionally wide tem-

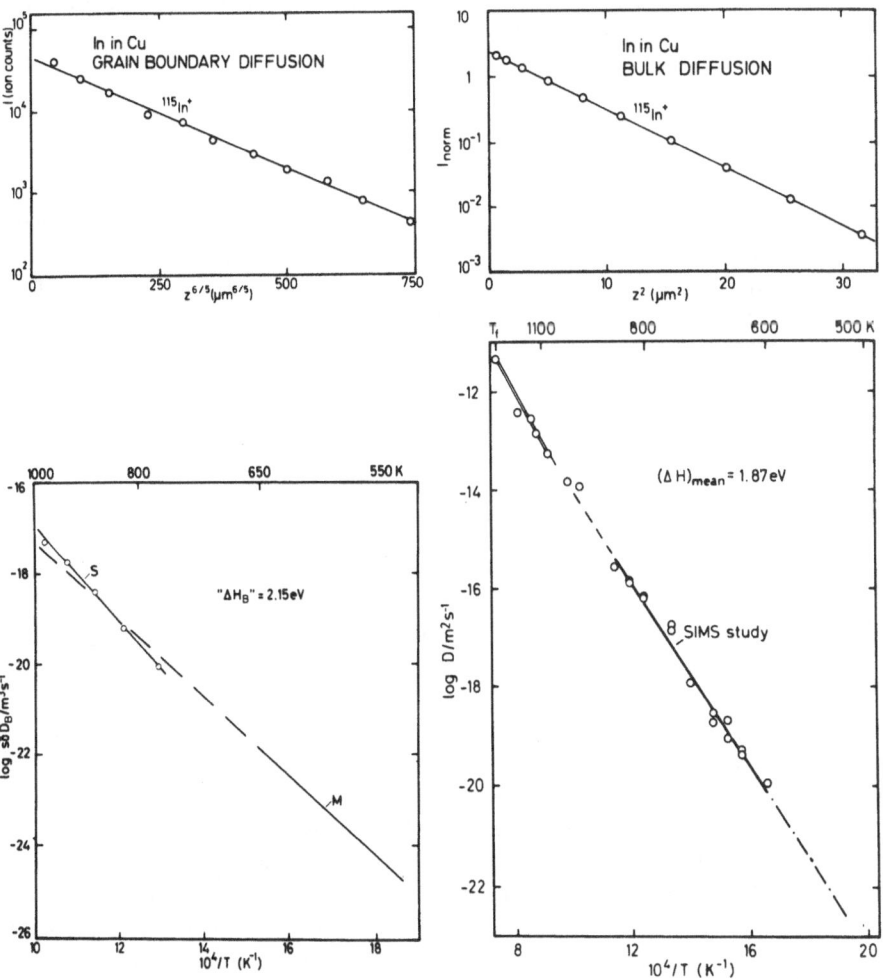

Fig. 11. Bulk and grain-boundary diffusion of In in copper. Above: corrected penetration
profiles of In tracer. Below: Arrhenius representations of the volume diffusion coefficient
and of the triple product of grain boundary diffusivity

perature range accessible for study by this technique makes it possible to distinguish a curvature, the slope in fact increasing by at least 3% throughout the range. Conclusions can be drawn regarding the contribution of divacancies to diffusion at high temperatures. Grain boundary diffusivity is represented by the triple product $s\delta D_g$, where the geometrical and distribution factors s and δ are likely to have a certain non-exponential temperature dependence. The effective activation energy of D_g therefore appears to be not very different from that of bulk diffusion. In the Arrhenius plot the line denoted by M is derived for mobile boundaries from experiments on discontinuous precipitation. The SIMS results suggest that in regard of diffusivity there is no essential difference between stationary and moving grain boundaries.

Remarks on Technique

The complementary uses of step-scan (along tapered surface) as well as head-on profiling make it possible to measure diffusivity quantitatively over exceptionally wide ranges of magnitude (here for bulk diffusion in principle between ca 10^{-7} and 10^{-19} cm^2/s). The good counting statistics achievable with step-scan also further the possibilities of isotope diffusion studies[46], but for this particular system the facility of high-voltage offset would be required.

Example: Glass Corrosion Study with Quantitative Profiling

System: Flat and polished block of borosilicate glass, containing i.a. simulated nuclear waste (see last row of Table 4). Corroded by 28 days' burial in Stripa[47] environment. The example is taken from a continuous study by SIMS of the corrosion and atom transport properties of glasses[48, 49].

Procedure

The corroded surface was provided with a 100 ÅU coating of Au to limit surface charge-up. As primary ions, O⁻ was used (ca 0.5 μA, beam diam. 50 μm, accelerated through 14.5 kV, scanned 80 x 80 μm). Positive secondary ions were collected from a central area 40 μm in diameter. The speed of sputtering varied between ca 5 and 15 ÅU/s, dependent on the degree of corrosion, and had to be determined separately for each depth layer. The profile was recorded by cyclic switching between the reference mass ^{29}Si and three other peaks. Thus by profiling in 4 adjacent craters, the following ions were studied: ^{7}Li, ^{11}B, ^{23}Na, ^{27}Al, ^{40}Ca, ^{56}Fe, ^{88}Sr, ^{133}Cs, ^{138}Ba, ^{139}La, ^{208}Pb,

Table 4. Concentrations (Atom % of Cations) at Three Depths in a Corroded Borosilicate Glass with Simulated Reactor Waste Products. Bulk Figures in Parentheses: Nominal Composition According to Ingoing Ingredients

	Si	B	Na	Fe	Al	Li	Cs	Ca	Ba	La	U
Surface	69.5	1.15	12.2	5.20	7.95	1.45	0.80	0.30	0.009	0.05	0.10
"Gel"	67.8	3.30	13.1	6.50	5.50	0.05	0.25	0.35	0.012	0.16	0.26
Bulk	41.2	26.1	20.8	3.85	3.25	0.16	0.32	0.01	0.10	0.14	0.24
	(42.5)	(24.5)	(21.9)	(3.75)	(3.20)	(---)	(0.33)	(---)	(0.15)	(0.20)	(0.30)

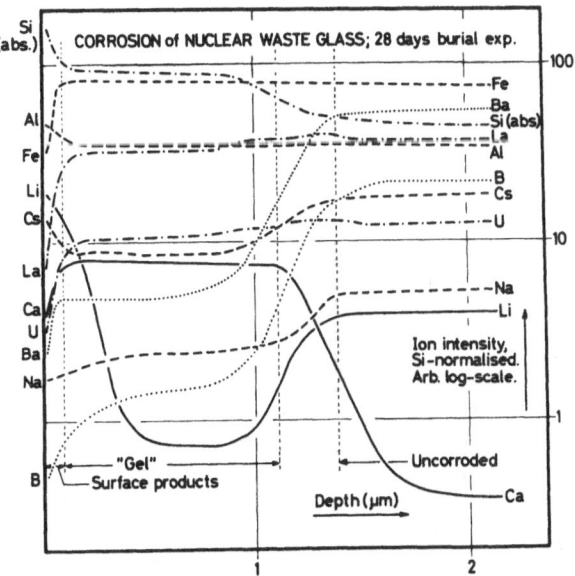

Fig. 12. SIMS in-depth profiles of components and of simulated nuclear waste in a leached borosilicate glass. Primary ions: O⁻. Secondary ions: positive. The ion currents of all elements except silicon are represented relative to Si^+

and ^{238}U. All peaks in the spectrum were checked to be well resolved and flat-topped. After profiling, the relative ion yields at selected depths were converted to the resp. element concentrations by means of established sensivity factors for SIMS of glasses.

Main Results

Fig. 12 shows the variations of the relative ion currents $I(E^+)/I(Si^+)$ with depth for 10 elements E. Not shown here are the profiles for Sr^+ (behaving similarly as Ba^+) and Pb^+ (present in concentrations less than ca 0.005 at.% throughout). The changes seen in the Si^+ signal reflect the sputtering speeds in the different zones. One may see the typical "gel" zone, here about 1 μm thick, where boron and the alkali metals are impoverished by atom exchange with the corrosive environment; on the other hand Ca is seen to have entered the gel from the surrounding clay. The outermost region, within ca 0.1 μm from the surface, contains partly non-glassy corrosion products. At depths beyond ca 2 μm all concentrations approach those of bulk glass, as is also seen in Table 4, which in addition to the quantitized results of the SIMS analysis also lists the composition as given by the proportions of different oxides used when making the glass. One may also see that the surface and gel regions are enriched in Al and Fe, which is thought to promote resistance to further corrosion.

Remarks on Technique

The procedure sketched above has during the last few years proved its worth as a sensitive and convenient tool in the study of glass corrosion. The total profiling time to 2 μm for the 13 elements in this example was about 90 minutes. The sensitivity for most elements is in the ppm range even in glasses of very complicated composition. The efficacy of the routine could be further enhanced by some improvements inherent in 2nd generation SIMS equipment, such as software permitting an unlimited number of peaks in each profiling cycle (cf. 4 peaks above), and electron flooding facility to neutralize surface charging and stabilize sample voltage at increasing depths.

The emission of positive ions from glasses is relatively abundant and stable due to the stoichiometric oxygen. Results obtained by the above procedure have shown good reproducibility. One prerequisite seems to be the wide energy pass band of the instrument; glass analysis in quadrupole type SIMS appears to entail some uncertainties in regard of sensitivity factors[50]. Notwithstanding problems of surface charging and other artifacts, calibration by means of external standards has been performed on a considerable range of glasses[49-51], and the results have been shown to conform rather well to the LTE formalism.

However, glass may exhibit considerable local fluctuations in composition. The surface of corroded glass, in particular, may contain inclusions or precipitates, rendering any quantitation based on glass standards more or less meaningless. As an example, Fig. 13 shows ion micrographs of a glass surface (imaged just after the penetration by the primary beam of the vapor-depos-

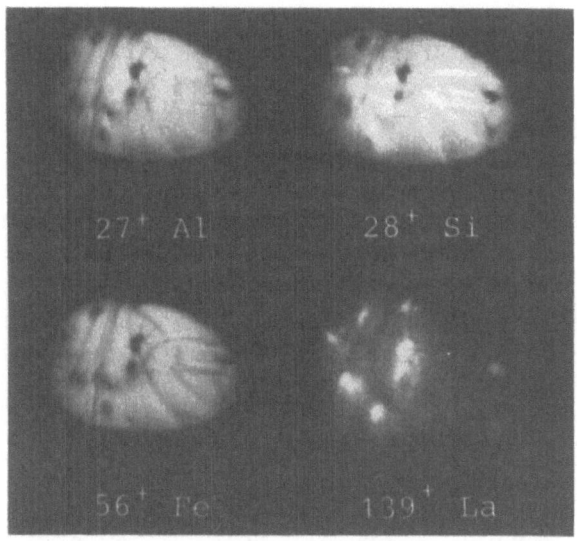

Fig. 13. Ion mass micrographs of the surface of a nuclear waste glass corroded in carbonated water. Length of imaged spot: ca 130 μm

ited Au layer), severely corroded by water in the presence of CO_2 [52]. The outermost glass is seen to contain mainly Al and Fe; the silicon is much impoverished, and the Si^+ signal is mainly due to the presence of cracks. As precipitates are some 10 μm in size, one may see a surface phase containing La (according to detailed study also Y and other rare earths). The quantitation of the surface zone is here hardly a routine task. The example shows some of the risks inherent in profiling of technical materials, and emphasizes the usefulness of the imaging mode of SIMS as a standard precaution and complement to in-depth studies.

Acknowledgements

We are indebted to the Swedish Board of Technical Development (STU) for financial support to several of the above-discusses applications. We wish to acknowledge the courtesy of the Cameca company for placing an IMS-3f instrument at our disposal for the silicon studies. The operational expertise of Mr H.-N. Migeon was invaluable in this context. We thank the Sandvik AB and AB Volvo for their permission to report some hitherto unpublished work. Thanks for valuable help and discussions are due to docent G. Dunlop, civ.ing. G. Andersson, Dr A. Salwén, civ.ing. U. Södervall, and Dr T. Thorvaldsson.

Summary

Applications of SIMS in Interdisciplinary Materials Characterization

The main assets and artifacts of secondary ion mass spectrometry are discussed mainly in relevance to practical microanalysis, profiling and imaging, as applied to different classes of technological materials. Limitations in element detection sensitivity, in topographic resolution and in apparatus development are illustrated by examples of different application of SIMS on steel, aluminium, copper, silicon and borosilicate glass, involving both surface mapping and in-depth profiling.

References

1. R. Castaing and G. Slodzian, J. microscopie 1, 395 (1962).
2. H. J. Liebl and R. F. K. Herzog, J. Appl. Phys. 34, 2893 (1963).
3. H. J. Liebl, J. Appl. Phys. 38, 5277 (1967).
4. J.-M. Rouberol, J. Guernet, P. Dechamps, J. Dagnot, and J.-M. Guyon de la Berge, Proc. 5th Int. Conf. X-Ray Optics & Microanalysis, 311. Berlin-Heidelberg-New York: Springer-Verlag. 1969.
5. I. W. Drummond and J. V. P. Long, Nature 215, 950 (1967).
6. C. A. Evans, Jr., Analyt. Chemistry 44, 67A (1972).
7. Secondary Ion Mass Spectrometry. K. F. J. Heinrich and D. E. Newbury (eds.), NBS Spec. Publ. 427, US Nat. Bureau of Standards. 1975.
8. H. W. Werner, in Applied Surface Analysis, ASTM STP 699. T. L. Barr and L. E. Davis (eds.), p. 81–110, Amer. Soc. Testing & Mat. (1980).
9. SIMS II. A. Benninghoven, C. A. Evans, R. A. Powell, R. Shimizu, H. A. Storms (eds.), Springer Ser. Chem. Phys. 9. Berlin-Heidelberg-New York: Springer-Verlag. 1979.
10. SIMS III. A. Benninghoven, J. Giber, J. László, M. Riedel, H. W. Werner (eds.), Springer Ser. Chem. Phys. 19. Berlin-Heidelberg-New York: Springer-Verlag. 1982.
11. A. Lodding, Rev. on Analyt. Chemistry, L. Niinistö (ed.). Budapest: Akademiai Kiadò. 1982.
12. A. Benninghoven, to be published.
13. H. W. Werner, Surf. & Interf. Analysis 2, 56 (1980).
14. C. A. Andersen and J. R. Hinthorne, Analyt. Chemistry 45, 1421 (1973).
15. A. Lodding, to be published.
16. M. Bernheim and G. Slodzian, J. Microsc. Spectrosc. Elec. 6, 141 (1981).
17. V. R. Deline, in Ref.[9], p. 48.
18. C. A. Andersen, in Ref.[7], p. 79.
19. D. S. Simons and D. E. Newbury, 3rd Int. SIMS Conf., Budapest, 1981.
20. H. Oechsner, in Ref.[10], p. 106.
21. G. Slodzian, in Ref.[10], p. 115.
22. J. A. McHugh, in Methods and Phenomena, Surface Analysis. S. P. Wolsky and A. W. Czanderna (eds.). Amsterdam: Elsevier. 1975.
23. J. C. Lorin, A. Havette, and G. Slodzian, in Ref.[10], p. 140.
24. V. Leroy, J. P. Servais, and L. Habraken, Centre Recherche Metallique, Liège 35, 69 (1973).

25. Ch.W. Magee and W.L. Harrinton, Appl. Phys. Lett. 47, 1232 (1976).
26. Ch.W. Magee, R.E. Honig, and C.A. Evans, Jr., Ref.[10], p. 172.
27. J. Hofmann, in Ref.[10], p. 186.
28. W.O. Hofer and U. Littmark, in Ref.[10], p. 201.
29. M.P. Macht and V. Naundorf, J. Appl. Phys. 11, 7551 (1982).
30. P. Dorner, W. Gust, M.B. Hintz, A. Lodding, H. Odelius, and B. Predel, Acta Metall. 28, 291 (1980).
31. H. Liebl, in Ref.[9], p. 176.
32. B.L. Bentz and H. Liebl, in Ref.[10], p. 30.
33. F.G. Rüdenauer, P. Pollinger, H. Studnicka, H. Gnaser, W. Steiger, and M.J. Higatsberger, in Ref.[10], p. 43.
34. G. Slodzian, in Ref.[7], p. 33.
35. G. Morrison, in Ref.[10], p. 233.
36. R. Seliger, J.W. Ward, V. Wang, and R.L. Kubena, Appl. Phys. Lett. 34, 310 (1979).
37. P.D. Prewett and D.K. Jefferies, Inst. Phys. Conf. Ser. 54, 316 (1980).
38. W. Steiger, F.G. Rüdenauer, H. Gnaser, P. Pollinger, and H. Studnicka, Mikrochim. Acta [Wien], Suppl. X, 1983, 111.
39. P. Dorner, W. Gust, A. Lodding, H. Odelius, and U. Roll, Acta Metall. 30, 941 (1982).
40. A. Lodding, H. Odelius, and U. Södervall, in Ref.[10], p. 351
41. E. Janzén, A. Lodding, H. Grimmeiss, and Ch. Deline, J. Appl. Phys. 11, 7367 (1982).
42. P. Dorner, W. Gust, A. Lodding, H. Odelius, B. Predel, and U. Roll, accepted for publ. in Philos. Mag. (1983).
43. U. Södervall, U. Roll, B. Predel, H. Odelius, A. Lodding, and W. Gust, Proc. DIMETA, Tihany (1982).
44. W. Gust, C. Ostertag, B. Predel, U. Roll, A. Lodding, and H. Odelius, Phil. Mag. 47, 395 (1983).
45. W. Gust, A. Lodding, H. Odelius, B. Predel, and U. Roll, Proc. DIMETA, Tihany (1982).
46. Th. Hehenkamp, A. Lodding, H. Odelius, and V. Schlett, Acta Metall. 27, 827 (1979).
47. L.L. Hench, L.O. Werme, and A. Lodding, in Scientific Basis for Radioactive Waste Management, Vol. 5, Berlin, 1982.
48. L.O. Werme, L.L. Hench, J.-L. Nogues, and A. Lodding, J. Nucl. Mat., in press.
49. L.L. Hench and D.E. Clark, J. Non-Cryst. Sol. 28, 83 (1978).
50. P.J. Hayward, E.V. Cacchetto, W.H. Hocking, and F.E. Doern, in Scientific Basis for Radioactive Waste Management, Vol. 5, Berlin, 1982.
51. A.E. Morgan, Surf. & Interf. Analysis 2, 123 (1980).
52. D.E. Clark, A. Lodding, and L.O. Werme, unpublished.

Mikrochimica Acta [Wien], Suppl. 10, 51–60 (1983)
© by Springer-Verlag 1983

Fachbereich Physik der Universität Kaiserslautern, Kaiserslautern, Federal Republic of Germany

Quantitative Secondary Neutral Mass Spectrometry Analysis of Alloys and Oxide-Metal-Interfaces

By

K. H. Müller and H. Oechsner

With 6 Figures

(Received January 19, 1983)

1. Introduction

For the chemical analysis of a solid sample a number of physical methods have been developed besides the classical, mostly time consuming methods of wet chemistry. Such analytical techniques are e.g. the optical emission and absorption spectroscopy or X-ray fluorescence, but also more recent techniques as X-ray Photoelectron Spectroscopy (XPS), Auger Electron Spectroscopy (AES), Secondary Ion Mass Spectroscopy (SIMS), Ion Scattering Spectroscopy (ISS) and others. This second group of analytical methods is of particular interest because of the possibility to determine the chemical composition at the very surface of the sample, and to obtain concentration depth profiles when the sample material is removed by sputtering.

Quite recently Secondary Neutral (or Sputtered Neutral) Mass Spectroscopy SNMS has been introduced as a novel analytical method by which the composition of the topmost atomic layers of a solid as well as its bulk composition can be determined quantitatively[1]. In the present work, a brief description of the principle of SNMS is followed by some examples for the application of SNMS which demonstrate the potential of this method for chemical bulk analysis and depth profiling.

2. Principle of SNMS

In SNMS neutral atoms and molecules being removed from a solid surface by
ion bombardment are ionized when passing a hot Maxwellian electron gas
before they are analyzed by a mass spectrometer[2]. The ionizing electron gas
is established by the electron component of a special resonant low pressure
high frequency plasma[3]. Using Argon, electron (or plasma) densities of
some $10^{10} cm^{-3}$ are achieved at Ar-pressures in the 10^{-4} mbar regime at electron temperatures corresponding to $10 - 15$ eV.
The ion bombardment of the sample is performed either by a separate ion
gun (Fig. 1)[4, 5], or by using directly the ion component of the plasma serving
for the space charge compensation in the ionizing chamber[6] (Fig. 2). In the
operation mode with the external ion gun it is possible to vary the bombarding current density at the sample over several orders of magnitude while stationary plasma conditions for optimum postionization efficiency are established. When the ion component of the postionizing plasma is used for the

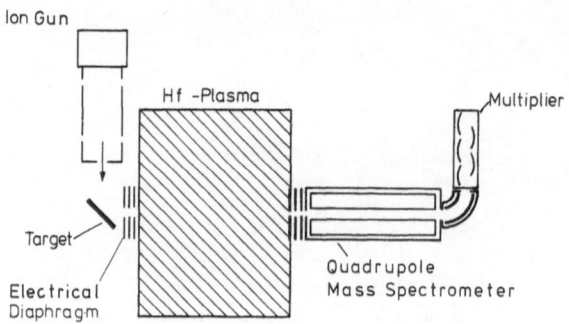

Fig. 1. Schematic diagram of the SNMS system with separate ion gun

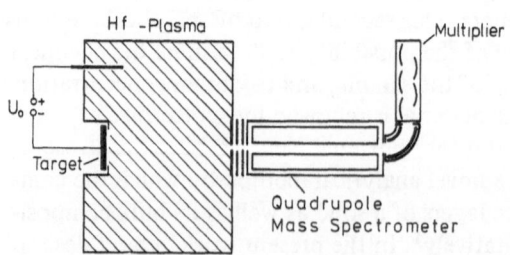

Fig. 2. Schematic diagram of the SNMS system for the "direct bombardment mode"
(target bombardment by ions extracted from the postionizing plasma)

bombardment of the analyzed surface, a high bombarding current density in the order several 10^{-3} A cm^{-2} is achieved. However, in this "direct bombardment mode", the bombarding energy can be varied from values as low as some 10 eV up to several keV. According to the energy dependence of the sputtering yield, the removal rate at the sample can consequently also be changed over a large range. At low bombarding energies, in particular bombardment induced interfering effects as cascade mixing are minimized which simplifies the evaluation of depth profiling measurements considerably. In the direct bombardment mode shown in Fig. 2 low energy ion bombardment with high lateral homogeneity is possible by means of an appropriate ion optical arrangement[7] producing extremely flat floors of the bombarding craters. In connection with the low bombarding energy this results in a very high depth resolution.

3. Experimental

The SNMS investigations presented in this work have been performed by the direct bombardment mode, schematically shown in Fig. 2, under ultra high vacuum conditions (base pressure in the 10^{-10} mbar regime). The specimen have been transported from normal atmosphere into the analysis chamber by a sample introduction system without breaking the UHV conditions in the equipment. During the SNMS analysis the plasma was operated in Argon at $8 \cdot 10^{-4}$ mbar. The current densities of the bombarding Ar^+-ions were around $2 \cdot 10^{-3}$ A/cm^2. The diameter of the analyzed circular area of the sample was 6 mm. Mass analysis was performed with a quadrupole mass spectrometer with attached ion optics in front. A conventional multiplier-rate meter or particle counting system was used for particle detection. These parts of the SNMS arrangement were identical with those of a commercial SIMS apparatus (SIMS 100 Leybold-Heraeus, Köln).

4. Quantification of SNMS

An SNMS signal $I(X^o)$ of a neutral sputtered species X is given by

$$I(X^o) = I_p \ Y_x \ \underbrace{\alpha_x^o \ \eta_x^o}_{\text{apparatus constant } D_x^o} \tag{1}$$

with Y_x partial sputtering yield of the species X

I_p bombarding ion current

α_x^o postionization coefficient for X

η_x^o geometry and transmission factor for X.

Since contrary to SIMS the emission and the ionization process of the analyzed surface particles are decoupled in SNMS, the postionization probability α_x^o is determined by the experimental conditions via the plasma parameters. Therefore, the product $\alpha_x^o \ \eta_x^o$ from Eq. (1) can be taken as one apparatus constant D_x^o, being the detecting coefficient for the species X in the particular SNMS system. For atomic sputtering of a nonelementary specimen we obtain under stationary conditions, i.e. for sputter equilibrium

$$Y_x = C_x^{bulk} \ Y_{tot} \tag{2}$$

with $$Y_{tot} = \Sigma \ Y_x. \tag{3}$$

Eqs. (1) and (2) can be combined to

$$\frac{C_i^{bulk}}{C_j^{bulk}} = \frac{I \ (X_i^o) \ D_j^o}{I \ (X_j^o) \ D_i^o} \tag{4}$$

Using

$$\Sigma \ C_x^{bulk} = 1 \tag{5}$$

the atomic bulk concentrations C_x^{bulk} of the sample can be readily obtained from the ratios of the SNMS signals when the relative apparatus constants D_i^o/D_j^o are known. These values can be easily determined experimentally by measuring the SNMS signals from elementary samples or nonelementary specimen with well known composition under the same operation parameters which are used during the SNMS analysis. A series of measurements with nonelementary reference samples of different composition have been performed in order to control if the ratios D_i^o/D_j^o depend on the individual matrix from which the particles indexed by i and j are ejected. For the relatively large number of nonelementary metallic examples investigated so far, no "matrix effects" have been found which had to be taken into account for the evaluation according to equ. 4. Also for large differences of the concentrations of the elements i and j in different matrices practically identical D_i^o/D_j^o ratios are obtained. An example is given in Table 1: The ratios of the apparatus constants D_x^o for Ni and Cu atoms which have been sputtered from an Fe matrix with 0.86 at% Ni and 0.6 at% Cu, and in a second case from an NiCu-alloy with 46 at% Ni and 54 at% Cu agree to better than 1.5%. Such deviations are within the accuracy of the composition of the reference sample.

For oxides of various materials such as high Z metals like Ta, W, Nb ..., the SNMS spectra contain peaks of the monoxide molecules, the intensity of

Table 1. Comparison of Relative Sensitivity Factors D^o_{Ni}/D^o_{Cu} of SNMS Obtained for Different Matrices

Sample	ratios of bulk concentrations C_{Ni}/C_{Cu} in the sample	SNMS peak ratios $I(Ni^o)/(Cu^o)$	relative sensitivity factors D^o_{Ni}/D^o_{Cu}
Fe-alloy with C_{Ni} = 0.86 at% C_{Cu} = 0.6 at%	1.43	2.05	1.43
Ni/Cu alloy C_{Ni} = 46 at% C_{Cu} = 54 at%	0.85	1.2	1.41

which is comparable to that of the atomic metal peaks[8]. In this case the assumption of purely atomic sputtering and therefore Eq. (2) is not valid. According to a model for the formation of sputtered metal monoxide molecules MeO ("direct emission model")[8, 9], the SNMS intensity I (MeO^o) is quantitatively related to the oxygen concentration at the surface of the sample[5, 9]. This allows an internal calibration of SNMS when, for instance, the oxygen surface concentrations at oxide samples have to be measured[5].

5. Quantitative Chemical Bulk Analysis of an FeCr-Alloy by SNMS

In Fig. 3 an SNMS spectrum is shown which has been measured at a metallic sample of unknown composition under normal bombardment by Ar$^+$-ions of 800 eV. From a gross overview the sample can immediately be identified qualitatively as an FeCr-alloy containing a number of other constituents within a wide concentration range. Several parts of the SNMS spectrum are additionally recorded with increased sensitivity. From the molecular peak intensities obviously the assumption of atomic sputtering, being made for the evaluation procedure in sect. 4, is well fulfilled in the present case.
Using a set of previously measured relative detection coefficients D^o_{Fe}/D^o_i, the absolute atomic concentrations C_i in the unknown sample have been determined according to Eqs. (4) and (5) from the normalized peak heights $I(X^o_i)/I(Fe^o)$ taken from the SNMS spectrum in Fig. 3. (The Fe^{54} peak was used in the present case, taking into account the natural isotope ratio of Fe.) The SNMS-results for C_i are given in Table 2. Simultaneously a comparison with the atomic concentrations C_i obtained by a subsequent wet chemistry

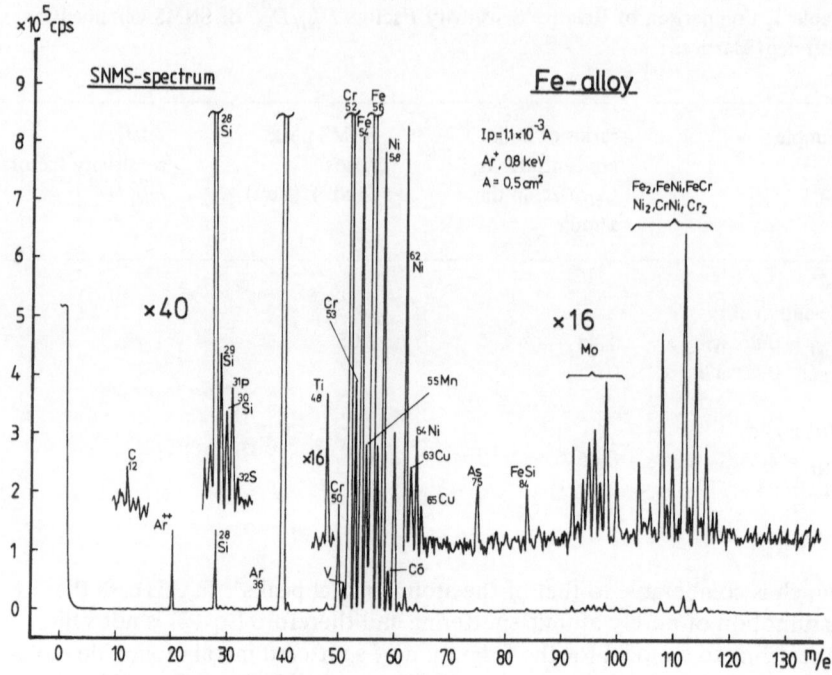

Fig. 3. SNMS spectrum of an Fe-alloy (with Cr as the second main component). The spectrum has been obtained by normal bombardment with Ar$^+$-ions of 0.8 keV in the direct bombardment mode shown in Fig. 2. (Bombarding Ar$^+$-current 1.1 mA, bombarded sample area 0.5 cm^2.)

analysis of the sample is presented. Both sets of C_i agree very satisfactorily, thus demonstrating the potential of SNMS for quantitative chemical analysis.

At the present technical state of SNMS equipments, concentrations in the ppm regime can be determined for all elements. Since all D_i^o/D_j^o ratios determined so far are in the order of unity, we conclude that the SNMS detection sensitivities are comparable for most of the elements.

In order to demonstrate the presently obtained detection limits of SNMS, the SNMS spectrum of an Fe-alloy containing 0.09 at% Ni is shown in Fig. 4. The mass range of the Ni-isotopes has been investigated with increased sensitivity (right hand section of Fig. 4). It is seen that the intensity of the Ni^{62} peak corresponding to an atomic concentration of 34 ppm amounts to $2 \cdot 10^4$ cps. Hence, even for a background level around 10^2 cps concentrations of 1 ppm are safely detectable.

Table 2. Composition of an FeCr Alloy in at% as Obtained by SNMS and by Wet Chemistry Analysis

Element	from SNMS spectrum	from wet chemistry
C	0.156	0.156
Si	0.76	0.765
Mn	1.37	1.54
P	0.051	0.043
S	0.0069	0.0086
Cr	18.2	18.9
Ni	8.73	8.91
Mo	0.13	0.16
Cu	0.128	0.104
Co	0.35	0.31
Ti	0.036	0.035
V	0.13	0.12
Fe	69.9	69.0
As	0.03	

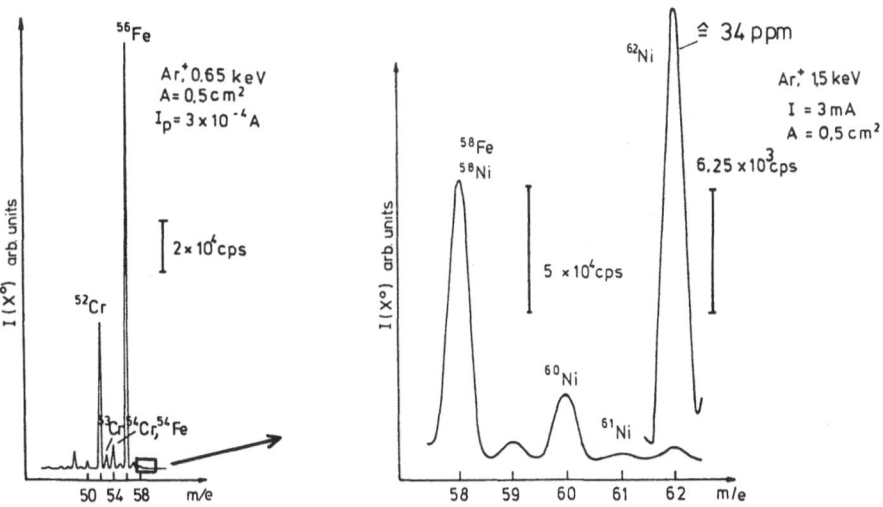

Fig. 4. Section of the SNMS spectrum of an Fe Cr-alloy. The mass range covering the Ni isotopes has been recorded with increased sensitivity for demonstrating the detection limit of SNMS to be $\leqslant 1$ ppm. (Direct bombardment mode with normally incident Ar^+-ions; bombarding parameters as shown in the figure.)

6. Depth Profiling by SNMS

Depth dependent concentration profiles of solid samples can be determined by SNMS with very high depth resolution because of the advantages mentioned in sect. 2, namely sputter removal by very low energy ions and high lateral homogeneity of the bombarding ion current[7].

As an example, an SNMS sputter depth profile of an anodic Ta_2O_5 layer (thickness 3360 Å) on polycrystalline Ta is shown in Fig. 5. The variations of the concentrations of tantalum and oxygen with depth are displayed by the behaviour of the SNMS signals for Ta and TaO plotted on a logarithmic scale versus the sputtering time. For a more precise determination of the Ta_2O_5/Ta interface the SNMS signals $I\,(TaO^0)$ and $I\,(Ta^0)$ in the transition interval are presented in Fig. 6 on a linear scale.

In comparison to previous SNMS depth profiles on the same samples[6] the bombarding energy for optimum depth resolution[7] has been reduced to a value of 202 eV for the normally incident Ar^+-ions. The 90 to 10% interface

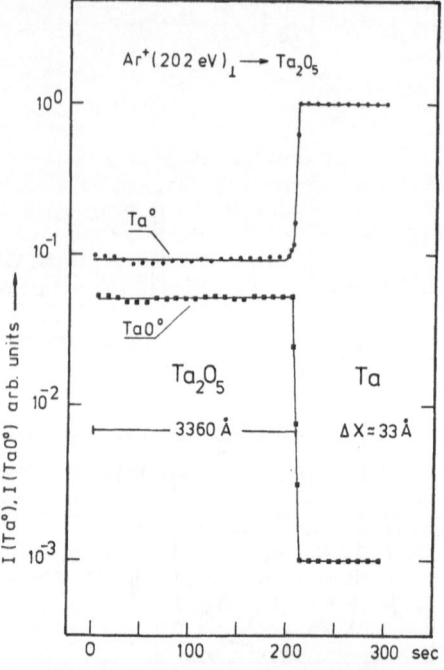

Fig. 5. SNMS depth profile of a Ta_2O_5 layer produced by anodic oxidation on polycrystalline Ta. The SNMS signals of the neutral Ta and TaO particles ejected under normal bombardment with Ar^+-ions of 202 eV are plotted on a logarithmic scale versus the sputtering time. (Thickness of the Ta_2O_5 layer 3360 Å.) The interface width Δx between Ta_2O_5 and Ta is determined to be 33 Å from the variation of $I\,(TaO^0)$ and $I\,(Ta^0)$

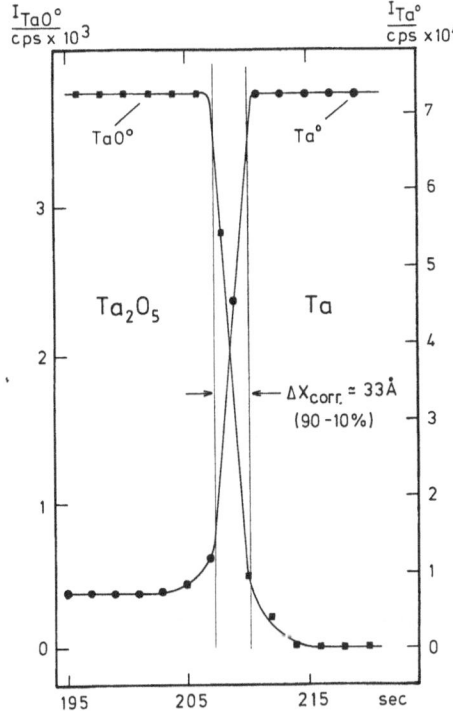

Fig. 6. Linear representation of the SNMS intensities from Fig. 5 in the transition range between the oxide film and the metallic substrate. Δx is taken as the width corresponding to the 90 to 10% decay or increase of the signals

width Δx has been found to be 33 Å which agrees with the best Δx values obtained by AES sputter depth profiling of the same interfaces[10]. For the evaluation of the SNMS profile in Figs. 5 and 6 the influence of the different sputtering yields[8] and the mean atomic volumes of Ta_2O_5 and Ta have been taken into account.

7. Conclusions

When comparing SNMS with other techniques for chemical bulk or surface analysis, some characteristic differences have to be mentioned. Contrary to SNMS the majority of the other methods used for bulk analysis are more or less matrix dependent thus requiring sometimes detailed reference measurements with special standards. This is particularly true for optical emission spectroscopy, where the detectors for different elements show different detection sensitivities which depend additionally on the individual technique. Information on compounds can be scarcely obtained in this case.

Amongst the methods for chemical surface analysis, SNMS and SIMS are close to each other not only from their principles. In both methods those particles which are removed from the surface are analysed, while electron spectroscopy techniques like XPS or AES, and ion backscattering are supplying with information about what is left at the surface. As soon as the surface is subjected to ion bombardment in connection with the analytical task or simply for sample cleaning, drastic changes of the surface composition may be induced by such effects as selective sputtering or cascade induced particle displacement. SNMS and SIMS are not influenced by such effects when the sample material is removed stoichiometrically, i.e. under sputter equilibrium. In SIMS, however, strong matrix effects influencing the secondary ion formation probabilities in a complicated manner are seriously hampering a quantitative evaluation of the secondary ion mass spectra.

From these arguments it may be concluded that in comparison to other techniques, SNMS fulfills very well the requirements for quantitative chemical surface and in-depth or bulk analysis.

Summary

Quantitative Secondary Neutral Mass Spectrometry Analysis of Alloys and Oxid-Metal-Interfaces

The characteristic properties and the operation modes of secondary neutral (or sputtered neutral) mass spectrometry (SNMS) are described briefly. Applications of SNMS for quantitative chemical bulk analysis and depth profiling are demonstrated for Fe-alloys and for anodic Ta_2O_5 films on Ta.

References

1. H. Oechsner, Appl. Phys. 8, 185 (1975).
2. H. Oechsner and W. Gerhard, Phys. Letters 40A, 211 (1972).
3. H. Oechsner, Plasma Physics 16, 835 (1974).
4. H. Oechsner, W. Rühe, and E. Stumpe, Surf. Sci. 85, 289 (1979).
5. H. Oechsner and A. Wucher, Appl. Surf. Sci. 10, 342 (1982).
6. H. Oechsner and E. Stumpe, Appl. Phys. 14, 43 (1977).
7. E. Stumpe, H. Oechsner, and H. Schoof, Appl. Phys. 20, 55 (1979).
8. H. Oechsner, H. Schoof, and E. Stumpe, Surf. Sci. 76, 343 (1978).
9. H. Oechsner, Molecule Formation in Oxide Sputtering. In: Secondary Ion Mass Spectrometry SIMS III, A. Benninghoven et al. (eds.). Berlin-Heidelberg-New York: Springer-Verlag. 1982. p. 106.
10. H. Schoof and H. Oechsner, Proc. 4th ICSS Cannes 1980, Vol. II, p. 1291 (Suppl. 201 of LeVide, le Conches Minces).

Correspondence and reprints: Prof. Dr. H. Oechsner, Fachbereich Physik der Universität Kaiserslautern, Technische Physik, Erwin-Schrödinger-Strasse, D-6750 Kaiserslautern, Federal Republic of Germany.

Mikrochimica Acta [Wien], Suppl. 10, 61–72 (1983)
© by Springer-Verlag 1983

Institut für Analytische Chemie der Technischen Universität Wien, Austria

Surface Analysis of Metals with SIMS

By

M. Pimminger and M. Grasserbauer

With 7 Figures

(Received January 19, 1983)

1. Introduction

Many properties of materials are strongly influenced by the surface state. Therefore a good analytical characterization of the surface layer could give support both for a direct improvement of technological processes and for tracing of fault sources, when failures in production are occurring.

As defined by IUPAC the surface or interfacial layer is "the region of space comprising and adjoining the phase boundary, within which the properties of matter are significantly different from the values in the adjoining bulk phases"[4]. In most cases these layers are very thin and so methods have to be employed, of which the analytical information originates from the utmost atomic layers. Most usual techniques are Auger Electron Spectroscopy (AES) and Photoelectron Spectroscopy (XPS, also called ESCA)[3]. Both of these methods have information depths up to 20 Å and their relative detection limit is in the range of about 0.X % in the analytical volume. With XPS also compound specific information is obtained but there is no possibility to carry out analyses in the microscale, whereas AES permits the representation of the lateral distribution of elements due to the microfocused exciting beam but only in some special cases information about compounds can be gained from the analytical signal.

Since also with Secondary Ion Mass Spectrometry (SIMS) the analytical signal is originating from the utmost layers, not only an additional method for surface characterization is at disposal, but due to the much lower detection limit the analytical capabilities are greatly enlarged. In this paper the techniques, the capability and also the limitations of applied surface analysis with SIMS will be presented. Investigations were carried out with the

CAMECA IMS 3*f* ion microanalyzer, an instrument which grants the most important combination of good in situ microanalysis with a high-standard mass spectrometric system (double focusing mass spectrometer with mass resolutions up to more than 10000).

2. Analytical Techniques

Dependent on the shape and dimension of the surface layer one has to choose the proper way of analysis. So surface layers are found on phases or crystals and on more or less even, in most cases polycrystalline and polyphase bulk materials (Fig. 1). If the thickness of the surface layer τ is in the range of X µm and if with phase boundaries the grain size d is larger than about 5 µm, lateral distribution analysis with sample cross sections would be preferred. Otherwise depth profiling is the only possible technique, also when phase boundaries with a low surface layer thickness and (or, respectively) with low grain sizes should be investigated.

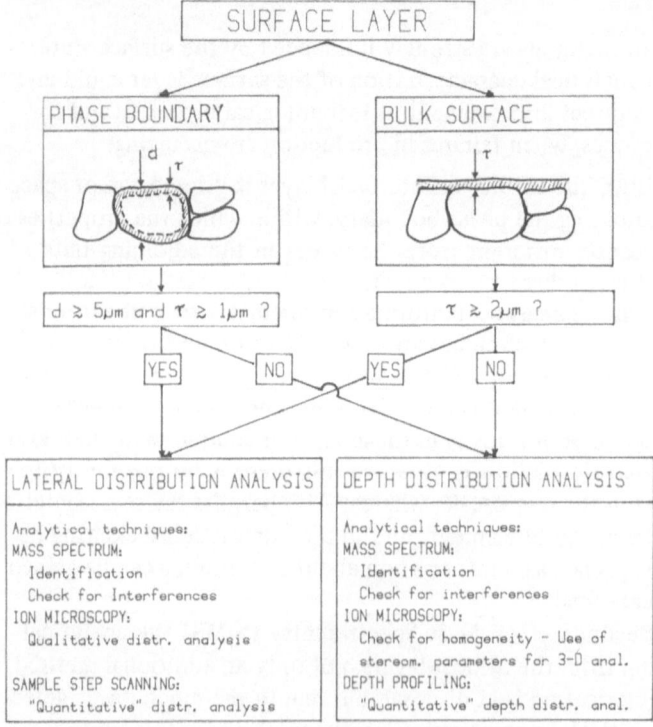

Fig. 1. Scheme of analytical strategy for surface analysis with SIMS

For lateral distribution analysis the limitations are set by the lateral resolving power which amounts to less than 1 μm for ion micrography, whereat for the step scanning mode the minimum analysed area of this instrument has a diameter of 1.4 μm. Two of the dominant advantages of lateral distribution analysis are that the sample can be precleaned from undesired contamination by ion etching and that measurements are only carried out, when already a steady state is reached.

With depth profiling the sample surface is eroded by sputtering with O_2, Ar or Cs primary ions and the removed material is analysed. In general primary ion current densities from 10^{-7}Acm^{-2} to 10^{-1}Acm^{-2} can be obtained which result in sputter rates from less than about 10^{-4}nms^{-1} to more than about 10 nms^{-1}. Since sputter depth profiling is a rather complex process some effects have to be taken into consideration, especially with technical samples and low primary current densities applied for thin surface films.

- Due to the enormous sensitivity in applied surface analysis all technological processes should be well known to get evidence of those analytical facts which are relevant for the solution of a given problem.
- Not quite plane as well as polycrystalline and polyphase sample surfaces cause a broadening of the real depth profile according to a nonuniform sputtering.
- As the depth calibration of very shallow craters is often impossible, the sputtering conditions should be kept very constantly because there is a linear relationship between sputtering time and layer thickness[2].
- One has to take into account that at interfaces considerable matrix effects can occur.
- According to a rule of thumb the erosion rate is much higher than the adsorption of molecules from the residual gas if Eq. (1) is fulfilled[8].

$$I_p(\mu A/cm^2)/p_b(mbar) \geq 10^8 \tag{1}$$

With a background pressure of 5.10^{-8} mbar during analysis contamination from residual gas is negligable if primary ion current densities are larger than about 5.10^{-6} Acm^{-2}.

3. Lateral Distribution Analysis: Mg in Spherulitic Graphite

The mechanical properties of cast iron are greatly influenced by the shape and extension of the precipitated carbon. For a long time it is well known that the usual monocrystalline and lamellar graphite can be modified to a spheroidal form by different techniques, like melting in vacuum or more easily by addition of reactive metals like Mg, Ca, Ce or Y[5]. It is assumed that Mg might not only cause the removal of harmful elements like O and S,

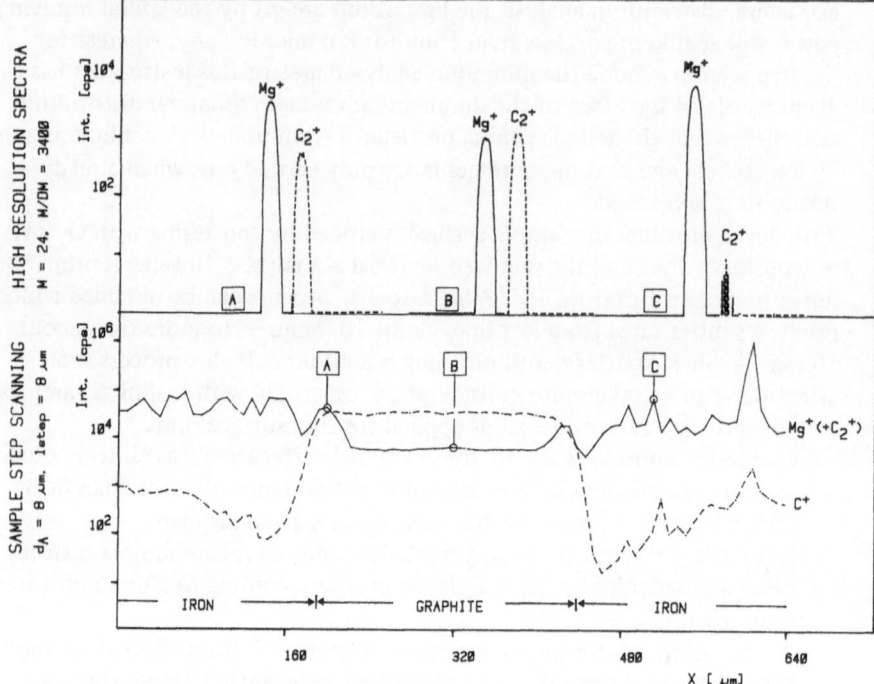

Fig. 2. Step scanning profile over a graphite spherulite and high mass resolution spectra of different microdomains: O_2^+, $E_0 = 12.5$ keV

but might have a specific influence upon the crystallization of graphite, propably by blocking the prism faces and so favouring the polycrystalline and spheroidal growth[6].

For supporting theoretical considerations the investigation of the distribution of Mg in carbon spherulites is of special interest. Different authors report average concentrations of 0.2 − 0.6 % Mg in the iron phase and 0.03 − 0.05% Mg in isolated graphite spheroids[5]. For analysis a focused primary beam of 1 μA O_2^+ was scanned over an area of 50 x 50 μm to clean the surface of the sample cross section. For the step scan technique the analysed area was limited to a diameter of 8 μm and a step width of 8 μm was chosen. The results in Fig. 2 show that besides an inhomogenous distribution in the iron matrix Mg is enriched at the surface of the spherulite relative to the background concentration in the inner parts of carbon. A better impression could be given by the ion micrographs of several elements in Fig. 3. Measurements in different spherulites always showed Mg-enriched layers with a thickness of 10 to 20 μm and only in one case a Mg-enrichment in the centre could be found. More often small iron inclusions in the carbon could be observed.

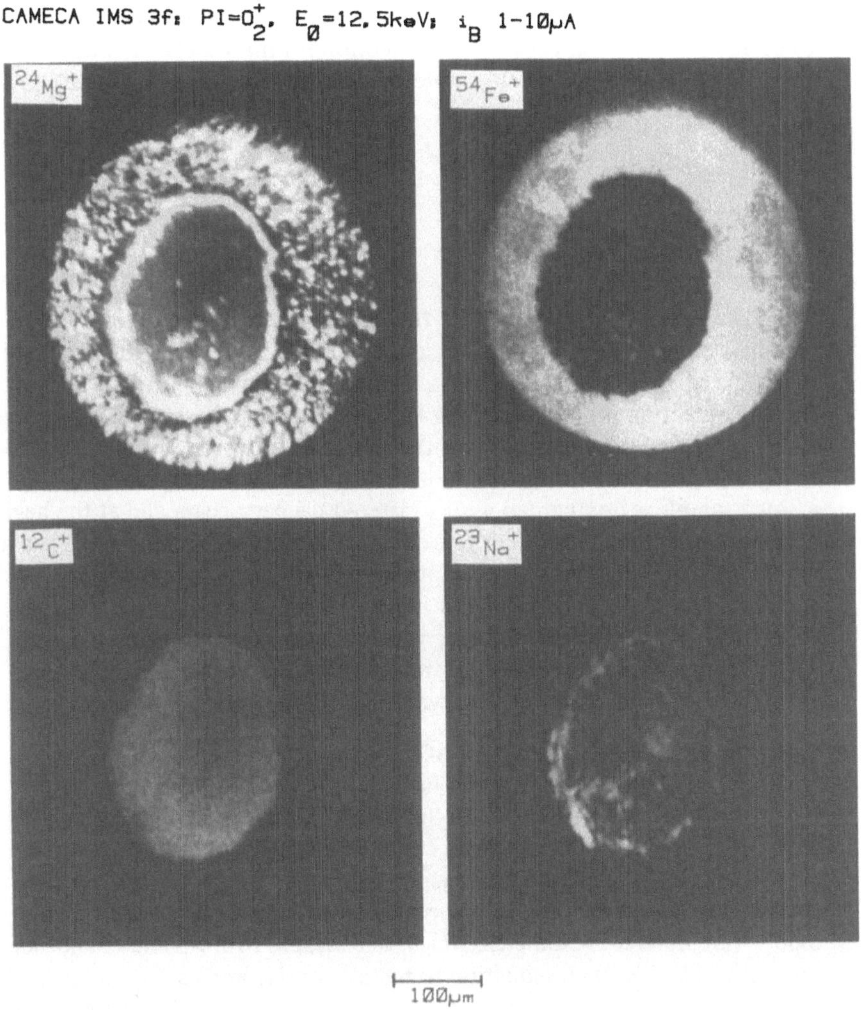

CAMECA IMS 3f: PI=O2+, E0=12.5keV; iB 1-10μA
24Mg+
54Fe+
12C+
23Na+
100μm

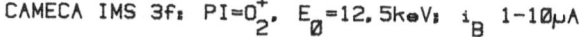

Fig. 3. Ion micrographs of spherulitic graphite in cast iron

As the mass 24 is subject to an interference with $^{12}C_2$, for a semiquantitative evaluation of the Mg-content high resolution mass spectra had to be taken at different sample sites (Fig. 2). A relative sensitivity factor was established by analysing pure Mg and C samples and so a Mg concentration of 0.03 % in bulk carbon and of 0.5 % at the surface layers of the graphite spherulites could be calculated. The relative error of these values is estimated to be about a factor 2, because no correction for matrix effect could be performed. More accurate results could be achieved by using implantation standards.

4. Depth Distribution Analysis

4.1 Trace Elements in WC Powder

When the grain size d and also the thickness of the altered layer τ is decreasing to micrometer dimensions and less, there is no possibility to characterize the surface of a single grain, because the lateral resolution is too low and the detection limit is also decreasing with extremely small analysed areas and very low primary current densities. So another way is proposed by analysing a large number of grains to obtain statistical information. A fine-grained sample is bombarded with a rastered primary beam and at the beginning only the utmost surface layers are analysed. By this method one has to expect a broadening of the true depth profile of elements and when a steady state is reached the secondary ion intensities are representing the average concentration, because also during sputtering in deeper regions the grain surface is detected to a certain extent. As there is no experience for this method first model investigations were carried out with WC powders.

Trace elements in WC powder are affecting the processes (especially the grain growth) during sintering and by that way also the mechanical properties of sintered materials. One of the first approaches for the systematic study of the role of trace elements could be clarifying the question, whether these dopants are dispersed homogeneously in the WC grains or enriched at the surface. The technological processes consisted of doping the "blue oxide" (an understoichiometric WO_3), one sample with Na and P, another one with Li and B, then reduction to metal with H_2 and carburization with graphite. These powders with an average grain diameter of $0.7 \mu m$ were then taken for the investigations (bulk analyses: Li, Na: flame emission spectroscopy, B, P: photometry) and also sintered with Co which brought about a much coarser grain size for the Na,P-doped samples.

In Fig. 4 the depth profiles of trace elements from both samples are represented. The matrix elements C^+, W^+, respectively, showed no significant intensity variations with sputter time, so the other secondary ion intensities were normalized to an equal C^+ intensity. Li and B are enriched at the sample surface by a factor of about 10 and are reaching a nearly steady state corresponding to average concentrations of $13 \mu g/g$ Li and $50 \mu g/g$ B after about ten minutes of sputtering. The erosion rate could not be measured but by a comparison with other experiments it was estimated to be about 0.5 nm/min. The sample which was not doped with Li and B gave a similar behaviour but

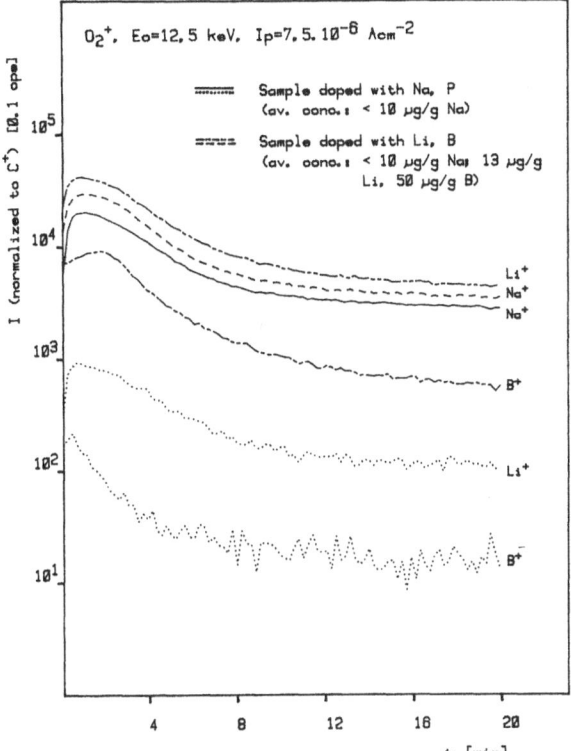

Fig. 4. Depth profiles of WC powder

with a decrease in intensity by a factor of about 35. As for industrial laboratories these contents are too low for determination, average concentrations
were calculated from SIMS depth profiles by a comparison of doped and
nondoped sample, and they amount in the latter case about 0.4 μg/g Li and
about 2 μg/g B. This example demonstrates the possiblity of determining
very low surface and also bulk concentrations with SIMS, even though low
primary current densities are applied.
Average Na concentrations were below the practical detection limit for
Flame Emission Spectroscopy in both samples. Depth profiling resulted in a
similar enrichment at the surface as Li and B, but the content of the non-
doped sample was higher than of the doped specimen. So it must be assumed
that the unavoidable background concentration in technological processes is
higher than the originally wanted doping. For example in graphite used for
carburization 380 μg/g Na was found. According to the coarser grain
structure of the Na, P doped WC powder this fact indicates that P must be
the active component during sintering.

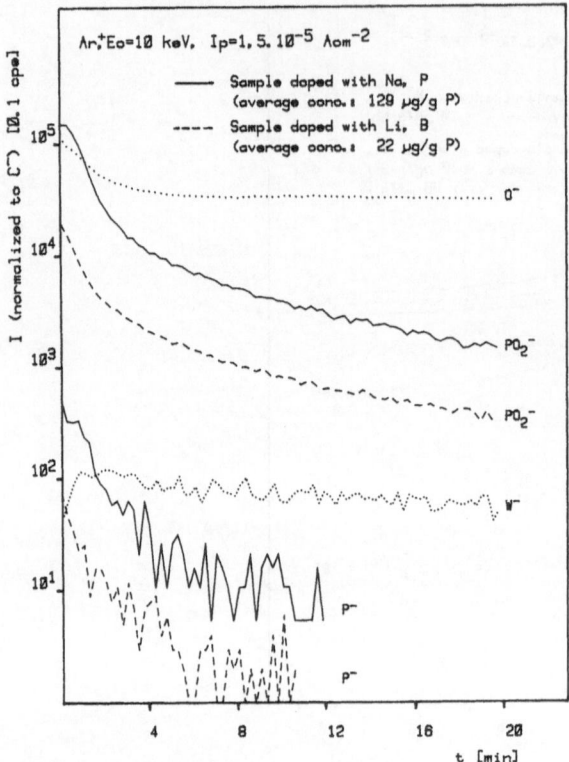

Fig. 5. Depth profiles of WC powder

As on account of interferences no clear results of the P^+ distribution could be obtained, in the negative secondary ion mass spectrum the peak at mass 31 (P^-) and the much more intensive peak at mass 63 (PO_2^-) were used for investigations (Fig. 5). Here a surface enrichment by a factor 100 can be seen as well for P^- as for PO_2^- and also for both samples. P seems to be localized more immediate at the surface than Na, Li and B. Also a good correlation between measured average concentrations and intensity ratios of doped and non-doped sample is found.

To estimate the influence of matrix effect for positive ions which is mainly caused by different O concentrations also the O^- profile was taken by using an Ar primary beam (Fig. 5). Only a slight enhancement at the surface and then a constant intensity of O^- could be noticed. Therefore the matrix effect for positive secondary ions is not significant in this case, especially as an O_2^+ primary beam was used for Na, Li and B depth profiles. The average concentrations amounts to 450 $\mu g/g$ O for the Li, B doped sample and 480 $\mu g/g$ O for the Na, P doped sample as determined by high temperature vacuum extraction.

These first investigations give encouraging results for the analysis of thin surface films on fine grains by gaining statistical information, a method which is also applicable for studies of fracture surfaces. The present findings are a starting point for further investigations like the measurement of the trace element distribution in different process stages.

4.2 Contamination Layers on Electrical Contacts

Similar to conventional mass spectrometry SIMS spectra contain molecular ion peaks. The use of fingerprint spectra for information about compounds had already been shown by Werner[8,9,10] and with similar methods by Benninghoven[1] and Oechsner[7]. Oechsner suggested two models for the formation of molecular ions. According to the atomic combination model (ACM) which is dominantly valid for samples with weak (metallic or ionic) bonds and comparable atomic masses, two or more not necessarily directly neighboured atoms which are individually sputtered in a local and temporal

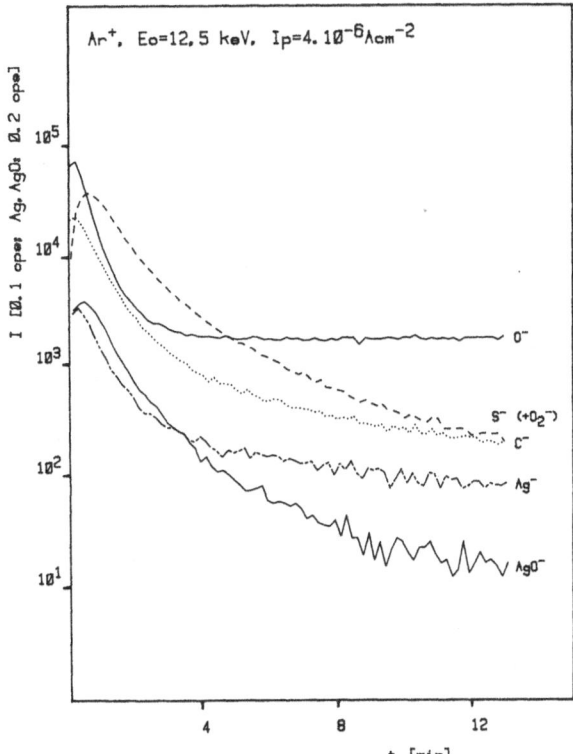

Fig. 6. Typical depth profile of contaminated Ag70Pd30 contact (Resistance = 1150 $m\Omega$)

correlation, recombine during ejection to a molecular ion. With the direct emission model (DEM) which is mainly valid for targets with strong (ionic) bonds and large relative mass differences the molecular ion is sputtered by an energy transfer to the heavier of next neighboured atoms[7]. As the primary ion bombardment causes a partial disintegration of the target surface, for each individual problem it has to be studied if the molecular ions represent the original arrangement of atoms. The generation of compound specific information was employed for the investigation of contamination layers.

Electrical contacts which are used in weak current relays in printed circuits fail due to an increase of the resistance from the usual value of less than 30 $m\Omega$ to more than 100 $m\Omega$. This is obviously caused by a building-up of thin contamination layers. Investigations should clear the reason for unusually high failures of relays with Ag70Pd30 contacts. Depth profiles show a surface enrichment of C, O and S (Fig. 6). As C is coming from an organic

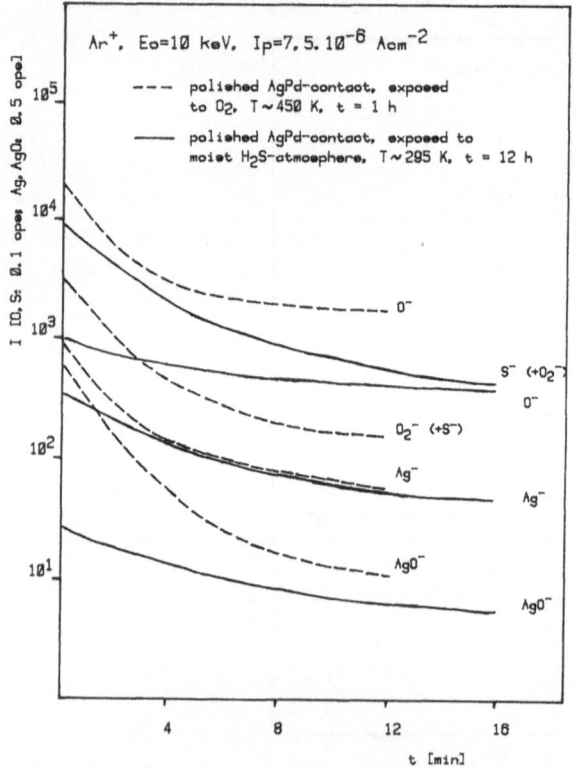

Fig. 7. Depth profiles of Ag70Pd30 contacts with oxidic and sulphidic surface layers (intensities are normalized to the Ag$^-$-intensity)

protective layer (a cinnamic acid – amine mixture), O and S could be associated with the increase of the resistance. S as a typical contamination element of Ag-Pd-contacts can be assumed to form a sulphidic layer. In order to gain information about the incorporation of O into the Ag-Pd-base material the AgO^--profile was measured. This enables a distinction of O bounded to Ag from O in the organic surface layer.

Experiments with oxidized and H_2S treated specimens and with Ag_2O and Ag_2S powders were carried out to evaluate the relative intensity ratios (Fig. 7). For Ag_2O a typical AgO^-/Ag^- ratio of 0.8 to 1 and a S^-/O^- ratio of 0.1 to 0.2, for Ag_2S AgO^-/Ag^- ratios of 0.1 to 0.2 and S^-/O^- ratios of 10 were found. Because of a considerable AgO_2^- interference the AgS^- peak is not significant and was therefore not evaluated. The AgO^-/Ag^- ratio proved to be a good indicator for the presence of a partly oxidized AgPd surface. A large series of investigations was carried out and the results confirmed the assumption of the additional insulating effect of AgO_x, because also in well protected contacts with low resistance values sulphidic layers were found.

Acknowledgement

The authors wish to thank Prof. Dr. B. Lux and Dr. W. Schubert from the Institute of Anorganic Technology of the Technical University of Vienna and Prof. Dr. E. Lassner, Woform Bergbau- und Hüttenges.m.b.H., St. Peter, for their cooperation in providing samples and the results of average analysis.

Summary

Surface Analysis of Metals with SIMS

Capabilities and limitations of surface analysis with SIMS are demonstrated by means of representative examples; the distribution of Mg in spherulitic graphite in cast iron for lateral distribution analysis, the distribution of trace element dopants in WC powders for depth profiling of very thin surface films on fine-grained samples and the investigation of contamination layers on electrical contacts as example for the use of molecular ions to gain information about compounds. It is shown that SIMS in its modern instrumental configuration has a high potential for surface analysis of metals, especially for trace elements and compounds.

References

1. A. Benninghoven, Surf. Sci. 28, 541 (1971).
2. F. Degreve and Ph. Ged, subm. to Surf. and Interf. Anal.

3. C.A. Evans and R.J. Blattner, Ann. Rev. Mater. Sci. 8, 181 (1978).
4. D.H. Everett, Pure and Appl. Chem. 31, 579 (1972).
5. B. Lux, Gießereiforsch. 22, 65 (1970).
6. B. Lux, Gießereiforsch. 22, 161 (1970).
7. H. Oechsner, Secondary Ion Mass Spectrometry SIMS III, A. Benninghoven, J. Giber, J. Laszlo, M. Riedel, and H. W. Werner (eds.). Berlin-Heidelberg-New York: Springer-Verlag. 1982. p. 106.
8. H.W. Werner, Vacuum 24, 493 (1974).
9. H.W. Werner, Appl. Phys. 7, 65 (1975).
10. H.W. Werner, Acta Electronica 18, 51 (1975).
11. E. Zinner, subm. to J. of Electrochem. Soc.

Correspondence and reprints: Prof. Dr. M. Grasserbauer, Institut für Analytische Chemie, Technische Universität Wien, Getreidemarkt 9, A-1060 Wien, Austria.

Mikrochimica Acta [Wien], Suppl. 10, 73–91 (1983)

Akademie der Wissenschaften der DDR, Zentralinstitut für Festkörperphysik und Werkstofforschung, Dresden, German Democratic Republic

Characterization of Metallic Glasses by Ion and Electron Microprobes

By

H. Mai

With 8 Figures

(Received January 19, 1983)

1. Introduction

The outstanding physical properties of metallic glasses like high mechanical strength, distinct corrosion resistivity, and a full range of magnetic behaviour of ferromagnetic, paramagnetic, diamagnetic and antiferromagnetic compositions[1] are decisive for new fields of application and improved conditions for conventional applications of metallic materials. This is true of the alloy as a self-contained material as well as of the vitreous phase created in the near surface region of a crystalline solid. Some typical properties of these materials[2,3] are to be mentioned:

- mechanical (substantial hardness, high tensile strength and durability against bending strain)
- magnetic (high saturation polarization, high permeabilies, and very low coercive forces)
- corrosion related (extremely high corrosion resistivity especially of alloys containing common metals like iron and nickel and small amounts of chromium).

Further economic advantages connected with the production of such alloys are energy saving, reduced consumption of high cost raw materials and the elimination of processing steps in comparison to conventional metallurgical methods for making wires and ribbons.

These features recommend as fields of application:

- fibre composite materials (e.g. reinforcing filaments in automobile tires,

transmission belts)
- cutting devices
- surface processed materials for application in corrosive environment (e.g. new biomaterials, cables that resist sea water)
- magnetic devices (e.g. inversion transformers, magnetic amplifiers, tape recording heads, transducers, magnetostrictive delay lines).

Mainly during the last ten years a diversity of metallic glasses has been prepared due to an extensive work in practice and theory. The preparation techniques comprise vapour deposition, cathode sputtering, chemical deposition from solution, ion implantation, LASER irradiation, and rapid quenching from the liquid melt. Ribbons — the todays predominating product — are obtained by the so-called melt spinning technique (Fig. 1). During this procedure a melt (3) of an already prepared alloy is formed in an inductively heated furnace (1, 8). The pressurized ejection gas (2) forces a stream of molten alloy (5) through the orifice (4) in the bottom of the crucible (1). This liquid alloy jet impinges onto the outer or inner surface of a rotating drum (7) of high thermal conductivity where it is rapidly solidified. A ribbon (6) of typically some 10 μm thick by some mm or cm wide can be taken from the drum. Minimum quenching rates greater than about 10^5 K s^{-1} and preferably about 10^6 K s^{-1} are necessary to suppress the creation of crystalline phases during the transition liquid $\rightarrow$ solid. Alloys of two, three and more components (transition metals, noble metals, metalloids, nonmetals, and elements of the Al-, Zn-groups) were formed. The composition usually approaches the eutectic composition where the liquid phase is most stable, and the melting temperature is relatively low. Ratios of the atomic radii of the major elements $\gamma_1/\gamma_2 < 0.88$ and > 1.12 resp. tend to improve the glass-forming tendency.

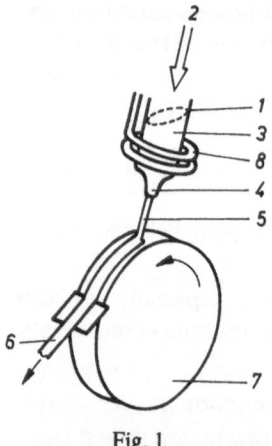

Fig. 1

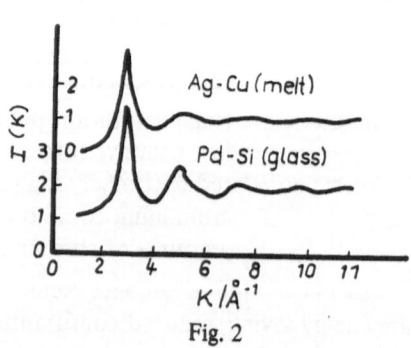

Fig. 2

Fig. 1. Schematic representation of melt spinning arrangement

Fig. 2. Comparison of the interference functions of a melt and a metallic glass

The structure of these metallic glasses shows some short-range order only whereas they do not posses any long-range order that is typical for crystalline structures. The X-ray and electron diffraction patterns (see e.g. Fig. 4) therefore show only a few diffuse rings, whereas the interference functions obtained from appropriate diffraction patterns of the melt and the metallic glass[4] show a considerable similarity (Fig. 2). This allows the metallic glass to be considered as an undercooled liquid. It exists in a metastable state which is characterized by a free energy that represents a relative minimum of the function of state and is located above the free energy of possible crystalline phases. The schematic diagram of Fig. 3 shows the behaviour of the free

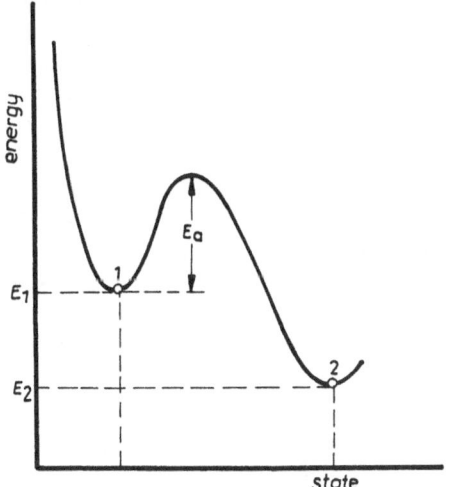

Fig. 3. Schematic diagram of the function of state involving the metastable conditions for metallic glasses

energy as a function of state transitions. The reaction amorphous → crystalline is represented by a transition from level E_1 to level E_2 when the activation energy $E_a \sim 80 \dots 400$ kJ mol^{-1} is supplied to the system[2]. This behaviour is especially of interest in relation to the problems discussed in section 4. There the conditions for the alloys $Fe_{81}Si_{3.5}C_2B_{13.5}$ and $Fe_{40}Ni_{40}P_{14}B_6$ – as the matter of our experimental investigations – are to be examined.

2. Problems of Characterization

The peculiarity of metallic glasses is caused by their structural order or disorder on a nm-scale. The investigation of these short-range structures as well as their chemical inhomogeneities in correlation to their outstanding physical features must be reserved to analytical methods like transmission electron microscopy (TEM), X-ray diffraction (XRD) and electron diffraction (ED)[5 -8], atom probe techniques[9] etc.

The analytical characterization of samples of this kind by methods of local microanalysis can only provide information about the mean chemical composition of sample parts of comparatively large dimensions. It can be used preferably for the confirmation of sample homogeneity or the explanation of local deviations of the physical properties on a μm- or cm-scale (i.e. across ribbon cross section or length). Additional possibilities of application can be found in the investigation of major component diffusion (self-diffusion of boron in $Fe_{40}Ni_{40}B_{20}$ glass[10]) or of segregations resulting in embrittlement or crystallization (Auger electron spectroscopy of the $Fe_{40}Ni_{40}P_{14}B_6$ alloy[11]).

The qualitative and quantitative investigations by electron probe microanalysis (EPMA) and secondary ion microanalysis (SIMA) to be described here were carried out to deal with the following problems:

– determination of major component and major impurity concentrations and comparison with nominal concentrations for estimation of consequences of the uncertainties of the metallurgical preparation process on the physical properties
– evaluation of major component spatial distributions for the deduction of possible dissolution or segregation processes during solidification resulting in local deviations of mechanical and magnetic features
– evaluation of impurity element spatial distributions for conclusions concerning the origin and solution behaviour of these elements during the preparation process, their contributions to property fluctuations, and the required purity of raw materials
– investigation of a particular behaviour of the alloys under influence of particle bombardment with regard to their possible application as standard reference materials of sub-μm-scale homogeneity.

The results have been completed by chemical analyses (see Table 1), TEM and ED experiments.

Table 1. Chemical Techniques Used for Determination of Major Components (s_{rel} – relative standard deviation)

element	method	s_{rel}
Fe	titration with Ce(IV)-solution	0.003
Ni	precipitation with dimethylglyoxime	0.0016
P	precipitation with ammonium-molybdate, solution in NaOH, backtitration in H_2SO_4	0.005
B	spectrophotometric determination with methylenblue	0.004

3. Sample Preparation

Usually amorphous ribbon prepared by the melt spinning technique shows a surface roughness that is inadequate for microanalytical purposes[12]. The top side shows surface undulations caused by material flow and solidification processes whereas the reverse side is an approximate copy of the surface topography of the substrate drum (a rather clean but, usually unpolished copper block to provide good adhesion of the melt jet as well as thermal contact). Therefore even under favourable conditions a remaining roughness of some μm has to be taken into account. For measurements within the ribbon cross section and at its surface optically flat areas have to be prepared — especially with regard to very precise determinations of major component concentrations. The high mechanical strength (Vickers hardness VH = 750 ... 1100 for $Fe_{40}Ni_{40}P_{14}B_6$ and $Fe_{80}B_{20}$ resp.)[3] and the low thickness of the ribbons involve a rather careful sample preparation. For the investigations discussed in the following sections the described preparation procedures proved to be successful:

- transverse sections[13], samples were degreased in trichloroethylene and washed in deionized water, coated with an anodic copper layer typically 500 μm thick (copper bath composition per 1 H_2O: CuCN 22.5 g, NaCN 34 g, Na_2CO_3 15 g), enclosed in epoxy resin, pre-finished with 600, 400, 280, 80 grit emery paper and metallographically finished using alumina (finest grade, vibration polishing)
- top and reverse side of ribbon, after degreasing and washing the samples were cemented to a metal block, mechanically pre-finished by hand on 600 grit emery paper under frequent stereo microscopic inspection (magn.: 100 x), short time vibration polishing using alumina.

4. Interaction of Primary Particles and Metastable Solid

For the generation of the desired analytical signals the solid sample has to be bombarded by ions or electrons. These particles create by means of their appropriate kinetic energy the necessary interactions with the solid atoms within the sampling volume (per definitionem: this part of the sample that emits the total group of analytical signals required for one complete "point" analysis). The prerequisites for a precise quantitative microanalysis — as for any other correctly elaborated analytical technique — are:

- a chemically homogeneous sampling volume
- a sampling volume that is representative for the whole solid section to be characterized
- an interaction process of primary particle and solid that involves either no modification of the sampling volume or reproducible transformations only which lead to steady state conditions preserving the initial analytical information to be extracted.

The methods of local microanalysis proved to be successful for the investigation of the stable solid are usually applied also for the analysis of metastable metallic glasses. Before describing our analytical applications however, we want to scan some methodical points of view that are of importance for the outcome of the measurements.

4.1 Electron Bombardment

The following primary beam parameters can be accepted for conventional EPMA investigations:

Accelerating voltage U_e = 10 ... 20 kV
Probe current I_e = 10 ... 20 nA (on brass)
Counting time t_c = 10 and 100 s resp.

A sampling volume of 10^{-11} to 10^{-12} cm^3 is obtained for appropriate electron ranges R_e typically 2 to 5 μm (i.e. ranges for that characteristic X-ray generation is possible). The estimation of a mean density of energy E_e injected by one primary electron into the sampling volume gives $E_e \sim 6 \times 10^{-4}$ J cm^{-3}. This does not exclude the appearance of much higher differential E_e-values along the electron path in the solid. But, neither during extended TEM experiments electron induced crystallization has been observed nor – in agreement with investigations of compact metallic specimens[14] – an increase in sample temperature during EPMA measurements is to be hoped for. The second statement agrees well with the estimation that for an electron dwelling time within the sampling volume of typically 5×10^{-13} s, for the interval of its relaxation of a similar order of magnitude, and the rate of impinging electrons of 6×10^{10} s^{-1} (according to $I_e \sim 10$ nA) an extremely low probability for a temporary interference of the interactions of two or more different electrons is obtained. This allows the conclusion that for our experimental conditions of EPMA a modification of the sampling volume (except surface contamination by residual gas cracking) can be excluded. Therefore metallic glasses could be accepted as useful standard reference materials for EPMA after precise chemical analysis.

4.2 Bombardment by $^{16}O_2^+$ Ions

Whereas the electrical parameters of the primary ion (PI) beam (U_{PI} = 20 kV, I_{PI} = 5...40 nA) are equivalent to those of the electrons due to the different ranges within the solid in the case of SIMA an injection of a similar kinetic energy into much smaller volumina of the solid takes place. The ratio between PI range R_{PI} and the maximum extent of the sampling volume $d_{AV}(PI)$ is characterized by $R_{PI} \ll d_{AV}(PI)$ in contrary to the EPMA analogon $R_e \sim d_{AV}(e)$. Now the energy density E_{PI} per impinging PI has to be estimated for the collision cascade volume only and the total sampling volume is represented by the sum of spatially and temporarily independent collision cascades. The analytical signal on the other hand appears as the sum of independent secondary ion (SI) emission processes proceeding from these collision cascades. An energy balance equivalent to section 4.1 yields for $R_{PI} \sim 1...2 \times 10^{-6}$ cm the mean value of the energy density within the collision cascade volumina $E_{PI} \sim 10^3$ J cm^{-3}. The range of the E_{PI} data appropriate to the R_{PI} range is overlapping the E_a range explained in section 1 ($E_a \sim 140...$ 680 J cm^{-3}). Influences on structure and state of the sampling volume therefore must be hoped for.

Moreover a substantial oxygen concentration (propably larger than that of the metalloids) has been built up in the sampling volume as soon as steady state conditions are approached[15,16]. Thus chemical reactions created by the ion bombardment will involve besides the major elements of the metallic glass also oxygen as an important component. The heats of formation for oxides which are nearly one order of magnitude larger than those for carbides, phosphides or silicides[17] can cause a preferential formation of major element oxides. The investigation of the situation existing under steady state conditions within the sampling volume has been attempted by TEM and ED techniques. Specimens of the alloys $Fe_{81}Si_{3.5}C_2B_{13.5}$ (sample I) and $Fe_{40}Ni_{40}P_{14}B_6$ (sample II) have been electrolytically thinned (acetic acid/perchloric acid ~ 10 : 1) for TEM studies and ion

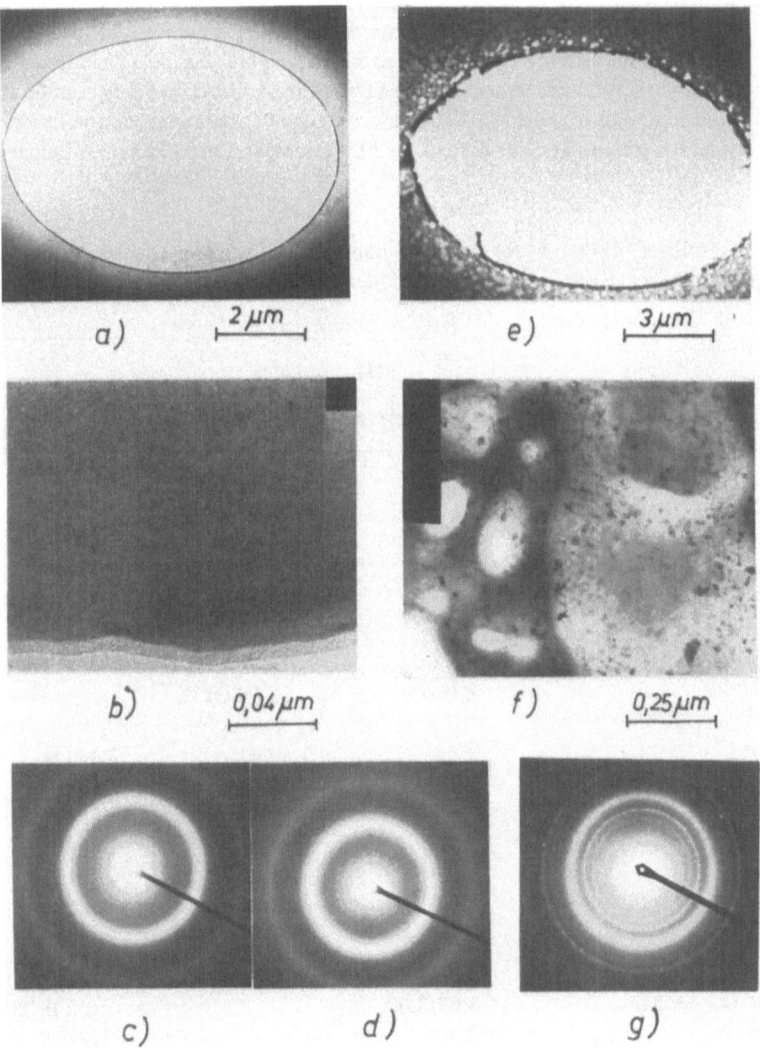

Fig. 4. TEM micrographs and ED pattern of metallic glasses; initial states − a, b, c, (sample I), d (sample II); oxygen bombarded − e, f, g (sample I)

bombarded in the SIMA apparatus IMMA (ARL, Sunland/California). The parameters of the primary beam were:

PI species   – $^{16}O_2^+$
PI accel. voltage U_{PI} = 20 kV
PI current I_{PI} = 40 nA
PI density i_{PI} = 4 x 10^{-4} A cm^{-2} (raster 100 x 100 μm^2, see Fig. 5a)
PI dose N_{PI} = 10^{17} ...10^{18} cm^{-2}

The amorphous state of the samples has been confirmed by TEM and ED photos (200 kV – TEMSCAN, JEOL Tokyo/Japan) taken prior to the ion exposure (Fig. 4a–d). After the ion bombardment the electron transparent regions of sample I (Fig. 4e, f) and sample II (Fig. 5e, f) exhibit crystalline volumina. The ED pattern of that (Fig. 4g, 5g) can be attributed to the structures of $\gamma – Fe_2O_3$ and Fe_3O_4 (sample I) and to mixed iron-nickel-oxide crystallites (sample II). The data of the ED pattern is given in Table 2. Whereas for

Table 2. Results of ED Pattern for Oxygen Bombarded (O_2^+) and Spontaneously Crystallized (s.c.) Sample Regions (d – interplanar spacing in Å)

| alloy | d (meas.) | ASTM-card index | | (hkl) |
		d (γ-Fe_2O_3)	d (Fe_3O_4)	
FeSiCB	2.96	2.95	2.97	(220)
(O_2^+)	2.51	2.52	2.53	(311)
	2.10	2.08	2.097	(400)
	1.70	1.70	1.714	(422)
	1.60	1.61	1.615	(333)
	1.48	1.48	1.484	(440)
		d (FeO)	d (NiO)	
FeNiPB	2.44	2.486	2.41	(111)
(O_2^+)	2.10	2.153	2.089	(200)
	1.79			
	1.49	1.523	1.476	(220)
	1.27	1.299	1.259	(311)
	1.08	1.077	1.044	(400)
		d (Fe_3B)		
FeSiCB	3.06	3.05/3.07		(220)/(201)
FeNiPB	2.51	2.50		(221)
(s.c.)	2.10	2.09/2.10		(140)/(321)
	1.86	1.88		(411)
	1.53	1.526/1.535		(440)/(402)
	1.31	1.31		(223)

sample I mostly irregular crystallite boundaries are observed (Fig. 4f) shows sample II (Fig. 5e, f) preferably crystallites of rather regular shape and a diameter of typically 10 to 20 nm (diameter of collision cascades $d_K \sim 10 \ldots 20$ nm!). These crystallites are surrounded by a border of a phase of higher electron transparency.

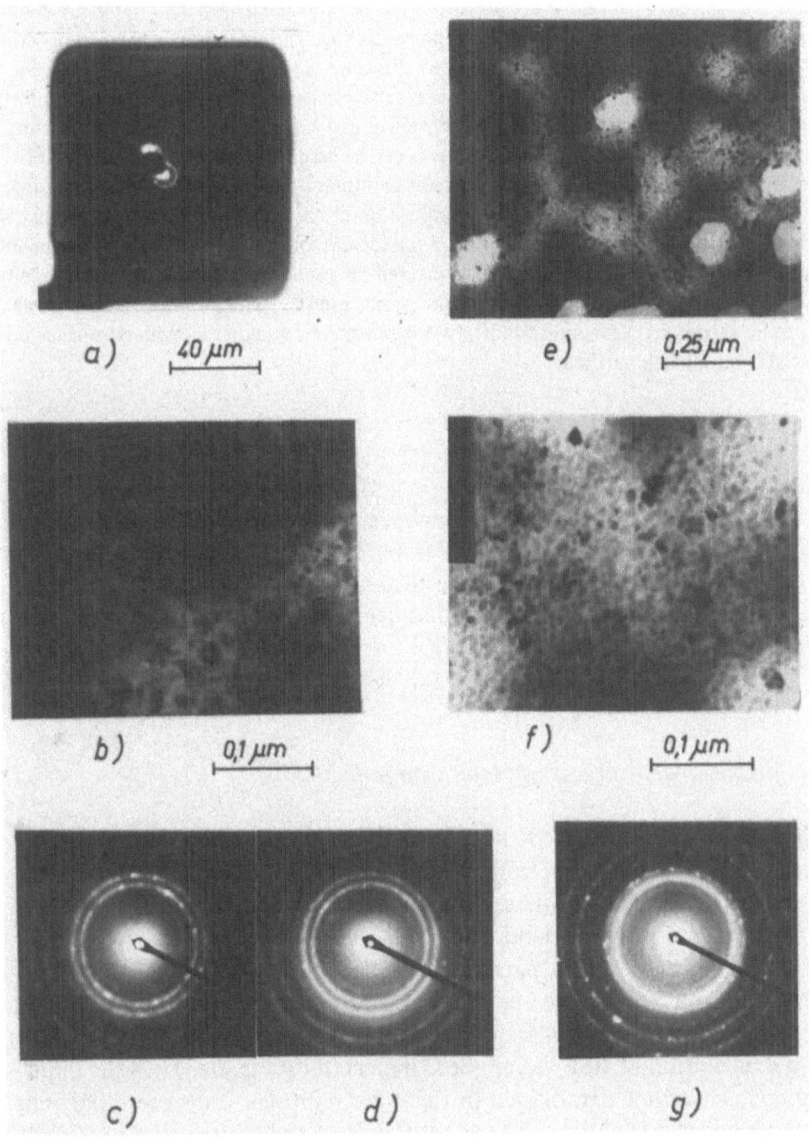

Fig. 5. Optical micrograph, TEM micrographs, and ED pattern of metallic glasses. PI raster – a (opt. micrograph); oxygen bombarded – e, f, g (sample II); spontaneous crystallization – b, c (sample I), d (sample II)

This dense and to a certain extent regular arrangement is quite different from regions of spontaneous crystallization which has been observed in sample I and sample II (Fig. 5b). The appropriate ED pattern are identical for both alloys (Fig. 5c, d). They can be attributed to the ED diagram of Fe_3B or $(Fe, Ni)_3(P,B)$.

Within every ED pattern the diffuse rings of a remaining component of the amorphous phase is observed. From these investigations can be concluded that the sampling volume has been modified during the interval of presputtering. The structure and the composition of the sampling volume under steady state conditions is rather different from that of a similar volume within the homogeneous as-received material. Now different phases exist and even for such small dimensions as mentioned above from a SIMA point of view no longer a solid solution of the major elements can be assumed. The creation of the SI is therefore determined by the different phases and their boundary regions[18] and no longer by a single matrix having the random composition of an amorphous state as has been discussed by other authors[19]. Above all the 5 μm local resolution of their SIMA instrument cannot reveal a crystallographic contrast created by ion bombardment. The homogeneity test within the sampling volume of metallic glasses used as standard reference materials has to be carried out on a nm-scale. Only this can answer whether a common surface potential[20] for all SI is verified.

5. EPMA Measurements

Beside the characterization of mean concentrations and distributions of the major components the behaviour of impurities having concentrations in the 0.1 ... 10 %-wt. range was to be investigated. A precision of 0.5 %-at. for metals and 0.3 %-at. for metalloids has been desired for major element determinations. This could be fulfilled with the exception of boron for which the EPMA determined concentration is taken as an estimate only.

5.1 Mean Concentrations of Major Components

The determination of precise mean concentrations has been carried out by wavelength dispersive spectrometers (WDS) and an energy dispersive spectrometer (EDS). Metallic glasses characterized by chemical analysis, gallium phosphide, and pure elements have been applied for standardization purposes. The electron beam parameters explained in section 4.1 were used. Sample regions not affected by inclusions or precipitates have been selected for the measurement from mappings of backscattered electrons (BSE). The spatial resolution of BSE technique is indicated by Fig. 6a. Thus the impurity concentrations determined in that way represent only those portions remaining in solid solution. The analytical results obtained from four ZFW-products, two products of BRD-factories, and one product of a US-factory are compared with the nominal concentrations in Table 3. The samples ZFW 1 through ZWF 4 indicate the development of our technology from the

Table 3. Results of EPMA Analyses of Alloy $Fe_{40}Ni_{40}P_{14}B_6$ (concentrations in %-wt., results in brackets are determined by chemical analysis, n.d. − not determined, x − not detected)

element sample	B	P	Ti	Cr	Mn	Fe	Ni	Fe/Ni
ZFW 1	(0.89)	9.1	1.6	1.1	1.8	41.6	43.9	0.948
ZFW 2	n.d.	9.1	1.1	1.2	1.6	41.7	43.8	0.952
ZFW 3	(0.89)	9.0	n.d.	n.d.	n.d.	41.8	44.2	0.946
ZFW 4	(1.3)	8.5	x	x	x	43.7	46.5	0.94
BRD 1	(1.85)	8.9	0.7	0.8	2.3	42.0	43.4	0.968
BRD 2	n.d.	8.8	x	x	0.4	43.2	43.0	1.004
US	n.d.	9.0	x	0.4	x	45.0	45.0	1.0
Nominal	1.28	8.53	x	x	x	43.97	46.22	0.951
s_{rel}		0.004				0.008	0.008	

early beginning (comparatively high impurity content) to today's state of the art. The concentrations of impurities detected by chemical analysis in sample ZFW 4 were generally found to be less than 0.23 %-wt. The relative standard deviations given in the lower row are typical for transverse section measurements. Whereas the iron-nickel ratio approaches the nominal value the phosphorus concentrations with the exception of sample ZFW 4 were found always as bit too high. The most serious problem was the adjustment of the boron concentration. Chemical analysis shows for this element nearly in every case the largest deviations from the nominal value.

5.2 Spatial Distribution of Major Components

The transverse section of sample II has been checked by a matrix of 66 point analyses (11 columns for ribbon width versus 6 rows for ribbon thickness). According to the rather small deviations hoped for the concentration changes have been calculated directly by a linear interpolation from the intensity changes of the appropriate $K\alpha$ lines. Counts accumulated for 20 s have been deadtime and drift corrected. A background correction has been omitted according to the assumption of an equal background contribution to every point analysis. The significance of concentration deviations across tape width and thickness has been assured by two-way analysis of variance. Nickel concentration shows a relative increase of 0.09 % across the tape width whereas phosphorus shows a relative decrease of 0.12 % in the same direction. Iron shows a similar behaviour like nickel but due to larger stochastical deviations its significance could not be assured. Across tape thickness no differences could be found for nickel. A relative increase of 0.3 % from both surface regions to the tape middle could be assured for iron and an opposite behaviour for the phosphorus concentration. But, a statement cannot be made about an equivalence of these phosphorus enriched near surface regions and the phosphorus segregations found by AES within a 0.36 μm thick surface layer[11].

5.3 Spatial Distribution of Impurity Elements

The inspection of the ribbons for inclusions and precipitates has been carried out first of all by BSE mapping. A surface region of typical inhomogeneity is shown in Fig. 6. The cluster of inclusions is mainly formed by aluminium oxide (Fig. 6b) which seems to be the only form of appearance of this main impurity. Aluminium has been found in the ferroboron typically 10 %-wt.

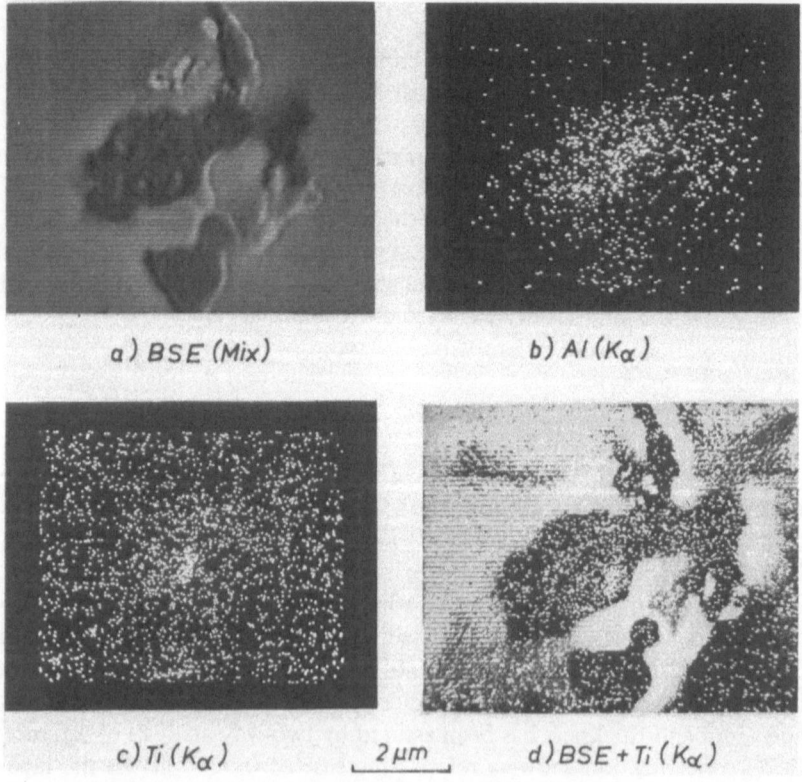

a) BSE (Mix)   b) Al (Kα)

c) Ti (Kα)   2 μm   d) BSE + Ti (Kα)

Fig. 6. EPMA representations of a cluster of inclusions in sample II

In the metallic glasses aluminium could not be found in solid solution. Titanium another important impurity element appears in solid solution and like chromium and manganese also in rather small inclusions (Fig. 6c). The one visible in the centre of the Ti ($K\alpha$) mapping can be easily localized within the cluster of Fig. 6a by a BSE-overlay (Fig. 6d).

6. SIMA Measurements

The IMMA has been applied for two different reasons, namely:

- for the complete survey of all detectable impurities and their distribution in the ribbon[21] and
- for investigations of the SI emission of metallic glasses and their applicability for quantitative test analyses involving LTE corrections[21] (LTE-local thermal equilibrium).

6.1 Spatial Distribution of Impurity Elements

Beside the major components a couple of impurities have been detected in the SI spectra. Those of particular interest were H, (C), O, Na, Al, (Si), Cl, K, Ca, Ti, V, Cr, Mn, and Cu (elements in brackets are major elements in sample I). Selected spatial distributions are shown as mappings of positive SI in Fig. 7. With the exception of V and Cu (always in solid solution) the others have been found in solid solution as well as precipitated or like Na, Al, K, and Ca precipitated only. Due to a PI beam diameter of 7 μm merely the larger inclusions show their structured shape (e.g. Fig. 7b, d, e and appropriate zero intensity areas in Fig. 7a, c, g, h). The smaller dots can rather be explained as an imaging of the PI beam cross section as has been confirmed by BSE and EPMA investigations.

The elements Na, Al, Si, K, and Ca appear preferably as complex oxidized inclusions whereas Ti rich inclusions seem to be a rather pure oxide. Due to the extremly high quenching rate it can be assumed that the identified inhomogeneities are real inclusions already present and detected by chemical analysis in appropriate concentrations in the premolten alloy. Precipitation during the melt spinning procedure can only be expected due to a residual gas oxygen pressure above the melt within the argon atmosphere of the inductively heated furnace.

Most of the impurities detected can be ascribed to components of the ferro-alloys of phosphorus and boron. The nature of these raw materials and their varying content of boron and phosphorus and the way of preparation of the premolten alloy could be found as the origin of major component variations and thus for variations of physical properties. The influence of the impurity elements seems to be of less importance within the concentration range in-

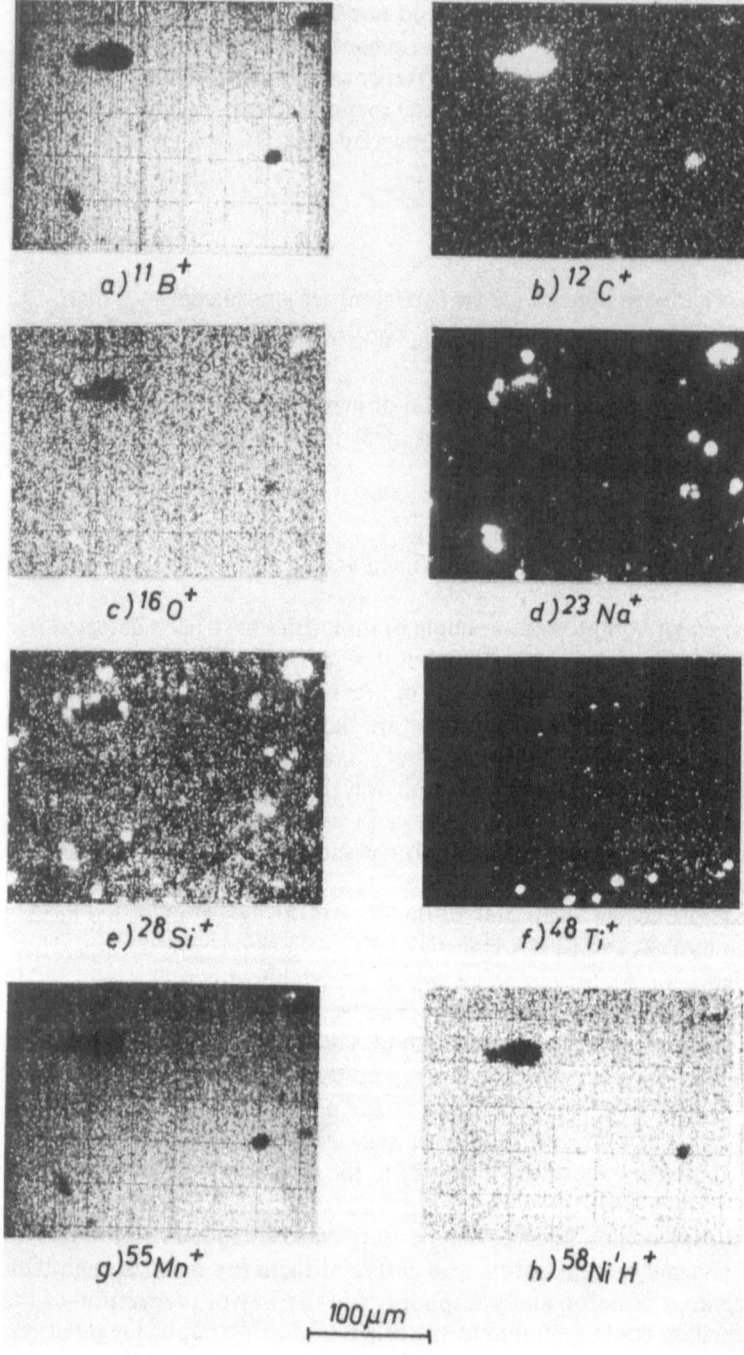

a) $^{11}B^+$

b) $^{12}C^+$

c) $^{16}O^+$

d) $^{23}Na^+$

e) $^{28}Si^+$

f) $^{48}Ti^+$

g) $^{55}Mn^+$

h) $^{58}Ni'H^+$

$\vdash\!\!\!-\!\!\!-\!\!\!-\!\!\!\dashv$ 100 μm

Fig. 7. SI intensity mappings of different SI species of sample II

vestigated here. Particles originating from the crucible material could not be detected.

6.2 SI-Emission from Metallic Glasses

An extensive study of the SI emission from binary alloy systems under $^{16}O_2^+$ bombardment[15] involves also the behaviour of pure iron and nickel-iron alloys. As has been found from XPS spectra in pure iron an unoxidized iron component Fe^0 and the ferric iron Fe^{3+} are present. Further support by valence-band spectra allowed the identification of γ-Fe_2O_3. For the $Fe_{60}Ni_{40}$ alloy the authors have found the disappearance of the Fe^0 state (i.e. the iron component is completely oxidized) and the appearance of Ni^0 and Ni^{2+} peaks. This has been attributed to a component of unreacted nickel metal and the formation of spinel like structures of mixed nickel-iron oxides. Taking into account the results of our own electron diffraction experiments we can suppose that the mass spectra obtained from sample I and sample II are mainly determined by iron- and nickel-oxides and unreacted metal components too. Fig. 8 shows the spectra of polycrystalline pure iron, sample I and sample II measured under steady state conditions of $^{16}O_2^+$ bombardment. The major differences between Fe and sample I are:

- higher intensities of $FeOH^+$, FeO_2H^+, $FeO_2H_2^+$, $FeO_2H_3^+$, $Fe_3O_2^+$, and $Fe_3O_3^+$ for pure iron
- appearance of atomic and atom oxide lines of the other alloying elements and major impurities (Al, Ti, V, Cr, Mn, and Cu) as well as lines of complex ions of different alloying elements in unoxidized and oxidized form for sample I.

The lines of FeO^+, FeO_2^+, Fe_2^+, Fe_2O^+, and $Fe_2O_2^+$ groups of these specimens do not show remarkable differences in height for comparable $^{56}Fe^+$ intensities.
The spectrum of sample II shows

- rather low intensities of NiO^+, Ni_2O^+, $Ni_2O_2^+$, $Ni_3O_2^+$, and $Ni_3O_3^+$ groups in comparison to those of iron (note equal atomic concentrations of Fe and Ni)
- high abundance of $Fe_xNi_y^+$ complex ions in unoxidized and oxidized form. $^{56}Fe^{58}Ni^+$ dominates in the appropriate line group and the $^{56}Fe^{58}Ni^{16}O^+$, $^{56}Fe_2^{58}Ni^{16}O^+$, and $^{56}Fe^{58}Ni_2^{16}O^+$ lines show comparable intensities to the appropriate iron oxide configurations but, in the $Me_2O_2^+$ and $Me_3O_3^+$ groups (Me for Fe and Ni resp.) a discrimination of complex oxides in comparison to the appropriate iron oxides is observed.

In comparison to sample I a similar spectrum of atomic and molecular iron and iron oxide lines is observed but, only those involving diatomic Fe groups reflect the concentration difference for iron between sample I and sample II.
The intensities of the atomic lines of B, C, Si, Ti, V, Cr, Fe and Ni have been used to test the CARISMA program[20] under diverse analytical conditions[22] (PI parameters: $^{16}O_2^+$ and $^{16}O^-$, $U_{PI} = 10$ and 20 kV, $i_{PI} = 6 \times 10^{-7} \dots 1.6 \times 10^{-3}$ A cm^{-2}). At the beginning the metallic glasses were supposed to be ideally homogeneous specimens for major elements and impurities in solid solution. The exchange of the internal standard elements (IS) with precisely known bulk concentrations was supposed to give information about the general applicability of the program. Table 4 shows the results of the same analysis corrected by different IS combinations for sample I. A comparison shows quite good agreement with true concentrations (results of chemical analysis, EPMA and carbon de-

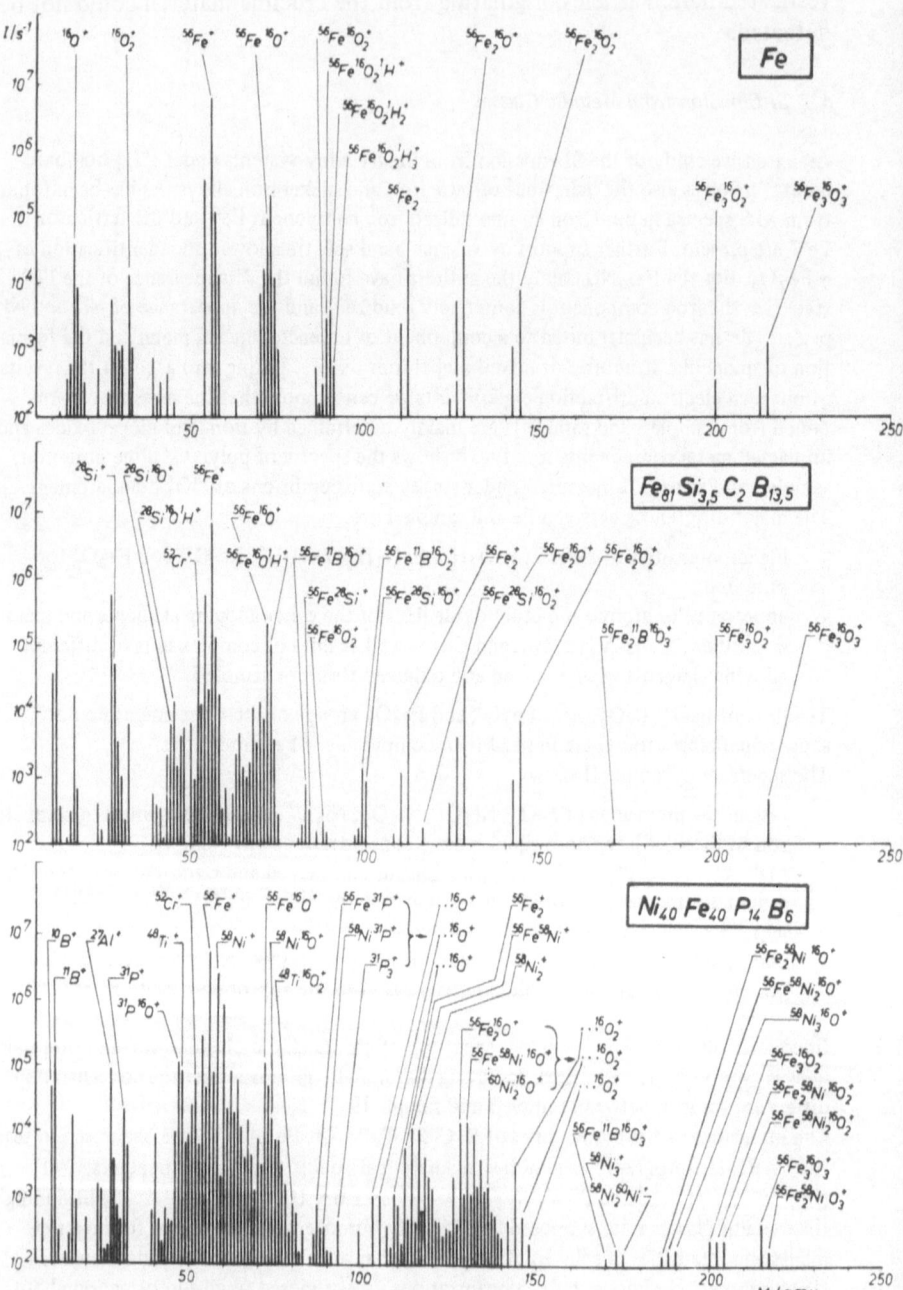

Fig. 8. SI mass spectra of iron, sample I, and sample II

Table 4. Influence of Internal Standard Combination (IS) on the Analytical Outcome; a – $Fe_{81}Si_{3.5}C_2B_{13.5}$; b – $Fe_{40}Ni_{40}P_{14}B_6$ (concentrations in %-at., n.d. – not determined)

	IS elements	B, C, Si, Fe	B, Fe	Si, Fe	C, Fe	B, C	true conc.
	B	14.3	14.3	15.6	16.1	14.3	14.3
	C	2.2	1.3	1.9	2.3	2.3	2.28
a	Si	4.4	3.8	4.3	2.1	1.9	4.33
	Ti	0.02	0.02	0.02	0.01	0.01	n.d.
	Cr	0.1	0.1	0.1	0.06	0.06	n.d.
	Fe	78.6	80.0	77.8	79.2	81.3	79.1

	IS elements	B, P, Fe, Ni	B, Fe	P, Fe	P, Ni	B, P	true conc.
	B	4.2	4.2	30.5	37.8	4.2	4.22
	P	0.8	0.9	14.8	14.8	1.3	14.9
b	Ti	1.9	1.9	1.2	0.2	2.5	1.74
	Cr	0.9	1.0	0.4	0.1	1.2	1.06
	Fe	62.5	61.9	36.5	27.1	66.5	38.2
	Ni	28.2	28.7	15.9	19.6	22.5	38.3

Table 5. Influence of PI Density i_{PI} (in mA cm^{-2}) on the Analytical Outcome (sample – $Fe_{40}Ni_{40}P_{14}B_6$, concentrations in %-at., true concentrations – see Table 4)

i_{PI} elements	6.25×10^{-4}	2.5×10^{-3}	2.5×10^{-2}	4×10^{-1}	1.6
B	4.3	4.3	4.2	4.2	4.2
P	0.8	0.9	0.8	0.7	0.8
Ti	1.0	1.3	1.9	2.2	2.0
Cr	0.6	0.7	0.9	1.1	1.0
Fe	48.4	53.2	62.5	68.8	66.9
Ni	44.1	38.6	28.2	21.5	23.5

termination). Serious deviations occur only for Ti, Cr and Si when C is used as IS and for C when it is not an IS. Complications really occured when the same procedure had been applied to sample II (Table 4b). Generally the phosphorus content is found to be low by more than one order of magnitude if it is not an IS in combination with iron or nickel. But, if it is used as an IS in these combinations B, Ti and Cr data show large deviations from the true concentrations. The major elements iron and nickel show unacceptable results. This indicates that in our experiments elements of high ionization energy show a particular behaviour and the consequences are so much greater as their concentrations in the matrix are. Another serious influence on the analytical results has been observed for increasing i_{PI}. Table 5 shows the behaviour of the CARISMA-results for the IS combination B, P, Fe, Ni. Again the results for phosphorus, iron and nickel are unsuitable. The different behaviour of Ti, Cr, Fe, and Ni (also observed for V and Mn) for change in i_{PI} is clearly demonstrated. This has also been described for crystalline Cr-Fe-Ni alloys[23]. A detailed explanation for this can not be given at the moment. It can be only said that different states of oxidation contribute to the SI creation process and the initial homogeneity of the sampling volume can no longer be supposed under steady state conditions.

Acknowledgements

The author would like to thank ZFW-members of the groups of metallurgical preparation, chemical analysis, spectroscopical methods, and analytical electron microscopy who supplied original data to this paper.

Summary

Characterization of Metallic Glasses by Ion and Electron Microprobes

Possibilities of characterizing metallic glasses are described for the alloys $Fe_{81}Si_{3.5}C_2B_{13.5}$ and $Fe_{40}Ni_{40}P_{14}B_6$. Mean concentrations and elemental distributions of major components and impurities have been determined by EPMA and SIMA. Aspects of the interaction of primary particles and solid atoms and possible interference to the analytical results are discussed on the basis of TEM and ED experiments. The influence of PI parameter variation is shown for different PI densities.

References

1. J.J. Gilman, Physics Today **28**, 46 (1975).
2. O. Henkel and J. Schneider, Proc. 14. Metalltagung. Dresden 1981.
3. W. Jaschinski, W. Wolf, U. König, and J. Hartwig, Tech. Mitt. Krupp, Forsch.-Ber. **39**, 1 (1981).
4. H. Warlimont, Z. Metallkunde **69**, 212 (1978).
5. J.C. Swartz, R. Kossowsky, J.J. Haugh, and R.F. Krause, J. Appl. Phys. **52**, 3324 (1981).

6. D.G. Morris, Acta Met. **29**, 1213 (1981).

7. J.L. Walter, P. Rao, E.F. Koch, and S.F. Bartram, Met. Transact. A **8A**, 1141 (1977).

8. T. Watanabe and M. Scott, J. Mat. Sci. **15**, 1131 (1980).

9. R. Wagner, Proc. 4. Tagung Mikrosonde, Dresden 1978, p. 173.

10. R.W. Calm, J.E. Evetts, J. Patterson, R.E. Somekh, and C.K. Jackson, J. Mat. Sci. **15**, 702 (1980).

11. J.L. Walter, F. Bacon, and F.E. Luborski, Mat. Sci. and Eng. **24**, 239 (1976).

12. H. Fiedler and L. Illgen, Proc. 14. Metalltagung, Dresden 1981.

13. M. Bitterlich, A. Drescher, H. Fiedler, J. Klosowski, W. Löschau, and H. Mai, Poster 14. Metalltagung, Dresden 1981.

14. O. Brümmer (ed.), Mikroanalyse mit Elektronen- und Ionensonden, 2. Aufl. Leipzig: Verlag Grundstoffindustrie. 1980. p. 50.

15. M.L. Yu and W. Reuter, J. Appl. Phys. **52**, 1478 (1981).

16. J.C.C. Tsai and J.M. Morabito, Surf. Sci. **44**, 247 (1974).

17. C.J. Smithells (ed.), Metals Reference Book, 5th ed. London: Butterworth. 1976.

18. G. Blaise, O. Lyon, and C. Roques-Carmes, Surf. Sci. **71**, 630 (1978).

19. M. Riedel, H. Gnaser, and F.G. Rüdenauer, Analyt. Chemistry **54**, 290 (1982).

20. C.A. Andersen and J.R. Hinthorne, Analyt. Chemistry **45**, 1421 (1973).

21. H. Fiedler, H. Mai, and R. Voigtmann, Poster 14. Metalltagung, Dresden 1981.

22. H. Mai, U. Seidenkranz, and R. Voigtmann, ZFW-Res. Rept. Quantitative Verfahren d. Verteilungsanalyse (SIMA), Dresden, Sept. 1980.

23. D.J. Grass and A.v. Rosenstiel, TNO, Apeldoorn, priv. communication.

Correspondence and reprints: Dr. rer. nat. Dipl.-Physiker H. Mai, Zentralinstitut für Festkörperphysik und Werkstofforschung, Postfach, DDR-8027 Dresden, German Democratic Republic.

Mikrochimica Acta [Wien], Suppl. 10, 93–102 (1983)

Institut für Analytische Chemie der Technischen Universität Wien and Metallwerk Plansee Ges.m.b.H., Reutte, Austria

Quantitative Surface Analysis of CVD-Hard Material Coatings with SIMS

By

G. Stingeder, M. Grasserbauer, H.M. Ortner, W. Schintlmeister, and W. Wallgram

With 5 Figures

(Received January 19, 1983)

1. Definition of the Problem

Protective coatings of hard materials are used on hard-metal (WC-Co) tools for improvement of wear resistance[1]. In a further development program, the properties of Al-B-mixed oxide and Ti(C,N,O) have been investigated with respect to wear resistance for cast iron machining and high speed cutting.

These coatings are produced by chemical vapour deposition (CVD) at approximately 1000 °C using $AlCl_3$, BCl_3, $TiCl_4$, CH_4, CO_2, H_2 and N_2.

Extensive analytical characterization of the coatings enables a purposive optimization of the various parameters (i.e. partial pressure of the reagents, temperature and flow rate) of the technical process.
The structure of the layers (using SEM, light microscopy, X-ray diffraction techniques etc.) and quantitative distribution analysis of elements of interest, i.e. N and B in Al-B-mixed oxides and O in Ti(C,N,O) have to be investigated.

2. Selection of Analytical Techniques

For the quantitative distribution of B in Al-B-mixed oxide and O in Ti(C,N,O)-layers only physical in-situ-techniques can be used since the coatings cannot be separated layer-by-layer for subsequent chemical analysis. Requirements for these techniques are:

i)   capability for surface and microstructures analysis – depth resolution of some nm and lateral resolution of a few μm
ii) high detection power in the range of μg/g
iii) capability for quantitative elemental analysis – required accuracy in the order of 25 rel %.

Evaluation of the in-situ-micro- and surface analysis technique available leads to the conclusion that only SIMS (Secondary Ion Mass Spectrometry) and AES (Auger Electron Spectrometry) offer a good chance to cope with the analytical problem.
Preliminary investigations showed that

i)   the detection power of AES was not sufficient for many samples, since the detection limits for the light elements B, N and O are in the order of 0.X %
ii) SIMS can only be applied successfully if the most modern instrumental configuration is used providing high useful yields[2] (e.g. CAMECA IMS 3*f* ion probe) and if the analytical technique is specifically developed for the given problem.

3. SIMS of Al-B-Mixed Oxides and Ti(C,N,O)-Coatings

Due to the different problems associated with the quantitative distribution analysis of B and N in Al-B-mixed oxides and O in Ti(C,N,O)-layers these topics are treated separately here.

3.1 B in Al-B-Mixed Oxide Layers

1–2 μm thick coatings were deposited on a WC-Co (90/10) substrate. SEM-micrographs showed a dense structure with a roughness of about 0.5 μm.

The in-depth profiling technique was used for distribution analysis. Optimization of the measurement parameters led to apply the following conditions for SIMS-depth profiles:

− Primary ions: O^-, 14.5 keV
− Primary current: $\sim$ 100 nA
− beam diameter: $\sim$ 30 μm
− raster: 250 x 250 μm
− analyzed area: 150 μm ϕ

To reduce charging effects on the extremely low conductive Al_2O_3 the surface was covered with a gold-layer of about 20–30 nm thickness. Negative primary ions (O^-) are commonly used in analysing insulating samples. The charging up could not be avoided completely and was about - 60 V. With a

modified ion gun implemented to the instrument after the investigation
of the CVD-coatings a primary beam current of 1 μA O⁻ was reached. A
charging up of -100 V was obtained with 500 nA O⁻ and it was not increasing significantly at higher currents.
The charging effect leads to a shift of the energy distribution and causes a
large loss in secondary ion intensity. Furthermore this change in the energy
distribution may cause a severe distortion of the shape of the depth profile.
It could be shown, however, that complete compensation of the effects
caused by changing is possible by careful adjustment of the energy slit (maximal acceptance 190 eV) of the double focussing mass spectrometer. This
possibility of compensating charge effects seems to be vital for accurate
quantitative analysis.
Under these conditions of analysis the sensitivity is sufficient (as indicated
by the intensity of B-peaks in Fig. 1 and high stability depth profiles can be
obtained. Fig. 2 shows the profiles of ^{11}B, ^{14}N and ^{27}Al during a sputter
time of about 1 $^1/_2$ hours. Such a high stability of the profiles is necessary
for quantification of the results. The sputterrate is extremely low ($\sim$ 200 nm/
hour) but with the new modified duoplasmatron it could be increased by a
factor of about 10.

Quantification

Standards were prepared by co-sputtering Al_2O_3/B_2O_3 mixtures of defined
composition on hard metal substrates. Homogenity (depth and lateral) was

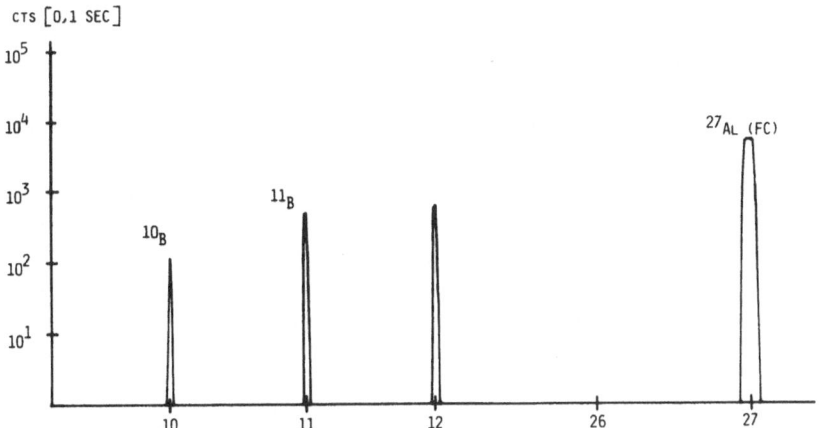

Fig. 1. Mass spectrum of B in Al-B-mixed-oxide coating. B measured with electron multiplier [EM], Al with Faraday cup [FC] (x 1200 less sensitive). B content in Al_2O_3 = 0.2
mol.% B_2O_3

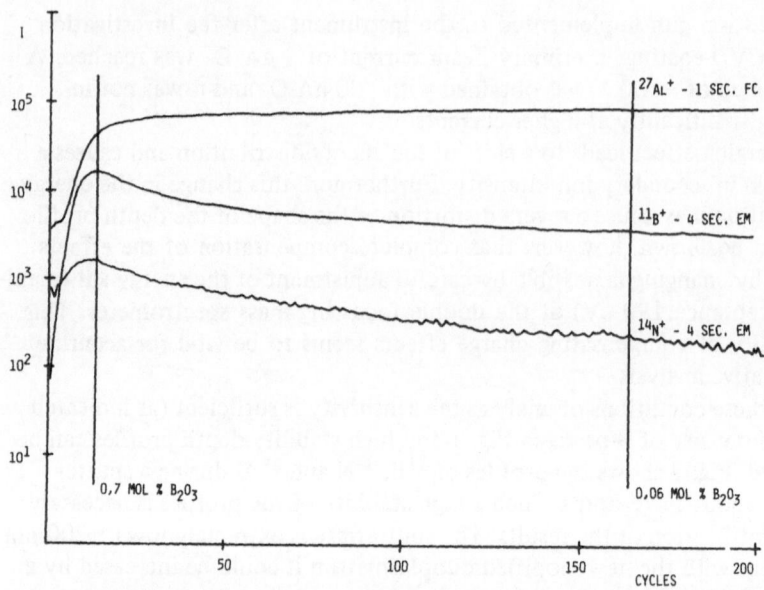

Fig. 2. Depth profile of B, N and Al in Al-B-mixed oxide coating. Quantitative boron-results obtained with calibration curve. Max. conc. of B_2O_3 = 0.7 mol.%, max. conc. of N = 0.5 at.%

tested with SIMS and found to be sufficient for quantitative analysis. The $^{11}B/^{27}Al$ intensity ratio was evaluated and a linearity over at least 3 decades was reached by using two Al_2O_3/B_2O_3 standards (0.7 resp. 7.5 mol.% B_2O_3 in Al_2O_3) and a well characterized glass standard (29 wt.% B_2O_3; 25.3 wt.% SiO_2; 13.7 wt.% Al_2O_3; 6.2 wt.% CaO; 13.9 wt.% BaO; 11.8 wt.% ZnO) to establish a calibration curve (Fig. 3).

The precision of the measurements of each standard is in the order of 1–5% when performed at short time intervals (e.g. during one day). The long time fluctuations of the measurements of the standards are in the order of 15% (rel. standard deviation of $^{11}B/^{27}Al$ values taken during a 6-month period after complete readjustment), but the standards can be implanted easily into the measurement cycle of the samples.

The accuracy can only be estimated from the deviation of the calibration points from the calibration curve and a careful evaluation of all sources for systematic errors since highly accurate reference methods for these concentrations (0.0X–X % B_2O_3 in Al_2O_3) in thin film samples are not available. The influence of surface roughness on the analytical result was investigated by comparative measurements of polished standards (0.3 μm roughness). The difference in the ratios $^{11}B^+/^{27}Al$ is in the order of 1–2 rel % indicating that this influence is negligible.

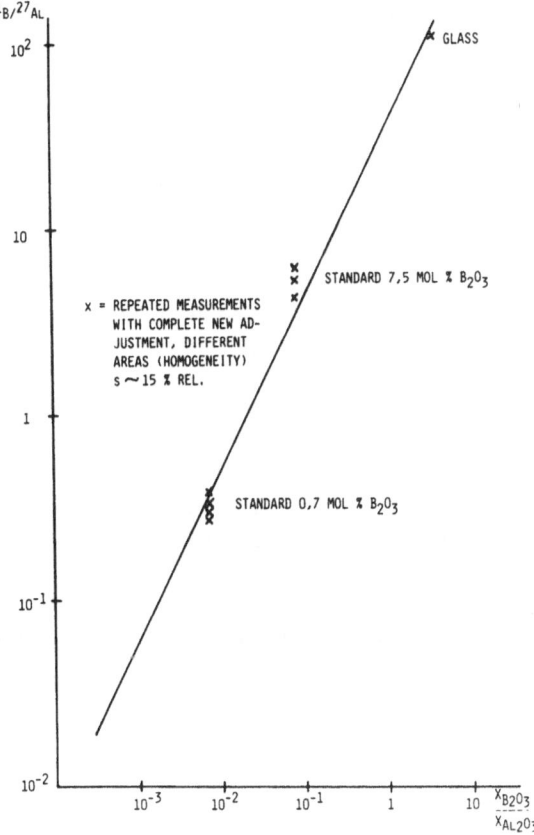

Fig. 3. Calibration curve for determination of B in Al-B-mixed oxide

Detection Power

The detection power could be calculated directly from the mass spectra since there are no interferences at masses 10 and 11. Applying the $3s$-criterion a practical detection limit of 1 μg B_2O_3/g Al_2O_3 was obtained under the analytical conditions used. This limit can be pushed to about 0.05 μg B_2O_3/g Al_2O_3 by optimizing the instrument for maximal detection power and using higher primary ion currents.

3.2 N in Al-B-Mixed Oxide Layers

The analysis was performed with the depth profile technique together with boron (Fig. 2). Only semiquantitative concentration values could be obtained for several reasons:

- well defined standards were not available
- the ionization yield (sensitivity) of nitrogen is low
- nitrogen was implanted by the primary beam and analyzed together with the nitrogen of the sample
- mass 14 shows an interference of CH_2^+. CH_2^+ gives a background of approximately 0.1 % N in Al_2O_3.

Better results would be obtained by using a nitrogen-free mass filtered primary beam and high mass resolution ($M/\Delta M \geqslant 1100$) to eliminate the CH_2^+ interference.

With a filtered O_2^+ primary beam in pure silicon as used in the electronics industry a detection limit of 20 μg/g could be achieved.

3.3 Oxygen in Ti(C,N,O)-Layers

Since the thickness of the Ti(C,N,O)-layers to be characterized was in the order of 5 μm depth profiling could not be used to obtain the distribution of oxygen across the whole Ti(C,N,O)-layer. Therefore distribution analysis was carried out on anglelapped cross sections (angel approximately 2° yielding a width of the layers between 100–250 μm) using step-scanning with a finely focussed ion beam.

The analysis conditions were chosen as follows:
- primary ions: Ar^+, 14.5 keV
- primary current: $\sim$ 30 nA
- beam diameter: $\sim$ 25 μm
- analysed area: 8 μm ϕ

The choice of the primary ions was dictated by the necessity to analyse oxygen. The angle lapped surface was coated with 20–30 nm gold. The charging effects were eliminated completely because a positive charge in small holes in the gold film is easily compensated by secondary electrons emitted from the gold layer.

Quantitative analysis with positive secondary ions. For quantitative characterization of the Ti(C,N,O)-layers positive secondary ions have to be used since all the other major elements (Ti, C, N and O) have to be measured and N forms only positive atomic ions.

Fig. 4 shows the step-scanning profiles of O, N and Ti in a sample of following structure: WC-TiC-TaC-Co-hard metal/TiC/Ti(C,N,O)/TiN. The incorporation of oxygen into a part of the Ti(C,N,O)-layer by the CVD-process could be proved.

For quantification the combination of electron probe microanalysis (EPMA) and SIMS was used. The $^{16}O^+/^{48}Ti^+$ signal-ratio was evaluated with external

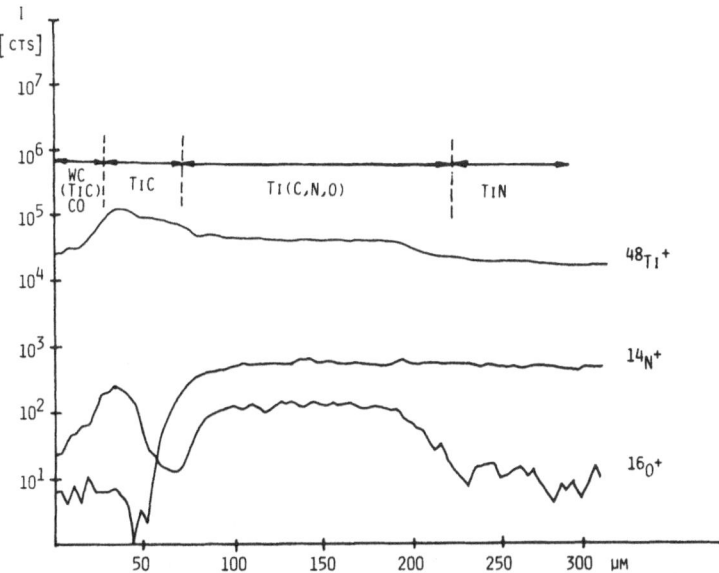

Fig. 4. Step-scan profiles of O, N and Ti across TiC, Ti(C,N,O) and TiN-layers (angle-lapped specimen). Lateral resolution = 8 μm. Oxygen concentration in Ti(C,N,O) = 0.9 wt.%

standards. The CVD-Ti(C,N,O) standards had a higher concentration (1–2 wt.%) and were characterized with EPMA with an accuracy of ± 10–20 rel % (using mixed TiN/TiO$_2$ sputter standards and TiO$_2$).

The comparison of quantitative results of EPMA and SIMS showed an agreement in the order of 25 rel % for samples of 1–2 wt.% oxygen. Since the two techniques applied exhibit completely different systematic errors this value may serve to characterize the accuracy of SIMS.

The great advantage of SIMS over EPMA is its surface sensitivity and its much higher detection power (see Table 1).

Compound specific distribution analysis with negative secondary ions. Compound specific distribution analysis of hardmaterial coatings is of interest because it should enable to get information about the incorporation of various elements into the deposited compounds (like Ti(C,N) or TiN), especially at interface zones where allocation on the basis of elemental analysis may be not sufficient.

As systematically shown by Werner[4] compound specific information can be extracted from the molecular ions. In a CVD-TiN sample containing small amounts of O and C the following molecular ions were detected in the negative ion mass spectrum: CN$^-$, CO$^-$, NO$^-$, TiN$^-$, TiO$^-$.

The molecular ions indicate that there is a nearest-neighbour relationship

Table 1. Detection Power of SIMS for Second Period Elements Calculated from Mass Spectra and Profiles over TiC/TiN [μg/g]

EPMA Pract. DL.	Element	Ion	SIMS Theoret. DL.	Pract. DL.
2000	C	$^{12}C^+$	100	
		$^{12}C^-$	50	
		$^{26}CN^-$	0.1	
8000	N	$^{14}N^+$	1000	
		$^{62}TiN^-$	100	
1000	O	$^{16}O^+$	200	
		$^{16}O^-$	0.5	50
		$^{30}NO^-$	20	

Conditions: SIMS: Ar^+, excitation energy 14.5 keV; beam current $\sim$ 20 nA; diameter of analyzed area 8 μm. EPMA: Excitation energy = 8–10 keV; beam current $\sim$ 200 nA; diameter of analyzed area 2 μm

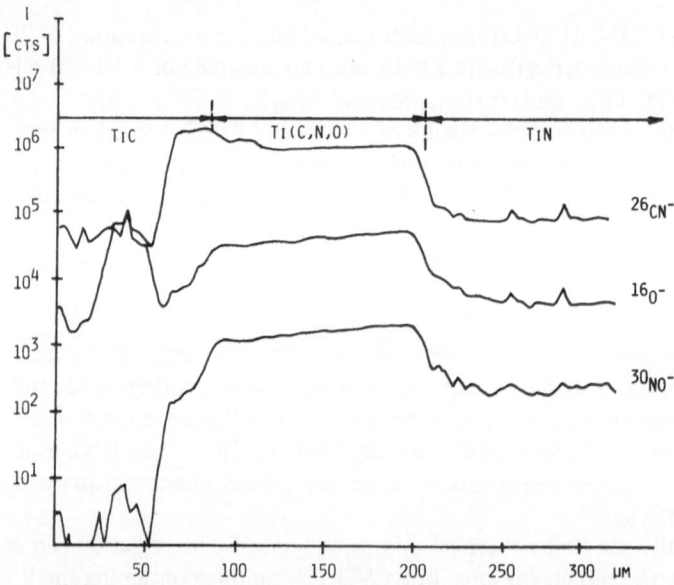

Fig. 5. Step-scan profiles of O^-, CN^- and NO^- across TiC, Ti(C,N,O) and TiN-layers (angle-lapped specimen). Lateral resolution = 8 μm. Oxygen concentration in Ti(C,N,O) = 0.9 wt.%

between these elements, or in other words, that C and O are incorporated into the TiN-lattice (most likely on lattice sites). Fig. 5 shows distribution profiles of NO^- and CN^- across TiC/Ti(C,N,O) and TiN-layers. NO^- enables the direct characterization of the oxygen incorporated into the TiN and allows a clear distinction from the oxygen in TiC (Al_2O_3-particles originating from sand blasting). The dynamic range of NO^- is higher than that of O^- CN^- enables to characterize the TiC/TiN diffusion zone and the region of C-deposition into the TiN.

It is not clear yet if those compound specific profiles can be calibrated for quantitative compound specific analysis with SIMS. It is evident, however, that – despite of the possibilities of the occurence of artefacts due to atomic mixing – SIMS may be a very powerful technique for compound specific surface analysis. Its main area of application may be seen for problems for which high lateral resolution is required since XPS up to now is not able to cope with the resolution problem and the capability of AES for compound specific analysis is very limited.

Detection Power

The theoretical detection limits extrapolated from the profiles and mass spectra vary between 0.1 μg/g for CN^- and 1000 μg/g for N^+ (see Table 1). The practical detection power (which is determined by spectral interferences, contamination from primary beam and residual gas etc.[5]) can be much lower. The practical limit of detection for oxygen with $^{16}O^-$ is by a factor of 100 higher than the theoretical detection limit because H_2O is the major residual gas compound (estimated partial pressure of about 10^{-9} mbar) and it is deposited to the sample surface even during sputtering. Theoretical and practical detection power for C, N, O can be increased mainly by use of Cs^+ as primary ions.

4. Conclusion

SIMS could be applied successfully to characterize the structure of coatings on hard metals and to assist the purposive optimization of the CVD-process.

It enables quantitative distribution analysis also of low concentration elements with reasonable accuracy and high sensitivity. Spatial resolution can be achieved either through direct depth profiling of surface layers or step-scan-microanalysis of anglelapped specimens. Compound specific analysis by evaluation of the molecular spectrum may yield important information on the incorporation of minor elements into the various matrices. A critical evaluation of SIMS for these problems leads to the conclusion that it is the most powerful technique – and therefore apt to become a standard method – but combination with other physical methods like AES or EPMA is often necessary, mainly for calibration purposes.

Summary

Quantitative Surface Analysis of CVD-Hard Material Coatings with SIMS

In the development of highly wear resistant coatings on hard metals information on the quantitative distribution of trace elements in hard material layers of only a few μm thickness is necessary.
The distribution of boron and nitrogen in Al-B-mixed-oxide and oxygen in Ti(C,N,O)-respectively Ti(N,O)-layers should be determined quantitatively.

After evaluation of different surface techniques the capabilities of SIMS using the ion microprobe CAMECA IMS 3f for this problem were investigated systematically.
The Al_2O_3 coatings were characterized with the depth profiling and the Ti(C,N,O)-layers with the step-scanning technique.
For the quantitative analysis of boron and oxygen and accuracy of better than ± 25 rel % could be achieved, but the analysis of nitrogen was only semiquantitative. Spatial resolution obtainable is a few nm for depth profiling and several μm for step-scanning profiles (corresponding to about 20–50 nm on the depth scale).
The high detection power of SIMS for depth profiling and microanalysis makes this method superior to other surface techniques.
Additional information can be gained by using molecular ions. The measurements indicate that SIMS is a very powerful technique for compound-specific microanalysis at high lateral resolution.

References

1. W. Schintlmeister, O. Pacher, W. Wallgram, and J. Kanz, Metall 34, 905 (1980).
2. F.G. Rüdenauer, Instrumental Aspects of Spatially 3-Dimensional SIMS-Analysis, in: Secondary Ion Mass Spectrometry II. A. Benninghoven, J. Giber, J. Laszlo, M. Riedel, and H. W. Werner (eds.). Berlin-Heidelberg-New York: Springer-Verlag. 1982.
3. E. Zinner, private communication.
4. H.W. Werner, Mikrochim. Acta [Wien], Suppl. VII, 1977, 63.
5. E. Zinner, Electrochem. Soc., in press.

Correspondence and reprints: Dipl.-Ing. Gerhard Stingeder, Institut für Analytische Chemie, Technische Universität Wien, Getreidemarkt 9, A-1060 Wien, Austria.

Mikrochimica Acta [Wien], Suppl. 10, 103–110 (1983)

Analytical Laboratories of Henkel KGaA, Düsseldorf, Federal Republic of Germany

ISS- and SIMS-Analysis of Thin Organic Layers on Metal Surfaces*

By

Herbert Puderbach

With 5 Figures

(Received January 19, 1983)

1. Introduction

Important for the development of industrial degreasing and cleaning products are

- good knowledge of the special chemical and physical reactions on the surface
- good knowledge of the materials used and their properties and last but not least
- experience with plant technology and procedure.

Different test methods can be used to probe the cleaning efficiency. These test methods differ enormously with regard to rating and procedure.

With the help of electron microscopy and X-ray microanalysis we have been able to assist in the development of products for special requirements, in the research of basic questions and in the analysis of defects[1-9].

Furthermore we did ISS- and SIMS-analysis to study the adsorption behaviour of surfactants and anticorrosion oils on metal surfaces. With these experiments we wanted to examine the adsorption behaviour of various substances on different metal surfaces and test whether it is possible to identify these substances on technical surfaces.

Some information on the composition of industrial cleaning products are of

---

* ISS   = Ion Scattering Spectrometry
  SIMS = Secondary Ion Mass Spectrometry

interest for this problem range. A degreasing in aqueous solutions is only possible in the presence of surfactants which displace the hydrophobe grease film on the surface. The cleaning efficiency of the surfactants is increased many times by combining them with suitable builders and complexing agents.

As builders alkaline cleaners contain alkaline hydroxides, carbonates, silicates, orthophosphates, condensed phosphates and borates. Non-ionic and anionic compounds are employed as surfactants.

The neutral cleaners are divided into builder containing and builder-free products. Hydrogen carbonates, borates and expecially ortho- and condensed phosphates are used as builders. These products have a high content of surfactants which are, depending on the application field, anionic and/or nonionic compounds.

The builder-free cleaning products consist of surfactant combinations and may additionally contain inhibitors, solubilizer and anticorrosion oils.

2. Experimental

The instrumentation used was a combined ISS/SIMS equipment of 3 M-Brand Company, Minnesota. The ISS-analysis were done with ^{3}He as probe ion and an acceleration voltage of 2.000 eV. The beam current ranged from 50 to 150 nA. The scattered ions were detected by using an electrostatic analyzer of cylindrical mirror type and a channel electron multiplier detector.

The secondary ions were generated with ^{20}Ne and mass analyzed with a quadrupole mass filter. It had a mass range capability of zero to 300 Atomic Mass Units with a resolution capability of one AMU and a transmission of 2.2×10^{-5}. Acceleration voltage and beam current were similar to the ISS-examinations.

The metal samples (aluminium sheets, cold rolled steel strip) were taken from production-line plants. The surfactants and anticorrosion oils were manufactured in our company. In this form they are also employed in industrial cleaning products.

3. Results

3.1 ISS-Examinations for the Adsorption of Surfactants and Anticorrosion Oils on Metal Surfaces

As far as we know up to now a direct proof of surfactants on the surfaces of technical products has not been possible. Till now we analysed with transmission electron microscopy as an indirect method[2]. The comparison of an untreated sample with a sample treated with a wetting agent, is the basis of this method. For this purpose both samples are dipped for a short time into sulfuric acid (15% H_2SO_4, 45 °C, 2 sec). The untreated sample shows an obvious pickling attack on the surface, while the sample treated with the wetting agent is only affected on the grain boundaries (Fig. 1). From that it can be concluded that wetting agents produce on the metal surface an inhibitory layer against acid attacks. This way different surfactants can be compared.

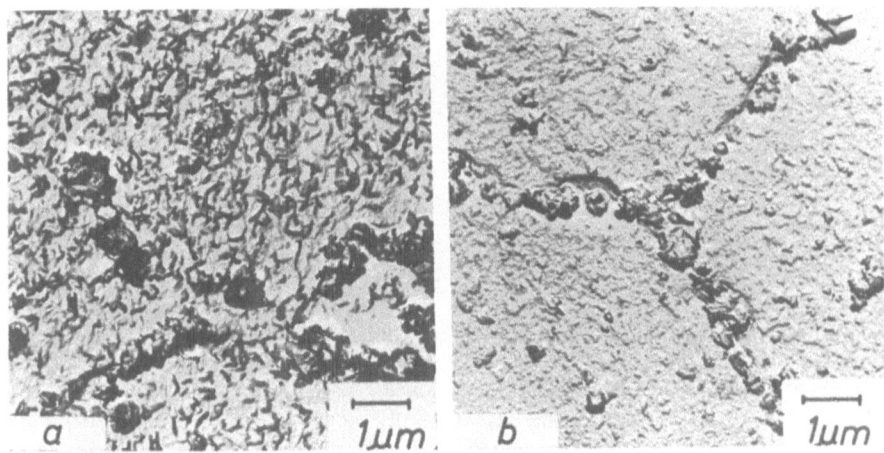

Fig. 1. Acid-pickling on steel strips; (a) untreated surface, (b) surface treated with wetting agent

In a first test series we used 10% aqueous solutions of the non-ionic surfactant nonylphenol with 9.5 moles EO, the anionic surfactants alkyl sulfonate and linear alkyl benzene sulfonate (LAS) and the anticorrosion oil triethanolamine. Aluminum and steel strips were dipped into the solutions for 30 sec and air-dried.

Fig. 2 shows the results of the ISS-analysis on the aluminum surfaces. Sodium and calcium found on the untreated strip are due to product and procedural conditions, carbon may be a residue from a degreasing bath, fluorine is apparently a residue remaining from a flux during aluminum electrolysis in the dry way. The surfactant alkylaryl polyglycol ether produces an inhomogenous cover as a strong aluminum peak is observed. From the absence of an aluminum peak in the case of alkyl sulfonate and linear alkyl benzene sulfonate we conclude, that they deposite as even layer on the surface. The carbon, oxygen, sodium and sulfur peaks are correlated with the molecular structure. The anticorrosion oil triethanolamine forms

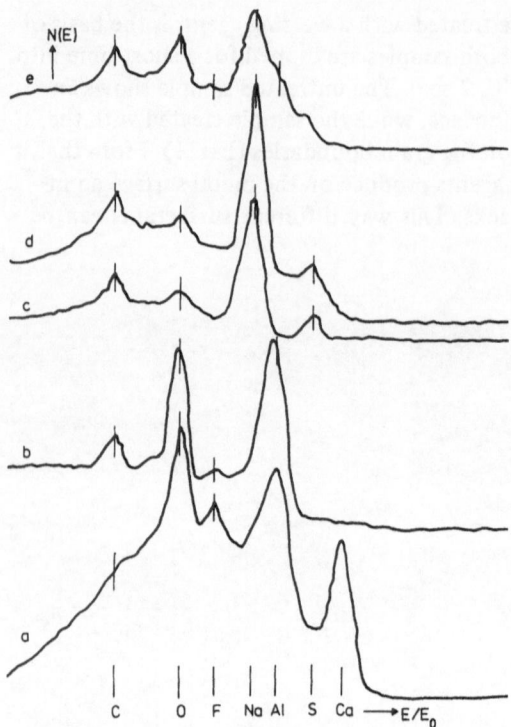

Fig. 2. ISS-spectra of treated aluminum strips (10% aqueous solutions); (a) untreated, (b) treated with alkylaryl polyglycol ether, (c) treated with alkyl sulfonate, (d) treated with linear alkyl benzene sulfonate, (e) treated with triethanolamine

a relatively homogenous layer, as only a small aluminum shoulder is observed.

Fig. 3 shows the results of the steel surface analysis. Each of the three surfactants produces a homogeneous cover on the steel surface. Only in the spectrum of the anticorrosion oil a strong iron peak is observed, so triethanolamine does not cover the steel surface completely. This is unusual for a good inactivating agent like triethanolamine. A possible explanation is, that the strips have been cleaned with alkali containing cleaners in the production line and a thin silicate film built up during this procedure, inhibits the surface. Therefore the steel strip was immersed in hydrochloric acid to remove this silicate layer. After rinsing with water and air-drying it was dipped into the triethanolamine solution and analysed. This time the anticorrosion oil covered the surface homogenously. The poor adsorption behaviour observed before was in fact due to the silicate layer.

In other test series carried out with 10% and 2% aqueous solutions of the surfactants and the anticorrosion oil the metal strips were rinsed thoroughly

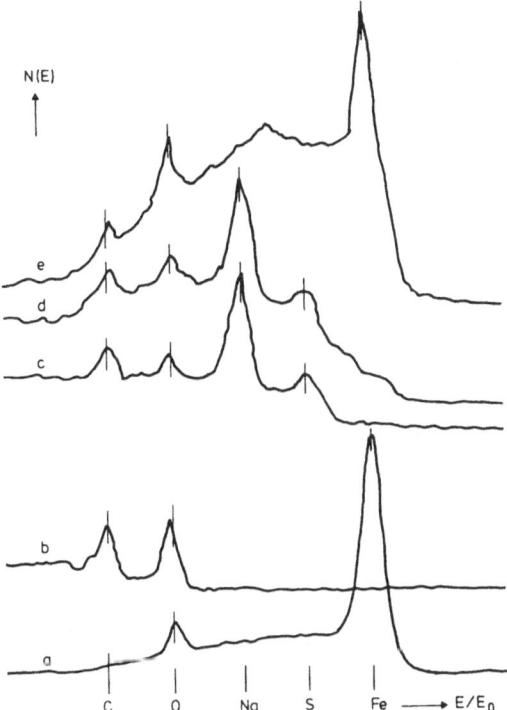

Fig. 3. ISS-spectra of treated steel strips (10% aqueous solutions); (a) untreated, (b) treated with alkylaryl polyglycol ether, (c) treated with alkyl sulfonate, (d) treated with linear alkyl benzene sulfonate, (e) treated with triethanolamine

after treatment. These experiments, which were based on conditions found in practice – that means decreased product concentration and thorough rinsing – confirmed the results obtained in the first test. An example is shown in Fig. 4. These investigations show that ISS provides a good method to prove the adsorption behaviour of cleaning agent components. The influence of the surfactant and the metal surface itself for the adsorption on the surface can be well demonstrated, even for practical product concentrations.

3.2 SIMS-Analysis of Organic Adsorption Layers on Metal Surfaces

Some reports about SIMS-analysis of organic compounds on metal surfaces have been published. Benninghoven and coworkers worked out excellent applications of metal supported organic compounds such as amino acids, peptides, drugs and vitamins[10−21]. In all cases parentlike molecular ions of

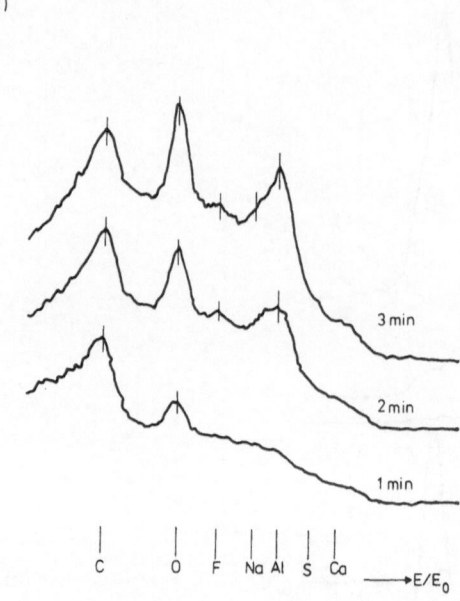

N(E)

3 min

2 min

1 min

C   O   F   Na Al   S   Ca E/E_0

Fig. 4. ISS-spectra of aluminum treated with linear alkyl benzene sulfonate (2% aqueous solution, water rinsed)

the general type $(M + H)^+$ and/or $(M - H)^-$ were found as well as characteristically large fragment ions correlated with functional groups. As supporting metal clean silver foils were used to avoid the influence of hydroxy groups and oxygen of the surface layer on the fragmentation.

Our experiments aimed to show the possibility of identifying surfactants on technical surfaces.

In the positive mass spectrum of the untreated steel strip mainly Fe^+, Na^+, Al^+, Si^+, K^+, Ca^+, O^+ and traces of SiO^+, $SiOH^+$, Cu^+ and FeO^+ were found. Within the mass range 100 to 200 only the mass 112 (Fe_2^+) was observed. In the negative spectrum O^-, Cl^-, SiO_2^-, SiO_3^- and organic fragments are indicated. The organic fragments are residues of rolling oils which can never be removed completely by a single cleaning procedure.

Fig. 5 shows the positive SIMS-spectrum of the steel surface treated with alkylaryl polyglycol ether. Various new fragments are observed, which must be correlated with the surfactant.

This result is confirmed by comparing this spectrum to a gas phase mass spectrum of alkylaryl polyglycol ether. Our proposal for an identification of the detected fragments is also given in Fig. 5.

In test series with 10% and 2% solutions and following intensive rinsing with water less surfactant signals were found. The influence of the metal surface (oxygen and hydroxy groups) was dominating. However a clear

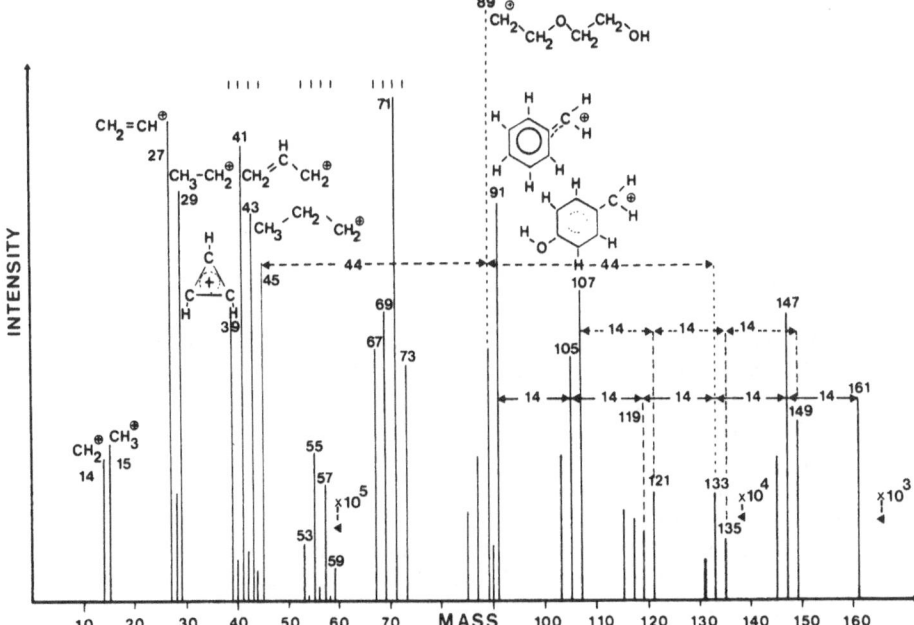

Fig. 5. Positive SIMS-spectrum of alkylaryl polyglycol ether on steeel strip (10% aqueous solution)

distinction between the treated and the untreated sample can be seen. These differences provide adequate information for practical problems. In the case of the other surfactants and the anticorrosion oil characteristic masses are proved. Inspection of the respective gas phase mass spectra and the SIMS spectrum of the untreated surface permit conclusions regarding the nature of the substance. According to the molecular structure of the surfactants S⁻ and SO⁻ fragments are found in the negative spectra.

The oxygen and hydroxy groups influence of the first surface layers was very strong in all test series with the treated aluminum strips. A fragment identification was hardly possible.

The experiments performed till now demonstrate that SIMS offers good opportunities for the proof of organic compounds on technical metal surfaces. Further examinations will be carried out with various other organic compounds and metal surfaces.

Summary

ISS- and SIMS-Analysis of Thin Organic Layers on Metal Surfaces

ISS offers a direct method for the analysis of the adsorption behaviour of surfactants and anticorrosion oils on metal surfaces. Even concentrations relevant under practical conditions can be tested. With SIMS it is possible to identify these organic layers on technical metal surfaces, if comparisons between test series with different concentrations and gas phase mass spectrometry are made.

References

1. H. Puderbach, Beitr. elektronenmikroskop. Direktabb. Oberfl. **5**, 671 (1972).
2. H. Puderbach and R. Schoenemann, Metalloberfläche **28**, 81 (1974).
3. H. Puderbach and R. Schoenemann, Metalloberfläche **29**, 107 (1975).
4. H. Puderbach, W. Friedemann, and T. Kamino, Metalloberfläche **30**, 401 (1976).
5. H. Puderbach and R. Schoenemann, Metalloberfläche **30**, 553 (1976).
6. W. Friedemann, H.J. Göhausen, and H. Puderbach, Aluminium **53**, 471 (1977).
7. H. Puderbach, H. Gotta, and K.H. Gottwald, Metall **31**, 1096 (1977).
8. H.J. Göhausen and H. Puderbach, Metall **33**, 250 (1979).
9. H.-G. Germscheid, Tenside-Detergents **19**, 3 (1982).
10. A. Benninghoven, D. Jaspers, and W. Sichtermann, Appl. Phys. **11**, 35 (1976).
11. A. Benninghoven, K.H. Müller, M. Schemmer, and P. Beckmann, Proc. 7th Int. Vac. Congr. **2**, 1063 (1977).
12. A. Benninghoven and W. Sichtermann, Org. Mass Spectrom. **12**, 595 (1977).
13. A. Benninghoven and W. Sichtermann, Analyt. Chemistry **50**, 1180 (1978).
14. A. Benninghoven, D. Jaspers, and W. Sichtermann, Adv. Mass Spectrom. **7B**, 1433 (1978).
15. S. Tanaki, A. Benninghoven, and W. Sichtermann, Springer Ser. Chem. Phys. **9**, 127 (1979).
16. A. Benninghoven, NBS Spec. Publ. **519**, 627 (1979).
17. A. Eicke, W. Sichtermann, and A. Benninghoven, Org. Mass Spectrom. **15**, 289 (1980).
18. W. Sichtermann, M. Junack, A. Eicke, and A. Benninghoven, Z. analyt. Chem. **301**, 115 (1980).
19. A. Benninghoven and W. Sichtermann, Int. J. Mass Spectrom. Ion Phys. **38**, 351 (1981).
20. W. Sichtermann and A. Benninghoven, Int. J. Mass Spectrom. Ion Phys. **40**, 177 (1981).
21. A. Benninghoven, SIMS III Proc. 3rd Int. Conf., 438 (1982).

Correspondence and reprints: Dipl.-Phys. Herbert Puderbach, ZR-FE Analytik - Phys. Methoden, Henkel KGaA, Postfach 1100, D-4000 Düsseldorf, Federal Republic of Germany.

Mikrochimica Acta [Wien], Suppl. 10, 111–117 (1983)

Österreichisches Forschungszentrum Seibersdorf and Institut für Experimentalphysik
der Universität Wien, Austria

New Developments in Spatially Multidimensional
Ion Microprobe Analysis

By

W. Steiger, F. Rüdenauer, H. Gnaser, P. Pollinger, and H. Studnicka

With 4 Figures

(Received January 19, 1983)

Due to the combination of lateral imaging properties, localized ion emission
and controlled removal of surface layers, ion microprobe mass spectrometry
inherently is capable of spatially 3-dimensional constitutional solids analysis.
Attempts to utilize this capability have been reported in the literature[1,2].
However, only the computerization of microprobe operation and the developments in on-line storage of a large number of digitized ion micrographs,
together with the developments of digital image processing and display
algorithms, made 3-dimensional distribution analysis a feasible mode of
solids characterization.
A sensitive problem area in secondary ion mass spectrometry is quantitation
of multidimensional elemental distributions. In addition to the problem of
computing time there is the problem of availability of internal standard
elements, needed by many quantitation algorithms. A promising development in this direction is the use of ion implanted internal standards[3]. In a
modified version of this technique the primary beam implanted atoms themselves are used as a continuously produced internal standard[4]. In this paper,
the use of an In-primary ion beam (produced in a liquid metal ion source)
for this purpose is described. This new type of ion source presently is considered to open up new possibilities for ion microprobe analysis of solids
with a lateral resolution limit in the 10-nm range[5].

Instrumental

The scanning ion microprobe used in these investigations has been previously described[6]. In addition to the standard duoplasmatron gas ion source (mass separated), an EHD ion source of the capillary type (not mass separated), operated with an indium charge[2], is mounted on the sample chamber. Up till now, almost exclusively Ga^+ ion sources were used for extreme microfocus applications[8]. For SIMS primary ion sources, however, indium is expected to have certain advantages compared to gallium (or, for that matter, oxygen), most of them being connected to the heavy mass of the indium atom: (a) increased sputtering rate; (b) short range in sample, reduced cascade mixing and increased resolution in depth profiling; (c) high secondary ion yields[4]; (d) potential use as internal implantation standard.

The computer configuration for instrument control and data evaluation was described previously[6]. Particularly for purposes of display of multidimensional ion images we have developed a 512 x 512 x 4 bit video memory (VIM) with a real-time 4 bit look-up table for contrast and pseudocolor manipulation. Also, the local PDP 11/34 has been coupled to a PDP 11/45 having access to a disk pack for mass storage of ion images. A software package (FORTRAN IV, RSX-11M) has been developed to handle multidimensional ion images:

VOLUME: acquisition of 3-dimensional elemental distributions in the form of a stack of 2-dimensional 256 x 256 x 4 bit-images covering n mass cycles of up to 8 elements.

TRANSX: selection of a "transaxial projection"[6] by intersection of the image stack along a selected line or column.

IPOLA: interpolates an image with arbitrary number of lines and columns to an image with equal number of lines and columns.

VIM: transfer of images from PDP 11/34 to video refresh memory.

IMPRO: communication between CPU and VIM; transfer of pixel data; read and write LUTs, read and write cursor; erase VIM; read in of single frame from TV-camera into VIM (frame-grab).

FRAME: transfer of single 512 x 512 x 4 bit image from VIM to magnetic disk with on-line compression to 256 x 256 pixel.

IDP: "image depth profiling"; computation of up to 16 local depth profiles in the 3-dimensional image stack; up to 16 square regions of interest (ROI) of arbitrary size and location are defined in a coaxial or transaxial image and overlayed onto the image(s) on the screen; image intensities inside each ROI are summed for every mass cycle and written as a regular depth profile file onto magnetic disk (Fig. 1).

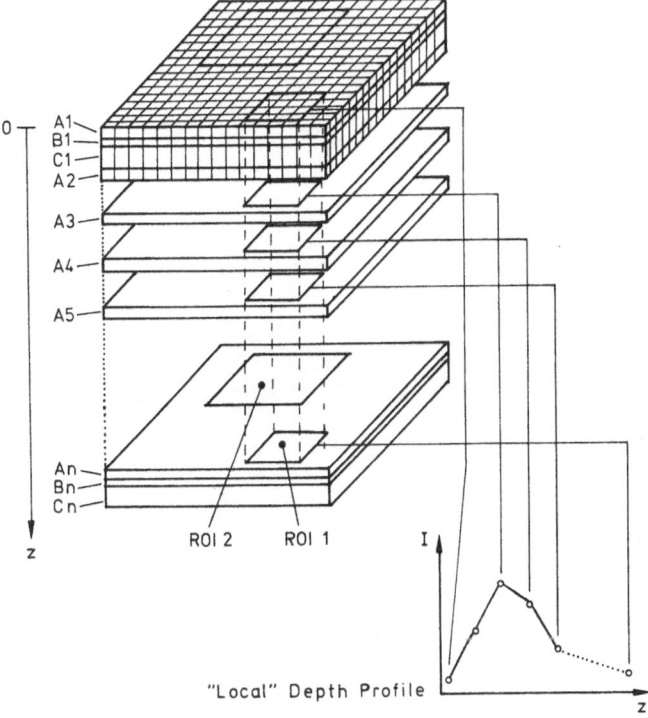

Fig. 1. Image depth profiling, schematic; n mass cycles shown for elements A, B, C: ROI ...region of interest; z...depth coordinate, I...ion count, integrated over ROI

Local Depth Profiles

Local depth profiling has been found to be a very useful mode of obtaining information concerning the depth distribution of an element[9]. Here, the original data base is a "volume" (stack of 2-dimensional images for multiple masses); the convenient thing about local depth profiling now is that the operator can decide on the optimum size and position of the profiling area(s) within the field of view *after* a preview of the elemental distribution across the full field of view. Fig. 2 shows the distribution of H and Cr across an INCONEL sample exposed to a hydrogen discharge in a TOKAMAK fusion device. These coaxial images were taken from a 3-dimensional image stack covering a total sputtered depth of ca. 230 nm. Note the strong enrichment of H^+ near the lower image boundary. The overlayed squares in the coaxial images mark 4 ROIs, in which local depth profiles are computed through the full image stack. Obviously, the depth distribution of hydrogen resembles an implantation profile in the enriched zone whereas it is constant with depth in the other ROIs (Fig. 3). The most likely interpretation is that

Fig. 2. Ion Images of H^+ (coax: u.l., transx: l.l.) and Cr^+ (u.r.) from INCONEL sample; coax projection along indicated column; ROIs superimposed

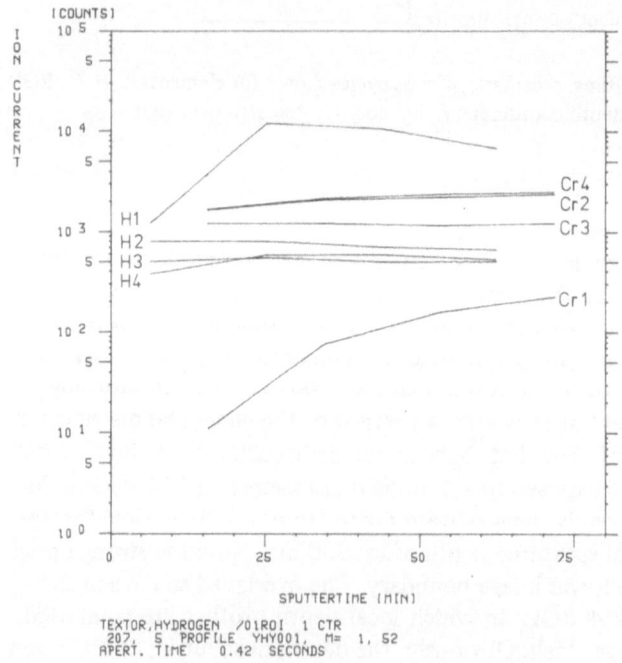

Fig. 3. Image depth profiles for Ni and Cr in ROI's of Fig. 2

the sample was locally contaminated (SiO_2-grain) previous to the plasma exposure. In the grain, the implanted hydrogen retained its original implant distribution (corresponding to an energy of 3 keV), in the base material hydrogen, due to its large diffusion coefficient in metals, very quickly became homogeneously distributed through the whole sample.

Quantitation of Depth Profiles Using Internal Implant Standards

The shape of a recorded SIMS-depth profile might strongly differ from the actual depth distribution of an element (due to the matrix effect) and even completely confuse the information to be extracted. Fig. 4 shows a profile of a sandwich sample consisting of 22 (probably partially interdiffused) alternate 10 nm Ni- and Cr-layers. Bombardment conditions were: 3 keV In-ions from the EHD source, 10 nA, 5 micron beam diameter. Certain artefacts can be observed: Cr shows almost no modulation, Ni and In are synchronously modulated. Obviously, all yields are enhanced in the Ni-layers, leading to an almost complete disappearance of the Cr modulation. The implanted equilibrium concentration, $\bar{c}$, of In, may be calculated[10] as

$$\bar{c} = (1/2Y) \cdot (1 + \mathrm{erf}\,(R_p/\sqrt{2}\sigma)) \tag{1}$$

where Y is the total sputtering yield of the (In-saturated) sample, R_p and σ are projected range and range straggling respectively of In in the sample. These parameters are almost identical in Cr and Ni, so that in this particular sample the surface concentration of In is almost independent on depth, with exception of the first 10 nm. Y is calculated from the time needed to sputter through one Ni- and Cr-layer, R_p and σ are taken from standard tables (Y = 8.3, R_p/σ = 2.45). Using now the equilibrium concentration of the primary beam species (1) as an internal standard for an LTE-type correction routine (to be performed at each incremental point of the profile), the concentration curves in the top portion of Fig. 4 are obtained (calibrated in ppm). The modulation of Ni now is retained, Cr-minima appear at the same depths as the Ni-maxima and In shows constant concentration in depth. The fact that the minimum Cr concentrations in the Ni layers are only slightly less than 50% could be ascribed to the inherent uncertainty of data obtained by LTE-type corrections or to initial interdiffusion of the layers. Auger measurements, done on the same sample, however yielded very similar results[11]. Concluding it may be said that the primary beam species, implanted in a sample during analysis, may serve as an internal standard for the correction of depth profiles even in situations where the matrix effect plays a dominant role.

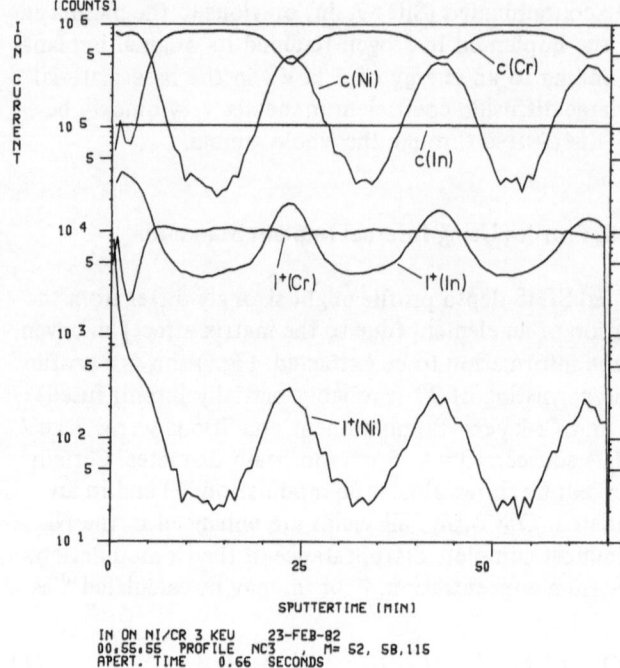

Fig. 4. Uncorrected (I^+) and corrected depth profile (c, in ppm) of same sandwich; primary beam: 3 keV In^+, $p = 2$. $E-8$ Torr

Acknowledgements

The authors thank Dr. D. Marton for supplying the Cr/Ni sandwich samples and Doz. Dr. W.O. Hofer for supplying the hydrogen-exposed INCONEL sample.

Summary

New Developments in Spatially Multidimensional Ion Microprobe Analysis

Requirements for the implementation of spatially 3-dimensional ion microprobe analysis are described and applications of this technique are demonstrated. A software package for computer control of the ion microprobe and display of digital secondary ion images has been written for a PDP 11/34 minicomputer. Novel developements are the measurements of an arbitrary number of local depth profiles within a scanned frame, and the use of the primary beam implanted ion as an internal standard for quantitation of depth profiles.

References

1. E. Berkey, in Microstructural Analysis, J. L. McCall and W. M. Mueller (eds.). New York: Plenum Press. 1973.
2. H.W. Werner, Acta Electron. **19**, 56 (1976).
3. D.M. Drummer and G.H. Morrison, Analyt. Chemistry **52**, 2147 (1980).
4. H. Gnaser, F. G. Rüdenauer, H. Studnicka, and P. Pollinger in Proc. 29th Int. Field Emission Symposion. Gothenburg 1982, in print.
5. R.L. Seliger, R.L. Kubena, R.D. Olney, J.W. Ward, and V. Wang, J. Vac. Sci. Tech. **16**, 1610 (1979).
6. F.G. Rüdenauer and W. Steiger, Mikrochim. Acta [Wien] **1981 II**, 375.
7. F.G. Rüdenauer, P. Pollinger, H. Studnicka, H. Gnaser, W. Steiger, and M.J. Higatsberger, in Secondary Ion Mass Spectrometry, SIMS III, A. Benninghoven, J. Giber, J. Laszlo, M. Riedel, and H. W. Werner (eds.). Berlin-Heidelberg-New York: Springer-Verlag. 1982.
8. A. Waugh, in Proc. 29th Int. Field Emission Symposium. Gothenburg 1982, in print.
9. A.J. Patkin and G.H. Morrison, Analyt. Chemistry **54**, 2 (1982).
10. F. Schulz and K. Wittmaack, Rad. Eff. **29**, 31 (1976).
11. P. Braun, unpublished results (1982).

Correspondence and reprints: Prof. Dr. F.G. Rüdenauer, Österreichisches Forschungszentrum Seibersdorf, Lenaugasse 10, A-1082 Wien, Austria.

Mikrochimica Acta [Wien], Suppl. 10, 119–134 (1983)
© by Springer-Verlag 1983

Max-Planck-Institut für Eisenforschung GmbH., Düsseldorf, Federal Republic of Germany

Investigation of Grain Boundary Segregation in Iron-Base Alloys by Auger Electron Spectroscopy

By

H. J. Grabke, H. Erhart, and R. Möller

With 10 Figures

(Received January 19, 1983)

1. Introduction

Most materials, metals and ceramics, are polycrystalline and contain more or less grain boundaries. The grain boundaries are of great importance for material properties. Dissolved atoms tend to segregate at grain boundaries and they can induce strong changes in the mechanical and corrosion behaviour. Some phenomena caused by grain boundary segregation of impurities in steels may be enumerated: temper embrittlement, creep embrittlement, intergranular corrosion, stress corrosion cracking and so on.

The equilibrium segregation of dissolved atoms in a metal is a reversible process, the grain boundary concentrations increase with the bulk concentration of the solute atoms and decrease with increasing temperature. The rate of equilibration depends on the diffusivities of the dissolved atoms. Thus equilibrium is easily established for carbon and nitrogen in iron even at temperatures down to 200 °C, whereas for phosphorus and sulfur and other substitutional atoms higher temperatures are necessary. Since the equilibrium segregation leads to an enrichment of the segregated atoms in only one or a few atomic layers at the grain boundary, until recently the grain boundary segregation could only be detected by its effects on materials properties. It is only for some years that surface analytical methods have been applied to detect the segregated atoms at grain boundaries, mainly Auger Electron Spectroscopy (AES) has been used. This paper will describe the method and some of its applications for the study of iron-base alloys and steels.

2. Auger Electron Spectroscopy at Grain Boundaries

2.1 Auger Electron Spectroscopy

The technique is performed in an UHV-system. The main components are an electron gun and an electron energy analyzer, which usually are combined in the CMA (cylindrical mirror analyzer). Electrons with a few thousand electron volts of energy are beamed at the surface of the specimen. These low energy electrons ionize the inner orbits of some of the atoms at the samples surfaces. When one of the atom's outer electrons drops from a higher energy level to the ionized level an Auger electron can be ejected with an energy which is characteristic of the excited atom. A competitive process is the emission of X-rays which prevails for the heavy elements. AES is particularly well suited for detecting lighter elements. The Auger electrons cause small peaks on the energy distribution curve of all secondary electrons, differentiation of this curve is usually employed to emphasize the Auger peaks. In this way the Auger spectra $dN(E)/dE$ vs. E are obtained. Numerous spectra of elements and compounds are listed[1] and tables are available of electron energies of the principal Auger peaks for each element. Thus qualitative analysis by AES of the surface of materials is no problem.

The sensitivity is different for different elements and their different Auger transitions, usually surface concentrations down to about 1 at% can be detected.

AES is a surface sensitive method since the Auger electrons cannot escape from the bulk without energy losses. The escape depth of the Auger electron depends on the matrix and the energy of the Auger electron. The collected data on the escape depth in dependence on the energy show that λ is nearly matrix-independent. There exists a range of energies about 50 to 500 eV where λ is only 5 to 10 Å, which is especially favourable for surface analysis.

2.2 Experimental Procedure[2-5]

For Auger analysis of the grain boundaries notched specimens are mounted in an UHV-system which should operate at a very low base pressure in the range $\leqslant 10^{-10}$ Torr. For the commercially available fracture unit the specimens should have a diameter of about 3 to 4 mm and a length of 20 to 40 mm. For fracture one specimen is introduced into the fracture stage which is cooled by liquid nitrogen. After reaching an appropriate temperature the sample is fractured by impact with a hammer. In case of insufficient vacuum conditions the fresh fracture surface of the cold sample would quickly adsorb gases from the residual atmosphere, mainly CO.

The fractured surface is moved in front of the CMA and analyzed. Different fracture modes can occur. At low temperature steels and iron-base alloys

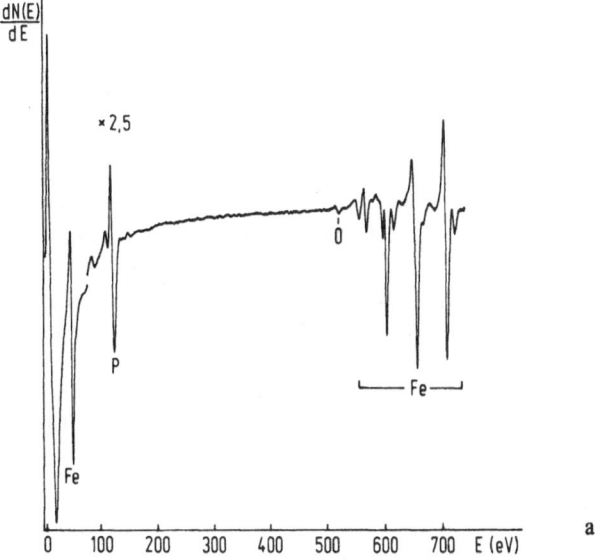

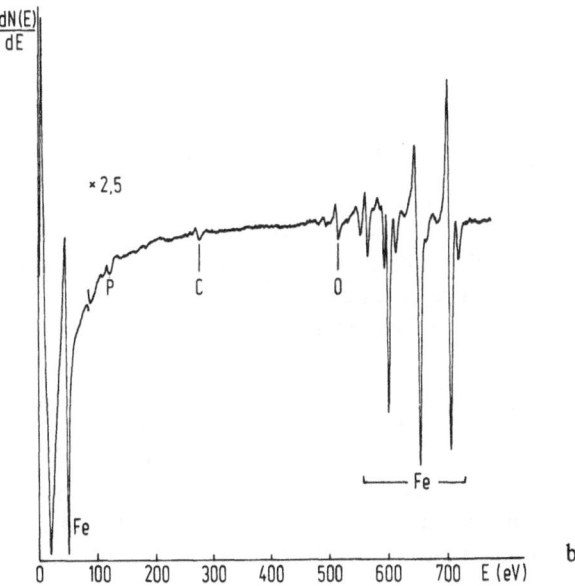

Fig. 1. Auger spectra of fracture surfaces of a Fe-0.17 P alloy, tempered at 450 °C; (a) intergranular fracture surface, (b) transgranular fracture surface

usually show intergranular and transgranular fracture. By using a scanning Auger microprobe (SAM) secondary electron images can be obtained which are helpful in characterizing the fracture mode. The surface can be observed on a TV-monitor. Grain boundary faces are identified where the grain boundary segregants can be detected. The grain boundary facets also are discernible by the Auger peaks of P, Sn, Sb or other impurities which enrich at the grain boundaries (Fig. 1a). For transgranular fracture the Auger spectrum shows the bulk composition (Fig. 1b). In order to obtain a maximum of intergranular fracture the optimum temperature is approximately 50° below the FATT (fracture appearance transition temperature) of the steel[3], i.e. 50° below the temperature where about 50% of the fracture surface is crystalline.

By SAM also Auger elemental maps are obtained fixing the analyzer energy at the Auger peak of a segregant's peak and recording the intensity as the electron beam rasters the specimen surface. Auger spectra are taken of several spots on intergranular fracture regions. The peak height for each element of interest is determined and normalized with respect to one Fe peak. The normalized peak height ratios serve as a measure of the concentration of the segregated elements. The normalisation serves to minimize the effects of instrumental and geometric factors.

2.3 Calibration Techniques

The Auger measurements can be quantified using sensitivity factors[1] but direct calibrations for the applied analyzer and the special studied systems are to be preferred.

One calibration technique which is appropriate for the determination of segregated solutes on surfaces involves the measurement of the peak height ratio as a function of coverage when one element is vapor deposited onto a substrate of a second element[6]. The coverage is measured separately by means of a quartz crystal oszillator, by ellipsometry or by LEED if the substrate is a single crystal.

Another calibration technique involves the analysis of cleavage fracture surfaces of homogeneous alloys[6, 7]. The samples must be homogeneous over the depth of analysis. The calibration can be applied for segregated atoms which are enriched at and near the surface only after correction. A method for that correction has been developed[7] for an idealized model in which the segregate is assumed to lie on the outermost atom layer. In order to derive the final calibration curve a value of the escape depth λ must be selected which is somewhat arbitrary.

Another calibration uses the surface segregation on doped single crystals. This method has been applied for P on iron[5]. Single crystals can be grown of Fe-1% P. Such crystals form a well-ordered, saturated $c(2 \times 2)$ structure of segregated phosphorus on Fe(100) upon heating the crystal at temperatures $\geqslant 800$ °C. This adsorption structure is observed by LEED and it indicates a coverage of $\theta = 0.5$, i.e. half a monolayer of phosphorus on the surface. The corresponding Auger peak height ratio of the phosphorus peak and an iron peak is determined on the same surface. Assuming a linear relation between surface concentration and peak height ratio the calibration is concluded. Similarly, by combination of LEED and AES observation of surface segregation on single crystals, calibrations were obtained for Sn, Si, C and other solute elements on iron[8].

The latter method has a great advantage for its application on grain boundary studies, since in both cases, surface segregation and grain boundary segregation, the segregated atoms are concentrated in only very few atomic layers near the interface or even only in one monolayer.

2.4 Distribution of Segregated Atoms

The value for the surface concentration on an intergranular fracture surface must be multiplied by the factor two, in order to account for the distribution of the segregated atoms on two fracture surfaces in the fracture process. The assumption that the fracture produces an equal division of segregated atoms was proved for an Fe-0.064% P alloy[9] and also for an antimony doped steel[10]. In both cases the analyses of the corresponding grain surfaces on both fracture surfaces showed about the same impurity concentrations.

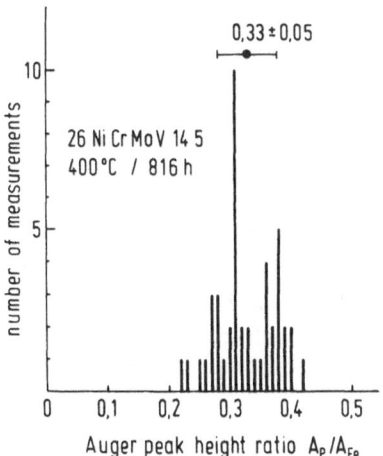

Fig. 2. Histogram of number of measurements vs. peak height ratio, for a measurement of fracture surface of a steel 26 NiCrMoV 14 5 with segregated phosphorus after 816 h annealing at 400 °C

The statement in the preceding chapter on the distribution of segregated atoms in only very few atomic layers near the grain boundary was confirmed by taking sputter profiles of fracture surfaces. The sputter profile for phosphorus on an intergranular fracture surface of Fe-P alloys[5] shows a fast removal of P while only about two monolayers had been sputtered away.

The extent of segregation seems to be different on different grain boundaries. This variation may be expected according to different grain boundary structures, as result from the orientations of the neighbouring grains. The variation in segregation was studied for phosphorus doped steels[10, 11] and antimony doped steels[10] and can be demonstrated by histograms of peak height ratios obtained from many individual grain surfaces analyzed on the sample (Fig. 2). Most boundaries have peak height ratios within about ± 20% of the average value. However, it is not quite evident that the variation only stems from differences in the extent of segregation. Some variation of the measured values is also to be expected according to the varying angle of incidence of the electron beam. The angle of the electron gun axis and the grain boundary faces varies and this fact also causes variations of the peak height ratio[9].

3. Studies of Grain Boundary Segregation in Iron-Base Alloys and Steels

3.1 Grain Boundary Segregation in Fe-P Alloys and Its Effects on Materials Properties[5, 11]

Iron base alloys with P contents between 0.009 to 0.33% were prepared and machined to specimens for Auger studies of fracture surfaces. All specimens were austenized at 1050 °C for 1 h, quenched and then held at 850 °C for 1 h to obtain a large grain size ($\geqslant$ 100 μm). The equilibrium of grain boundary segregation was established by annealing at temperatures in the range 400 to 800 °C for sufficient time. After that, the specimens were introduced into the UHV-chamber, fractured and analyzed as described. For each specimen about 10 spectra were recorded from different spots on intergranular fracture facets. The average values for the P/Fe-peak height ratio are plotted in Fig. 3 for all specimens. The grain boundary concentration increases with increasing bulk concentration of the phosphorus, with increasing equilibration temperature the grain boundary concentration decreases. This behaviour can be described by the Langmuir-McLean equation:

$$\log \frac{\theta}{1-\theta} = \frac{1791}{T} - 0.621 + \log\% \text{ P}$$

when θ is the grain boundary coverage in parts of a monolayer and %P is the bulk concentration in wt.%. With this equation the grain boundary concentration in equilibrium can be calculated for any Fe-P alloy in the α solid solution range.

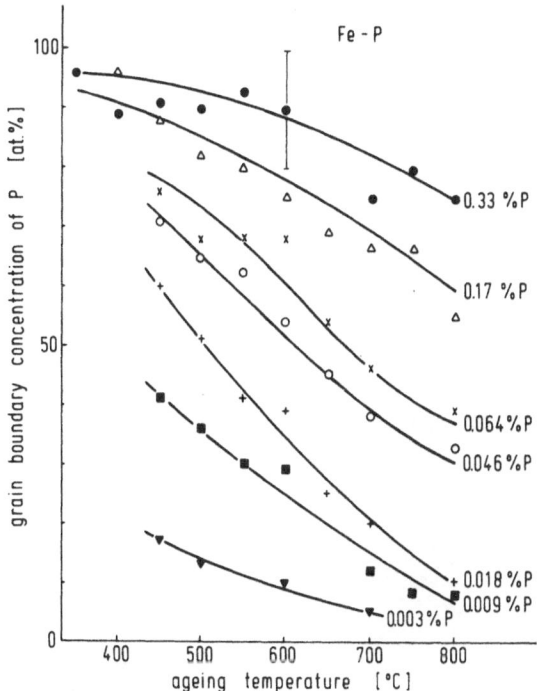

Fig. 3. Grain boundary concentration of phosphorus plotted as a function of equilibrium temperature for different Fe-P alloys

The presence of phosphorus in the grain boundaries induces grain-boundary embrittlement. With increasing grain boundary concentration the fracture mode at low temperature changes from transgranular to intergranular, see Fig. 4. Phosphorus has a weakening effect on the cohesion of the grain boundaries.

The presence of phosphorus at grain boundaries also induces intergranular corrosion in hot nitrate solutions[12]. The samples investigated by AES were also tested in calcium-nitrate at 60 °C, at a constant potential in the passive range, 1000 mV (SHE), the samples with P enriched grain boundaries showed strong intergranular corrosion. The average corrosion current density which was observed after some hours could be clearly correlated to the grain boundary concentration, see Fig. 5. The presence of phosphorus obviously prohibits the passivation of the iron surface at the intersections of grain boundaries and surface.

The effect of several alloying elements on the extent of phosphorus segregation was studied[5]. The elements carbon and nitrogen also tend to segregate at grain boundaries and displacement equilibria were observed:

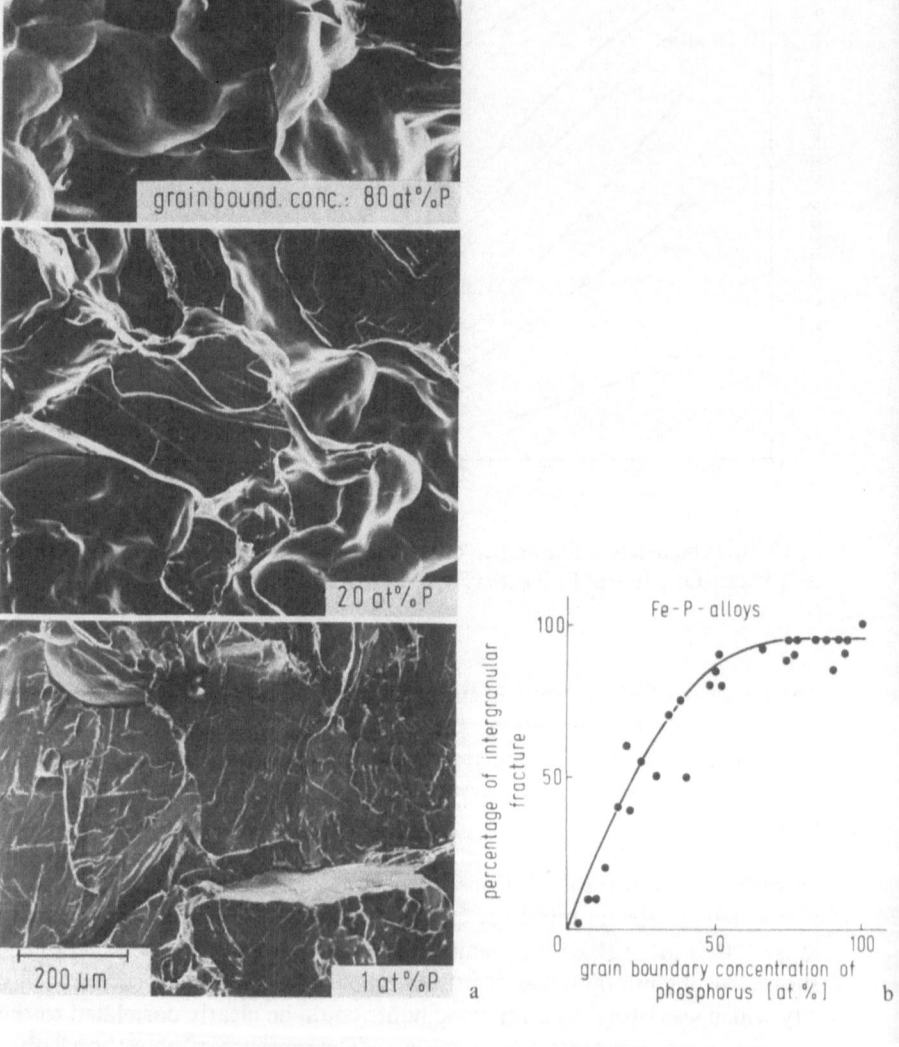

Fig. 4. Effect of phosphorus on the fracture mode of Fe-P alloys; (a) scanning electron micrographs of fractured samples with different grain boundary concentrations of P, (b) percentage of intergranular fracture vs. grain boundary concentration of P

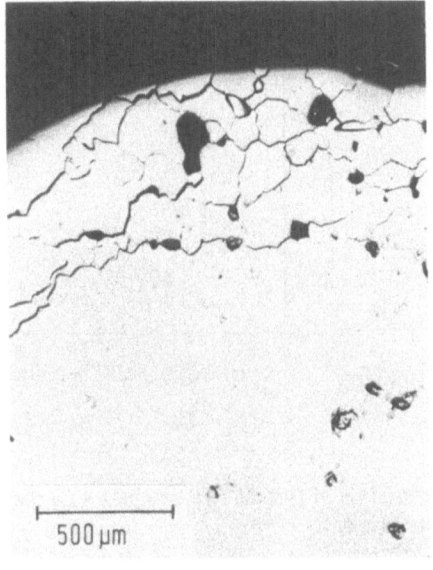
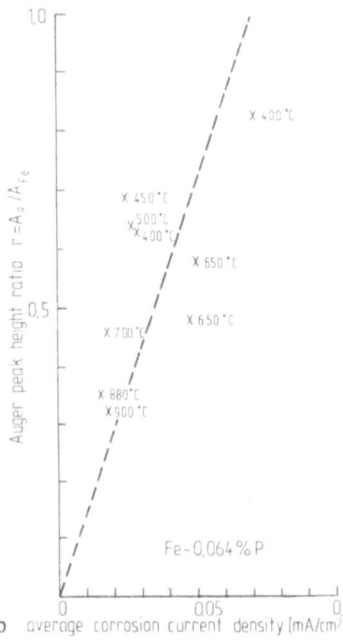

a   b   average corrosion current density [mA/cm²]

Fig. 5. Intergranular corrosion of Fe-P alloys in hot nitrate solution at constant potential +1000 mV (SHE); (a) metallographic cross section of corroded specimen (magnification 50 x), (b) correlation of grain boundary concentration and average corrosion current density for Fe-0.064 P

$$C \text{ (in the lattice)} + P \text{ (at a grain boundary)}$$
$$\rightleftharpoons C \text{ (at a grain boundary)} + P \text{ (in the lattice)}$$

which are determined by the relative bulk concentrations. The effect of carbon hinders phosphorus segregation in unalloyed carbon steels which, therefore, show no embrittlement by phosphorus.

Other alloying elements such as Cr, Ni, Mn, and Si have no direct effect on the P-segregation[5, 11, 13]. However, they affect the P-segregation by their effect on the carbon solubility and activity in iron-base alloys, resp. low alloy steels. Such effects have been misinterpreted by the theory of 'synergetic cosegregation' until recently.

3.2 Temper Embrittlement of a Low Alloy CrMo-Steel

The steel 34 CrMo 4 is used for steel tanks for gases. Specimens of this material investigated by AES showed embrittlement by phosphorus. The conditions for this embrittlement were elucidated[11].

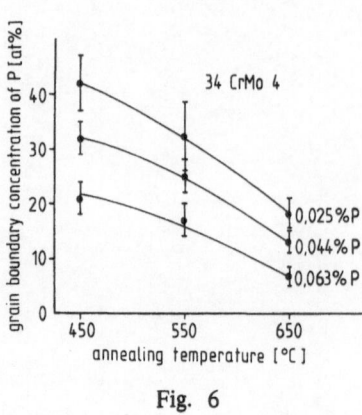

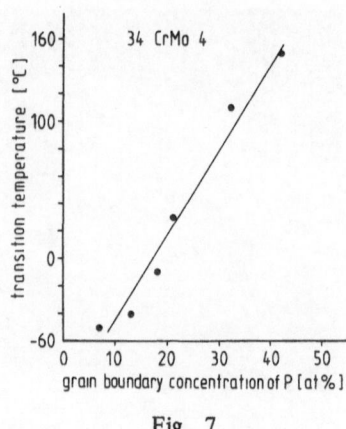

Fig. 6   Fig. 7

Fig. 6. Dependence of grain boundary concentration of P in the steel 34 CrMo 4 on the annealing temperature and the bulk concentration of P

Fig. 7. Dependence of the transition temperature for the steel 34 CrMo 4 on the grain boundary concentration of P

Starting material with the composition: 0.33% C, 1.09% Cr, 0.62% Mn, 0.18% Mo, 0.20% Si was doped in three ingots with 0.025, 0.044 and 0.063% P. After forging, samples were prepared for AES studies and notch impact tests the transition temperature was derived, for the transition between brittle fracture at low temperature and ductile fracture at higher temly establishes the grain boundary concentrations of P, see Fig. 6. From impact tests the transition temperature was derived, for the transition between brittle fracture at low temperature and ductile fracture at higher temperature. The transition temperature increases to higher values with increasing bulk concentration of P and decreasing tempering temperature. If the transition temperature is plotted vs. the grain boundary concentration of P, a linear relation results, see Fig. 7.

The given results allow predictions on the mechanical behaviour of the studied steel. For example, if a steel with 0.035% P (which is an admissable concentration) is tempered at 550 °C the grain boundary concentration will be about 20 at%. This grain boundary concentration leads to a transition temperature above 20 °C, thus brittle fracture can occur at ambient temperature. This problem can be avoided either by a lower P-concentration in the material or by higher tempering temperatures.

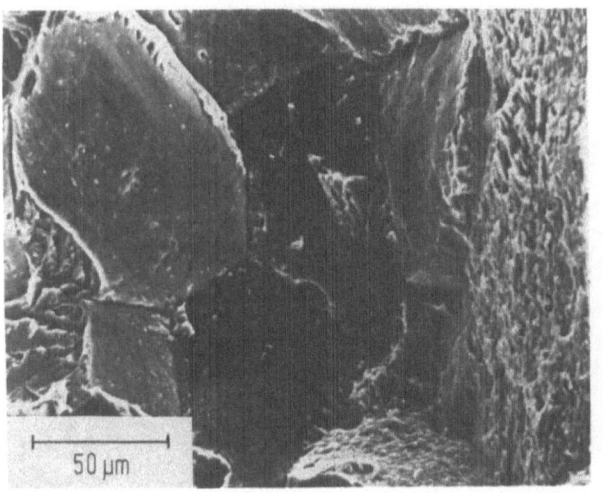

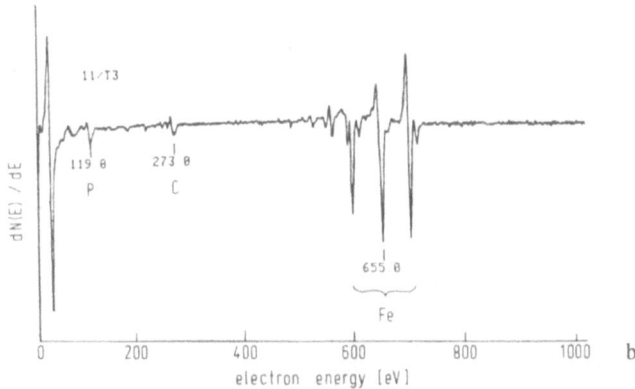

Fig. 8. Brittle failure of a 1% CrMoV-steel, sample fracture after creep test for 2500 h at 530 °C and 300 N/mm^2; (a) scanning electron micrograph of the fracture surface, (b) Auger spectrum taken on the fracture surface

3.3 Creep Embrittlement of Low Alloy CrMoV-Steels

Heat resistant CrMoV-steels are applied in the temperature range 400 to 600 °C under stress. Grain boundary segregation occurs in this temperature range and may affect the creep and rupture of such steels. Similarly as in temper embrittlement phosphorus segregates to grain boundaries, reduces the cohesive strength along grain boundaries and may induce intergranular

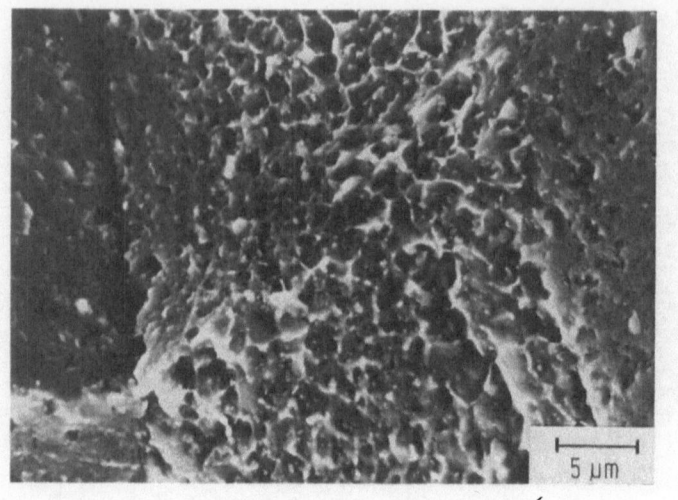

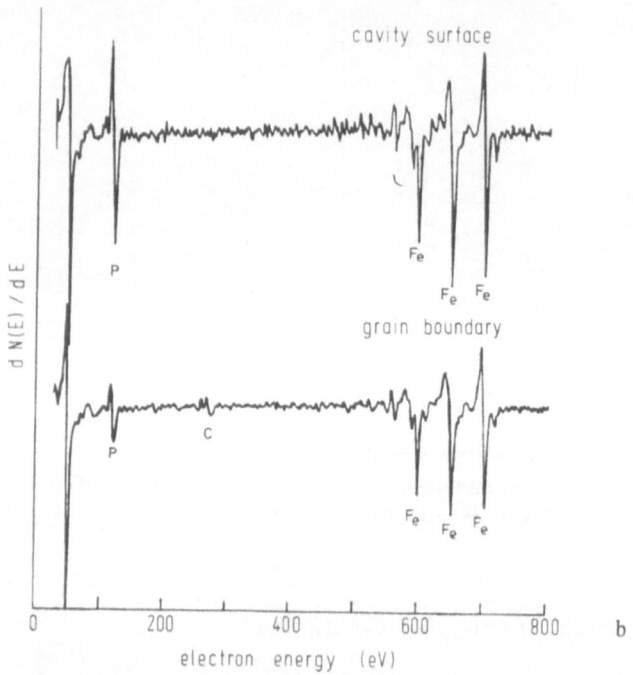

Fig. 9. Cavities at grain boundaries of 1% CrMoV doped with 0.045% P after creep at 546 °C and 265 N/mm² ; (a) scanning electron micrograph of a cavitated area, (b) Auger spectrum out of a cavity surface compared with the spectrum of an even grain boundary region

fracture, see Fig. 8, but only in specimens strongly embrittled by additional factors.

For the deformation and rupture under creep conditions other mechanisms are more important: nucleation and growth of cavities, dislocation mobility, grain boundary sliding etc. These processes certainly can be affected by impurity segregation. The rate of cavity formation at grain boundaries should be enhanced according to the reduced critical stress for the formation of cavities. On the other hand the rate of cavity growth may be retarded since in iron the grain boundary self-diffusivity is decreased by the presence of grain boundary segregates such as P, S, C, and N[14]. Many different effects of impurities in steels on the creep behaviour have been discussed[15-17] and favorable as well as negative influences have been observed.

One study on the effect of P on the creep of a 1% CrMoV-steel[18] showed an increase of ductility and creep rate with increasing P-content. However, this behaviour was mainly caused by an influence of P on the carbide morphology and composition. The presence of a high P-content favored the formation of coarse $(Fe,Cr)_3C$ carbides whereas the formation of fine Mo_2C and V_4C_3 was suppressed, caused by the competitive interaction of P with Mo and V.

Also the concentration of cavities at grain boundaries was increased by the presence of P but this effect was not decisive for the creep behaviour. By applying a small diameter of the primary electron beam (0.2 µm), AES anal-

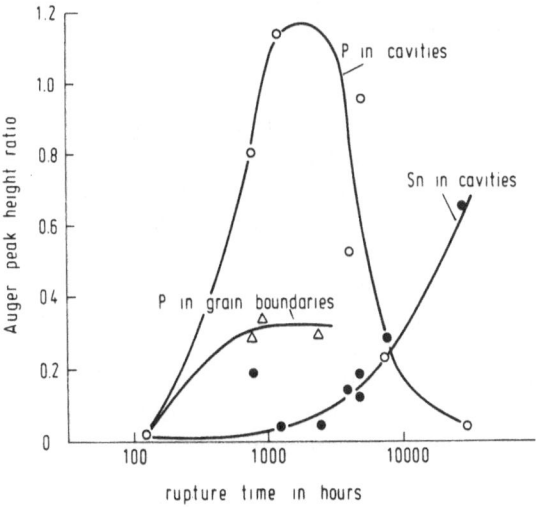

Fig. 10. Changes of the concentrations of P and Sn at the grain boundaries and in the cavities with creep time (time to rupture), data from a large selection of 1% CrMoV specimens crept at 550 °C to fracture

yses could be performed separately for the flat grain boundary areas and for the inside of the cavities. The cavity surfaces show much higher P-concentration than the grain boundaries, see Fig. 9. The free surface segregation on cavity surfaces leads to higher concentrations. Studies of creep specimens with long exposure times have shown that phosphorus in the cavities is slowly replaced by other elements with higher surface segregation enthalpies, such as Sn and Sb[19, 20], Fig. 10. No effect of the applied stress on the level of P-segregation at grain boundaries has been observed. This is in agreement with theoretical estimates.

Still many questions are open concerning the effect of impurities and grain boundary segregation on the creep of steels, however, careful studies on model alloys with the aid of AES should ensure future progress.

4. Conclusions

AES is a very useful technique for the study of grain boundaries in polycrystalline metals. After fracture inside an UHV-system the segregated layers at the grain boundaries can be analyzed, a qualitative and quantitative determination of the elemental composition is possible. The segregation is concentrated in one monolayer or only very few layers, therefore, the high surface-specificity of AES is very favorable.

AES in most cases only renders information about the elements present and conclusions about their chemical state and binding mode are difficult. Such information may be obtained by photoelectron-spectroscopy (XPS, UPS), however, in this method the much lower sensitivity is a disadvantage. Some other surface analysis methods cannot be applied, since ISS and LEED need a plane or even single crystalline surface.

A further advantage of AES is the high lateral resolution which is possible nowadays by using very small beam diameters (0.2 μm or better). In combination with the possibility to obtain secondary electron images of the surface one can analyze small grain surfaces or small details on larger grain boundary facets. In this way selective segregation in cavities on grain boundaries can be detected[19, 20] and also small inclusions in steels can be indentified[21].

A great drawback is the fact that only embrittled grain boundaries are accessible for analysis. Grain boundaries with a high cohesion cannot be opened by fracture. Thus it is not possible to study the segregation of C, N or B in iron-base alloys by AES. These elements do not embrittle the grain boundaries.

One method which can be applied in such cases is field-ion microscopy and mass spectrometry[22, 23]. A preliminary study[23] could be performed on the segregation of P in an iron alloy, but also on the segregation of Mn and C, which would not have been possible with AES. The method has unique capabilities for very sensitive compositional analyses and lateral resolution in the atomic range. However, it is difficult to prepare the specimens, the thin wire tips, to find a grain boundary in such tip and to avoid splitting of the tip in the high electric field. This method could give more information on the segregation of nonembrittling elements and on the structure of grain boundaries than AES.

Summary

Investigation of Grain Boundary Segregation in Iron-Base Alloys by Auger Electron Spectroscopy

By applying Auger electron spectroscopy to the analysis of intergranular fracture surfaces it has become possible to elucidate the grain boundary segregation of solute atoms in metals.
Segregation equilibria in binary pure alloys are studied, for example for Fe-P. The effects of the segregate on fracture mode and corrosion can be correlated to the measured grain boundary concentrations.
Grain boundary concentrations of P and other impurities in steels are measured in dependence on temperature and bulk concentrations and correlated to the ductile-brittle transition temperature. In this way temper embrittlement and long-time embrittlement behaviour can be predicted.
Effects of impurity segregation on the creep of heat resistant steels, on cavity nucleation and growth during creep, are studied.
One drawback of the technique is that the grain boundaries must be embrittled to be fractured. Grain boundaries with non-embrittling segregates can be studied by field ion microscopy and mass spectrometry.

References

1. L.E. Davis, N.C. MacDonald, P.W. Palmberg, E. Riach, and R.E. Weber. Handbook of Auger Electron Spectroscopy, 2nd ed. Edina, Minnesota: Physical Electronics Industries. 1976.
2. D.F. Stein, A. Joshi, and R.P. Laforce, Trans. ASM **62**, 776 (1969).
3. R. Viswanathan and A. Joshi, Metallurg. Trans. A **6A**, 2289 (1975).
4. G. Tauber and H.J. Grabke, Ber. Bunsenges. physikal. Chemie **82**, 298 (1978).
5. H. Erhart and H.J. Grabke, Metals Sci. **15**, 401 (1981).
6. M.P. Seah, Surface Sci. **40**, 595 (1973).
7. L. Marchut and Ch. J. McMahon Jr., Electron and Positron Spectroscopies in Materials Science and Engineering. Academic Press. 1979. p. 183.
8. H. Viefhaus, H. Erhart, and R. Möller, unpublished data.
9. H. Viefhaus, R. Möller, H. Erhart, and H.J. Grabke, Scripta Met. **17**, 165 (1983).
10. C.L. Briant, Acta Met. in press.

11. H. Erhart, H.J. Grabke, and R. Möller, Arch. Eisenhüttenwes. **52**, 451 (1981).
12. J. Küpper, H. Erhart, and H.J. Grabke, Corrosion Sci. **21**, 227 (1981).
13. C.L. Briant, Scripta Met. **15**, 1013 (1981).
14. L. Stratmann and H.J. Grabke, unpublished results.
15. H.R. Tipler and B.E. Hopkins, Metals Sci. **10**, 47 (1976).
16. D.S. Wilkinson, K. Akibo, N. Thyagarajan, and D.P. Pope, Metallurg. Trans. **11A**, 1829 (1980).
17. D.J. Gooch, Metals Sci. **15**, 45 (1981).
18. Jin Yu and H.J. Grabke, Metals Sci. in press.
19. W.G. Hartweck and H.J. Grabke, Scripta Met. **15**, 653 (1981).
20. U. Franzoni, H. Goretzki, and S. Sturlese, Scripta Met. **15**, 743 (1981).
21. H.J. Grabke, H. Viefhaus, and B. Egert, Stahl u. Eisen **99**, 1497 (1979).
22. A.R. Waugh and M.J. Southon, Surface Sci. **68**, 79 (1977).
23. T. Sakurai, Y. Kuk, A.K. Birchenall, H.W. Pickering, and H.J. Grabke, Scripta Met. **15**, 535 (1981).

Correspondence and reprints: Prof. Dr. H. J. Grabke, Max-Planck-Institut für Eisenforschung GmbH., Max-Planck-Strasse 1, D-4000 Düsseldorf, Federal Republic of Germany.

Mikrochimica Acta [Wien], Suppl. 10, 135–143 (1983)
© by Springer-Verlag 1983

Max-Planck-Institut für Metallforschung, Institut für Werkstoffwissenschaften, Stuttgart, Federal Republic of Germany

Characterization of Anodic Oxide Layers by Sputter Profiling with AES and XPS

By

S. Hofmann and J. M. Sanz*

With 5 Figures

(Received January 19, 1983)

1. Introduction

The chemical characterization of thin films can be accomplished by the combination of a surface analysis technique such as Auger electron spectroscopy (AES) or X-ray induced photoelectron spectroscopy (XPS) with ion sputtering[1,2]. The high in-depth resolution[3] of the electron spectroscopies of the order of 1 nm together with the gradual, slow erosion of the surface by the impinging primary ion beam is advantageous for the determination of composition profiles. Although AES spectra contain chemical information, XPS is ideally suited to obtain the chemical bond characteristics of an element in a compound by observation of the chemical shift of the respective elemental peaks[4]. On the other hand, AES generally gives better depth resolution than XPS because of the smaller analysis spot with respect to the sputtered area[1,2,5]. A combination of both techniques should reveal the elemental composition profile with optimum depth resolution (AES) and a chemical bonding profile (XPS). However, besides the many possible pitfalls due to incorrect experimental arrangement as discussed elsewhere[1,6], there is a fundamental limitation in obtaining the true profile: the change of elemental composition and chemical bonds induced by the interaction of the primary ion beam with the sample. The principal result of this effect is an atomic mixing of the first few surface layers[7] with subsequent formation of

* Present adress: Dpto Fisica Aplicada, Universidad Autonoma, Madrid, Spain.

new chemical bonds and a depletion of the elements which are preferentially sputtered.

It has been shown that many oxides suffer reduction induced by sputtering[8], but most results are only qualitative and sometimes contradictory[9]. However, a correct interpretation of oxide profiling results with respect to the original composition can only be made if the ion beam effects are quantitatively known.

The aim of this paper is to show the main problems encountered in the sputter profiling of some anodic oxides (Ta_2O_5, Nb_2O_5, HfO_2, ZrO_2, TiO_2, Al_2O_3) and to present the quantification of the analysis of the altered layer with respect to elemental composition and chemical state.

2. Experimental

Thin foils (100 μm – 300 μm thickness) of the base metal of 99.99% nominal purity and typically 5 x 10 mm^2 area were anodized in the usual arrangement[10] with appropriate electrolytes and thickness/voltage parameters to obtain an oxide layer of about 30 nm thickness[10, 11].

Before mounting the samples in the UHV analysis chamber, they were rinsed subsequently in acetone, ethanol and distilled water under supersonic excitation and dried.

Sputter profiling of the anodic oxide layers was done using a 3 keV Ar$^+$ ion beam at 45° incidence angle rastered over an area of 4 x 4 mm^2. AES and XPS analysis were performed with a double pass cylindrical mirror analyzer (DP-CMA, Perkin-Elmer, Phys. Electronics Div., Mod. 15–255 GAR) with

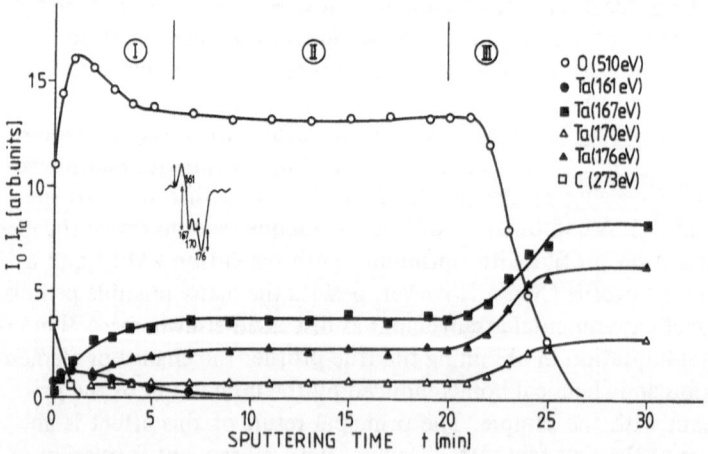

Fig. 1. AES depth profile of 30 nm thick Ta_2O_5/Ta sputtered with 3 keV Ar$^+$ ions. (Signal intensities in peak to peak heights)

30° between surface normal and CMA-axis. A rotable drum within this analyzer with a slit of 12° opening allows emission angle dependent analysis used for non destructive depth profiling for a thickness < 5 nm of the altered layer of Ta_2O_5 [12]. In AES a 2 keV, 15 μA electron beam of about 0.2 mm diameter was employed. For XPS, a Mg $K\alpha$ X-ray source was operated at 10 keV and 40 mA emission current. The CMA pass energy setting at 25 eV corresponds to a constant energy resolution of 0.15 eV. The analyzed area was about 2.5 mm in diameter.

3. Results

The profiling results for Ta_2O_5/Ta as a typical example for sputtering induced oxygen depletion are reported in Figs. 1 − 3. Fig. 4 shows the XPS depth profile for Al_2O_3 which is stable under ion bombardment. The results of all the oxides studied are compiled in Table 1.

3.1 AES Profiling

Fig. 1 shows the AES depth profile of a Ta_2O_5/Ta layer of 30 nm thickness. After sputtering through the contamination layer consisting mainly of hydrocarbons as disclosed by the rapidly decreasing C peak and the corresponding increase of the O peak from the oxide layer below. After about 1 min sputtering time corresponding to ca. 1.5 nm (the contamination layer thickness) the O peak begins to decrease. Contrary to the behavior of the O peak, the Ta (167, 170, 176 eV) peaks are increasing monotonously. Since their intensity ratio stays almost constant with eroded depth and is the same in the oxide layer and the metal (at $t > 25$ min), those peaks contain no chemical information. The behavior of the Ta 161 eV peak, however, is totally different. Its relative intensity/time relation follows that of the O 510 peak with sputtering time. After about 6 min sputtering time corresponding to a removed layer thickness of 7 nm, the oxygen peak stays constant (up to $t = 21$ min) and the Ta 161 eV peak vanishes. Although this indicates a chemical state information of that Auger peak related to the pentoxide (*cf.* sect. 3.2), the XPS profiling analysis shows a clear correlation to the different oxidation states induced by sputtering.

3.2 XPS Profiling

As an example, Fig. 2 shows the XPS depth profile of a 30 nm thick Ta_2O_5/Ta anodic oxide layer. It is rather similar to the AES profile (Fig. 1) with three regions characterized by a transition zone to the steady state region (II), the

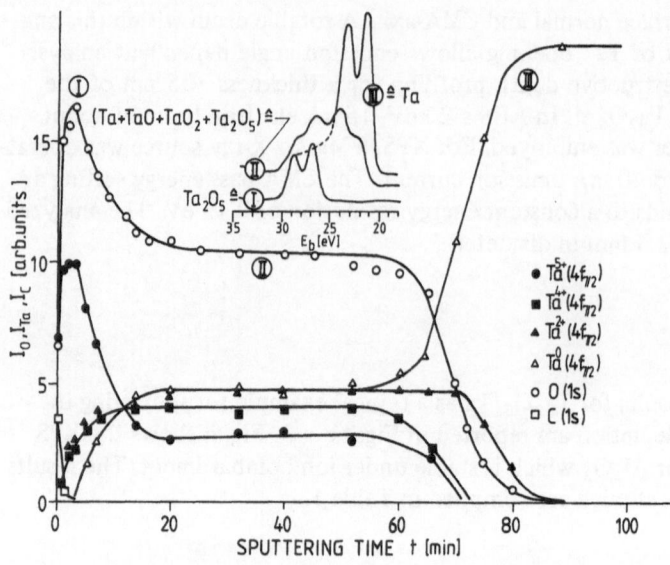

Fig. 2. XPS sputtering depth profile ("chemical profile") of a 30 nm thick Ta_2O_5/Ta layer. (Details see text)

interface region and the pure metal (as seen from the oxygen XPS peak intensity O1s, E_b = 530 eV). The peak intensities are given in the relative peak areas of the C1s (E_b = 285 eV), O1s, and Ta $4f_{7/2}$ peaks. The latter are obtained by the peak fitting (with Gaussian approximation) of the Ta $4f_{7/2, 5/2}$ doublet as shown in Fig. 3 for the steady state region (II). The insert in Fig. 2 indicates the shape of the Ta spectra obtained at the unsputtered surface (I) the steady state region (II) and after reaching the pure metal (III). The chemical shift due to the different valence states of Ta is clearly revealed[11, 12]. Thus, the four states obtained: Ta°, Ta^{2+}, Ta^{4+} and Ta^{5+} corresponding to metallic Ta and the lower oxides TaO, TaO_2, and Ta_2O_5 are plotted against the sputtering time in Fig. 2.

Similar XPS profiles showing depletion and the sputtering induced generation of lower oxides have been obtained for Nb_2O_5, HfO_2, ZrO_2, and TiO_2 [11, 12]. In contrast to these results, Al_2O_3 shows no reduction under sputtering as seen in the XPS profile in Fig. 4.

3.3 Quantitative Composition Analysis

A quantitative analysis of both the AES and XPS results requires a knowledge of the thickness of the altered layer[1, 13]. Assuming a thickness d of that layer of homogenous composition, it can be shown that the intensity of the metal (I_{Me}) peak in the surface layer is[1, 13]:

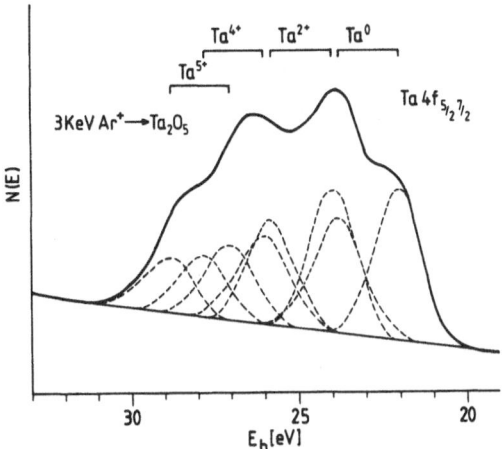

Fig. 3. Example of the peak fitting (broken line) of the Ta ($4f_{7/2, 5/2}$ peaks in region (II) of Fig. 2 with the simplified assumptions of Gaussian peaks of constant FWHM (1.6 eV), linear background substraction and ≈ 1 eV/valence state relative peak shift. After Ref.[12]

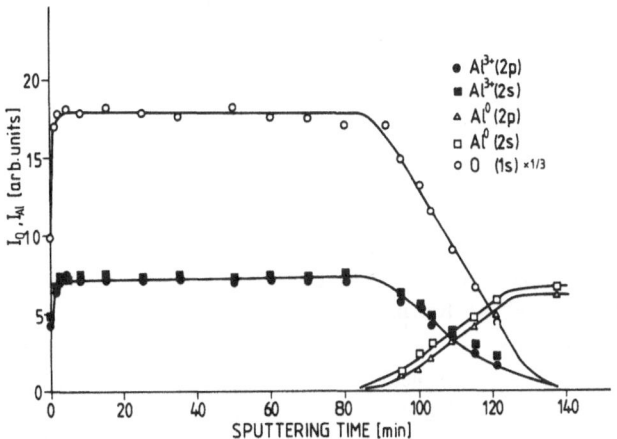

Fig. 4. XPS sputtering depth profile of a 30 nm thick Al_2O_3/Al anodic layer (as in Fig. 2)[12] showing no oxygen depletion

$$I^{s}_{Me} = I^{o}_{Me} \cdot X^{s}_{Me} \left[1 - e^{-d/\lambda_{Me}} \right] \tag{1}$$

and the corresponding intensity I^{b}_{Me} from the undistorted bulk oxide underneath:

$$I_{Me}^b = I_{Me}^o \cdot X_{Me}^b \cdot e^{-d/\lambda_{Me}} \qquad (2)$$

where I_{Me}^o is the standard intensity of the pure metal, X_{Me}^s and X_{Me}^b are the surface layer and bulk metal mole fractions, respectively, and λ_{Me} the effective escape depth[1, 3] of the respective metal Auger or photo electrons in the distorted oxide layer.

The ratio of Eqs. (1) and (2) gives:

$$\frac{I_{Me}^s}{I_{Me}^b} = \frac{X_{Me}^s}{X_{Me}^b} \cdot \left(e^{d/\lambda_{Me}} - 1 \right) \qquad (3)$$

The AES signal intensities cannot differentiate between the altered surface layer contribution and the bulk oxide contribution. Therefore, the solution of Eqs. (1), (2), (3) for the surface layer composition requires the knowledge of d (e.g. ion range[14]) or a second AES peak with different λ_{Me}[1].

In XPS profiling, we are in a much better position since we know the altered surface layer composition from the relative contributions of the deconvoluted metal peaks of the lower oxides in this layer and the bulk oxide below (cf. Figs. 2, 3). In region II, for the case of Ta_2O_5 we arrive at the surface layer composition by comparing the relative peak intensities with the molar fractions of the oxidation states from which they originate, i.e..

$$I_{Ta^0}^s : I_{Ta^{2+}}^s : I_{Ta^{4+}}^s = X_{Ta^0}^s : X_{TaO}^s : X_{TaO_2}^s \qquad (4)$$

Table 1. Composition and Thickness d of the Oxygen Depletion Layer of the Different Oxides After Sputtering with 3 keV Ar^+ Ions in the Steady State Region (II)

Oxide	Decomposition to	Depletion Layer Composition [mole fract]	Oxygen Loss [at.%]	Depletion Layer Thickness d [nm]
Al_2O_3	–	–	–	–
TiO_2	$Ti_2O + TiO$	$X_{Ti}^s = .43$	10	1.8 ± 0.6
ZrO_2	$ZrO + Zr$	$X_{Zr}^s = .62$	29	1.3 ± 0.5
HfO_2	$HfO + Hf$	$X_{Hf}^s = .62$	29	2.0 ± 0.6
Nb_2O_5	$NbO_2 + NbO$	$X_{Nb}^s = .47$	18	2.2 ± 0.7
Ta_2O_5	$TaO_2 + TaO + Ta$	$X_{Ta}^s = .50$	21	2.5 ± 0.7

Eq. (4) yields the values in region II of Fig. 2:

$X_{Ta}^s = 0.18$, $X_{TaO}^s = 0.35$, $X_{TaO_2}^s = 0.47$, which correspond to an average

composition of TaO ($\overline{X_{Ta}^s} = 0.5$). With $I_{Ta}^s/I_{Ta}^b = (I_{Ta^0} + I_{Ta^{2+}} + I_{Ta^{4+}})/I_{Ta^{5+}}$

$= 4.56$ and $X_{Ta}^b = X_{Ta^{5+}} = 0.29$ Eq. (3) can be solved giving $d/\lambda_{Ta} = 1.29$.

With $\lambda_{Ta} = 1.9$ nm we get $d = 2.5$ ($\pm$ 0.7) nm.

The procedure outlined above was applied to the XPS profiles of the other oxides studied. The results are compiled in Table 1.

(For Ta_2O_5, the thickness and composition homogeneity of the altered layer was independently obtained by the elegant method of emission angle variation[15] using a CMA with rotable acceptance slit[1, 12].)

4. Discussion

The characterization of oxides by sputter profiling is generally influenced by oxygen depletion due to preferential sputtering of oxygen. Thus, for Ta_2O_5, Nb_2O_5, HfO_2, ZrO_2, and TiO_2 an apparent oxygen and a corresponding metal intensitiy profile versus sputtering time is obtained, which is not present in the original oxide layer. This is revealed by a transition to a steady state characterized by its decreased AES oxygen intensity. The more powerful technique is XPS, which shows the lower oxide states generated, the quantity of the oxygen loss and the thickness of the altered layer (cf. Table 1) according to the primary ion range. However, there are exceptions like Al_2O_3, which shows no decomposition (Fig. 4), and SiO_2[16]. For these oxides, the measured profile corresponds directly to the original, true profile.

To predict the behavior of other oxides not studied in this paper a correlation to the oxide properties appears highly desirable. There exist thermodynamic models[17, 18] relating the observed reduction to the free binding energy of the oxide. Whereas Kim et al.[17] predict that oxides with $(-\Delta G_0)$ $\geqslant 472$ kJ/mole will remain stable, Naguib and Kelly[18] correlate the amount of reduction to a partial dissociation enthalpy ΔH_a of the suboxide with minimum ΔH_a which should be the only one final oxide phase. Both predictions, however, are not confirmed by our results (Table 1).

According to the Sigmund sputtering theory[19], for 3 keV Ar^+ bombardment we are presumably in the linear collision cascade regime, where a rather small influence of the atomic mass ratio is predicted. Haff[20], assuming equipartition of energy and diffusion-like motion of the atoms, derives for the sputtering yield ratio of alloy components A, B with atomic masses M_A, M_B : $Y_A/Y_B = (M_B/M_A)^{1/4}$. Although it is not yet clear whether spike sputtering conditions exist for the 3 keV Ar^+ bombardment, for the sputtering of oxides they were assumed by Kelly[18], who pointed out the importance of bonding. However, the estimation of surface binding energies of metal and oxygen for an oxide in the collisional spike sputtering regime appears to be highly speculative at present. Restricting our considerations to the mass ratio effect, according to Sigmund[19] $Y_A/Y_B \propto (M_B/M_A)^{1/2}$, therefore, we try a correlation of the altered layer composition $(X_{Me}/X_O)_s$ given by:

$$(X_{Me}/X_O)_s \cong (X_{Me}/X_O)_b \cdot (M_{Me}/M_O)^{1/2} \qquad (5)$$

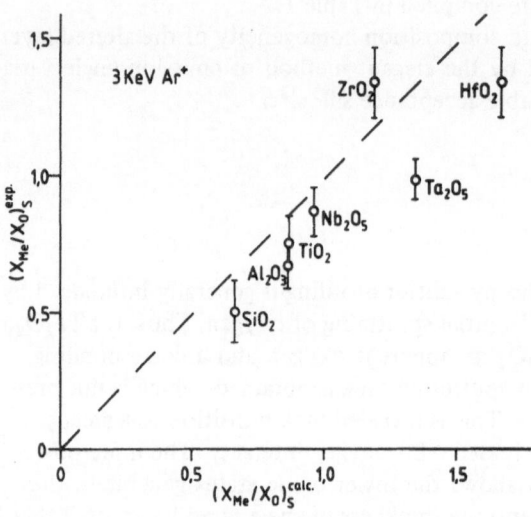

Fig. 5. Experimentally determined composition of the altered surface layer, $(X_{Me}/X_O)_s^{exp}$, according to Table 1 plotted against $(X_{Me}/X_O)_s^{calc}$, predicted by Eq. (5). (SiO$_2$ was studied in Ref.[16])

In Fig. 5, the ratio $(X_{Me}/X_O)_s^{exp}$ for the oxides studied (*cf.* Table 1) is plotted against the ratio $(X_{Me}/X_O)_s^{calc}$ calculated after Eq. (5).

A correlation is obtained which indicates that the mass ratio argument according to Eq. (5) is more decisive in predicting the sputtering induced reduction than the formation and/or binding energies. A slightly better correlation is obtained for a mass ratio exponent 0.4. Nevertheless, Eq. (5) can only be considered as a crude approximation.

In conclusion, it has been shown that the amount of oxygen depletion during sputtering is negligible for light metals, increases with the cation mass and can be at least qualitatively explained by predictions from the Sigmund sputtering theory. The sputtering induced compositional and phase changes have to be taken into account in AES and XPS depth profiling, and quantitative data as compiled here allow to correct the measured profile of an oxide to disclose the original oxide phase present in the sample before sputtering.

Summary

Characterization of Anodic Oxide Layers by Sputter Profiling with AES and XPS

Quantitative depth profiling results obtained by 3 keV Ar^+ bombardment in combination with AES and XPS are presented for the anodic oxides Ta_2O_5, Nb_2O_5, HfO_2, ZrO_2, TiO_2, Al_2O_3. The interpretational difficulty is a sputtering induced oxygen depletion and the generation of new oxide phases in a thin surface layer, which is observed in all the oxides except Al_2O_3 and is discussed in detail for Ta_2O_5. A comparison of the different amounts of oxygen loss shows a clear correlation with the atomic mass of the metallic element according to the spike regime of sputtering. The procedures of quantification of the altered surface layers are outlined which allow a correction of the oxygen loss.

References

1. S. Hofmann, Analusis **9**, 181 (1981).
2. M. Pijolat and G. Hollinger, Analusis **10**, 8 (1982).
3. M.P. Seah and W. Dench, Surf. Interface Anal. **1**, 2 (1979).
4. D. Briggs (ed.), Handbook of X-Ray and UV Photoelectron Spectroscopy. London: Heyden. 1978.
5. J.B. Malherbe, J.M. Sanz, and S. Hofmann, Surf. Interface Anal. **3**, 235 (1981).
6. S. Hofmann, in Proc. 3rd Int. Conf. on SIMS (III), A. Benninghoven, J. Giber, J. Laszlo, M. Riedel, and H.W. Werner (eds.). Berlin-Heidelberg-New York: Springer-Verlag. 1982. p. 186.
7. J. Kirschner and H.W. Etzkorn, Appl. Surf. Sci. **3**, 251 (1979).
8. R. Holm and S. Storp, Appl. Phys. **12**, 101 (1977).
9. J.W. Coburn, Thin Solid Films **64**, 371 (1979).
10. L. Young, Anodic Oxide Films. London: Academic Press. 1961.
11. J.M. Sanz, PHD Thesis. Stuttgart University, 1982.
12. S. Hofmann and J.M. Sanz, to be published in J. Trace and Microprobe Techniques.
13. M.P. Seah, Analusis **9**, 171 (1981).
14. P.H. Holloway and G.C. Nelson, J. Vac. Sci. Technol. **16**, 793 (1979).
15. M.F. Ebel, J. Electr. Spectr. Rel. Phen. **14**, 287 (1978).
16. S. Hofmann and J.H. Thomas III, J. Vac. Sci. Technol. **B1**, 43 (1983).
17. K.S. Kim, W.E. Baitinger, J.W. Amy, and N. Winograd, J. Electr. Spectr. Rel. Phen. **5**, 351 (1974).
18. R. Kelly, Surf. Sci. **100**, 85 (1980).
19. P. Sigmund, in Sputtering by Particle Bombardment, I.R. Behrisch (ed.). Berlin-Heidelberg-New York: Springer-Verlag. 1981. p. 9.
20. P. Haff, Appl. Phys. Lett. **31**, 259 (1977).

Correspondence and reprints: Priv.-Doz. Dr. S. Hofmann, Max-Planck-Institut für Metallforschung, Institut für Werkstoffwissenschaften, Seestrasse 92, D-7000 Stuttgart 1, Federal Republic of Germany.

Mikrochimica Acta [Wien], Suppl. 10, 145–158 (1983)
© by Springer-Verlag 1983

Institut für Spektrochemie und angewandte Spektroskopie, Dortmund, and
Henkel KGaA, Düsseldorf, Federal Republic of Germany

Investigations of Phosphate Coatings of Galvanized Steel Sheets by a Surface-Analytical Multi-Method Approach

By

H. Bubert, R. Garten, R. Klockenkämper, and H. Puderbach

With 13 Figures

(Received January 19, 1983)

1. Introduction

Year after year, monetary amounts reaching thousands of millions are lost
by corrosion. In the Federal Republic of Germany alone, costs due to corrosion rose to 4 x 10^{10} DM during the past year[1]. Great exertions are made
by manufacturers, all of which with one solitary aim, to develop improved
protection methods against corrosion. Widely used protection coatings for
steel sheets are produced by electro-galvanizing, phosphatizing and passivation by chromatizing of the surface[2]. Steel sheets coated in this manner may
be converted, soldered and welded without any difficulty. The phosphate
coating makes for an excellent adhering base for varnishing and enameling
paints which can be applied without any additional treatment. They find use
in household and electrical supplies, in automobile and building industries.
The macroscopic properties of the corrosion-protective coatings depend to a
great extent upon the details of the layer structure. In order to optimize the
specific kind and the homogeneity of this layer structure, and in order to
avoid faults, the demand for analytical investigations is stringent. Such investigations have been carried out with several surface-analytical methods.
Our strategy here is to carry out synergetic investigations according to a
multi-method concept. Further multi-method approaches to the analysis of
surface layers of steel makings have been presented by Ebel and
Wernisch[3], Gettings and Kinloch[4], Janssen[5], Kool, Mittemeijer, and
Schalkoord[6], Vetters, Hoffmann, and Schaaber[7], and Wakano and Usuki[8].
SCHAABER[7], and WAKANO and USUKI[8].

2. Experimental

2.1 Technical Process and Sampling

Fig. 1 shows schematically the production process for surface refinement of the steel sheets. The steel sheet is degreased, acid-pickled and electro-galvanized; it then passes another acid-pickling solution and a rinse using tap water. The galvanized surface is pre-activated to prepare it for a uniform deposition of phosphate. The effect of pre-activation upon the size of phosphate crystals has been studied formerly[9]. The steel sheet is then phosphatized and rinsed. Finally, the phosphate coating is post-passivated in a chromate rinse, air-dried and rolled.

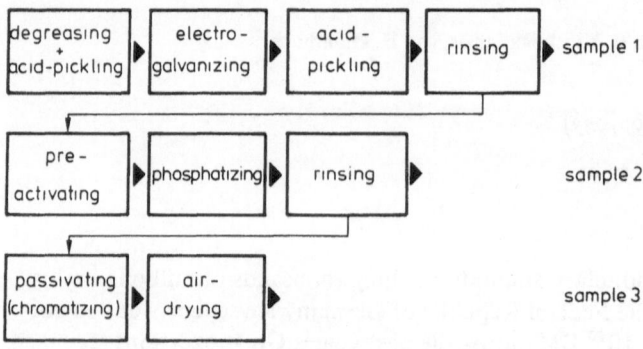

Fig. 1. Schematic process of production

Fixodine C16 was used as agent for the pre-activation, Granodine® C 5804 was used for phosphatizing and Deoxylite 41 was used for post-passivating; all are products of Gerhard Collardin GmbH, Köln, FRG.

For the purpose of our investigations, the refined steel sheet was taken from a production-line plant. The first analytical sample (Sample 1) was taken after the first four steps of the preparation process (see Fig. 1) had been accomplished. This sheet then passed the remaining stages of the production plant, but without any further treatment, just into the air-drying stage. Further treatments of this sheet were carried out on a laboratory scale. A phosphatized sample (Sample 2) and an additionally post-passivated sample (Sample 3) were obtained as indicated in Fig. 1.

2.2 Analytical Instrumentation

The following methods were applied so far in our analytical investigations:

Scanning electron microscopy (SEM);
energy-dispersive X-ray spectrometry (EDX) combined with SEM,
Auger electron spectrometry (AES), secondary ion mass spectrometry
(SIMS), and (low-energy) ion scattering spectrometry (ISS).

Apparatus:

SEM, EDX: Jeol JSM-35
             electron energy E_p = 5 ... 30 keV
             beam current i_e = 0.1 ... 10 nA
             magnification typical f = 2000
             EDAX-Spectrometer Model 707 A Si(Li)-Detector;
             spectral resolution ΔE_x = 155 eV at E_x = 5.9 keV.

AES: Physical Electronics Industries (PHI) SAM 590,
             primary electron energy E_p = 3 keV,
             beam current i_e = 120 μA, beam diameter d_e = 2.5 μm,
             analyzer resolution $\Delta E/E$ = 0.3 %,
             argon sputtering, ion energy E_i = 1 keV,
             ion current density j_i = 60 μA $\cdot$ cm^{-2}.
             Point analyses as well as areal scanning analyses (typical scanned area
             q = 0.25 mm^2) have been performed.

SIMS, ISS: 3M – Brand 525
             working gas and primary ion energy
             SIMS: ^{20}Ne, E_i = 4 keV
             ISS: ^{3}He, E_i = 2 keV
             ion current i_i $\cong$ 100 nA
             beam diameter d_i $\cong$ 100 μm, areal scanning
             analyses (scanned area q = 2 mm^2)

Table 1. Instrumental Parameters of the Methods Used

method	lateral resolution μm	information depth nm	detection limit μg/g
SEM	0,01	10	
EDX (SEM)	0,5	500	3000
AES	2,5	2	3000
SIMS	100	1	1
ISS	100	0,5	3000

Table 1 summarizes important specifications of the methods used. For each
method, the Table emphasizes one particular capability: with respect to

SEM, EDX the lateral resolution, with respect to AES and ISS the depth resolution, with respect to SIMS the power of detection. During our investigations our attention was centered on these specially efficient charcteristics, at times at the expense of other parameters. But in this way, the combination included the individual performance in a very effective way.

3. Results

3.1 Zn-Coated Steel Sheet

A SEM-image of the surface of a steel sheet treated with a zinc coating only (Sample 1) is presented in Fig. 2. The zinc coating displays a dense, polycrystalline structure. Fig. 3 shows typical spectra of the original zinc coated surface. The AES-spectrum indicates, apart from Zn and O, also the adventitious C, as well as S and Cl. The presence of these elements is explainable by spread residues of acid-pickling (2 % H_2SO_4) and rinsing bath (tap water). The ISS-spectrum completely confirms the result of AES. − Both of the SIMS spectra, due to their high power of detection, additionally display traces of e.g. Na, K, Ca and Fe, as well as traces of organic fragments which originate from chemical bath solutions and roller oils. The occurrence of Cr can be explained by spreading via running the steel sheet through squeezing

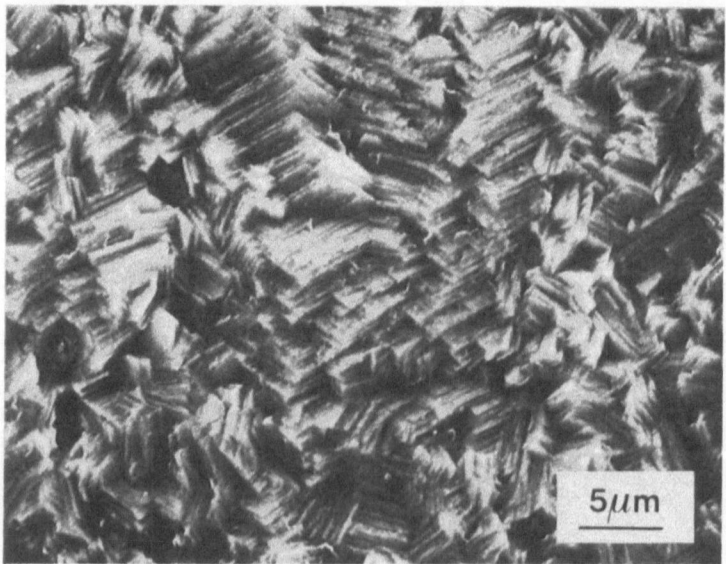

Fig. 2. SEM-image of a Zn-coated steel sheet (Sample 1 as indicated in Fig. 1)

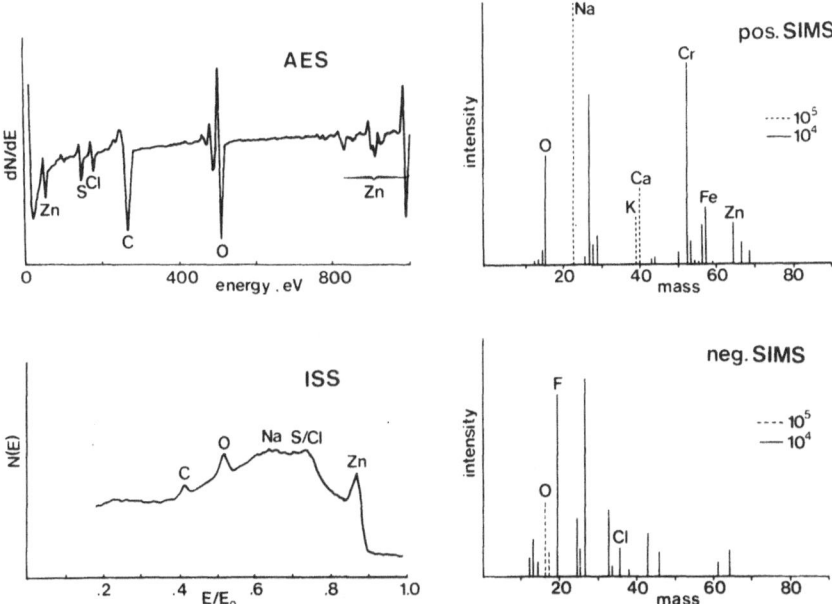

Fig. 3. Typical spectra of the surface of the original Zn-coating (Sample 1) obtained by different methods

rolls which are adhered with Cr from previous materials passing through the production-line. The appearance of the element F could not be explained satisfactorily as yet. (F is present in numerous samples, using SIMS, it is detected with particularly high sensitivity.)

To study the concentration depth profile, Sample 1 was treated by successive ion sputtering. Fig. 4 shows AES-depth profiles of the elements C, O, Zn and Fe. This quantitative analysis has been carried out by use of sensitivity factors[10]. The depth scale is based on the measurement of Zn by EDX-line scans of a polished cross-section; an average Zn-layer thickness of 5 μm was obtained. The elements C and O are found in the uppermost layers, the bordering zones from Zn to Fe are considerably indistinct. – This is mainly a consequence of the roughness and the complicated structure of the layer as well as of the broadening effects due to sputtering.

3.2 Phosphatized Zn-Coating

The SEM-micrograph of a phosphatized Zn-coating is presented in Fig. 5. The phosphate layer consists of flake-like crystals which are found to be partly upright, and partly lying, as revealed by stereo-micrographs. On the strength of their arrangement, some crystals appear like stars, in the top

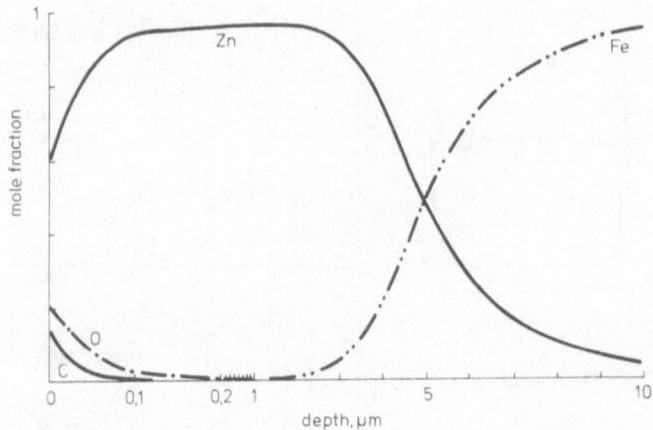

Fig. 4. AES-depth profiles of C, O, Zn, and Fe for a Zn-coated steel sheet (Sample 1)

Fig. 5. SEM-image of the spray-phosphatized, Zn-coated steel sheet (Sample 2 as indicated in Fig. 1) showing hopeite crystals

view. Their edge lengths measure about 5 μm and their thickness about 0.2 μm. The layer thickness averages at approx. 0.5 μm. By X-ray diffraction, the crystallites of the phosphate layers were identified as hopeite $(Zn_3(PO_4)_2 \cdot 4 H_2O)$.

Fig. 6 shows typical spectra of the phosphatized surface: in addition to the elements detected so far, the AES spectrum shows a P-signal. Due to the C-overlayer, this signal is only weak, but appears very much more distinct after short, cautious sputtering. The C-overlayer covering the phosphatized surface may be desorbed not only by ion sputtering but also, to a great extent, by the primary electron beam, if current densities $j_e \geqslant 2 \cdot 10^{-2}$ Am^{-2} are applied. For example, in Fig. 7 a mapping for P is presented from a sample location, at which beforehand, a line scan and a point analysis have been carried out in the centre of the displayed area. Along the trace of this line scan the P-signal is intense due to the electron bombardment-induced desorption of the adventitious C-overlayer.

In Fig. 6 also the element Cr appears in the AES-, ISS- and positive SIMS-spectra, due to spreading within the production-line, as already mentioned for the original galvanized steel sheet (Sect. 3.1). In the positive SIMS-spectrum, organic fragments additionally occur which may originate from the organic buffer substances in the pre-activating solution.

An AES-depth profile is shown in Fig. 8. The scale of depth is based on the assumption that Zn and hopeite have the same sputtering rate. The flat maxima of P and Cr occurring here are the consequence of a surface covering by C and O.

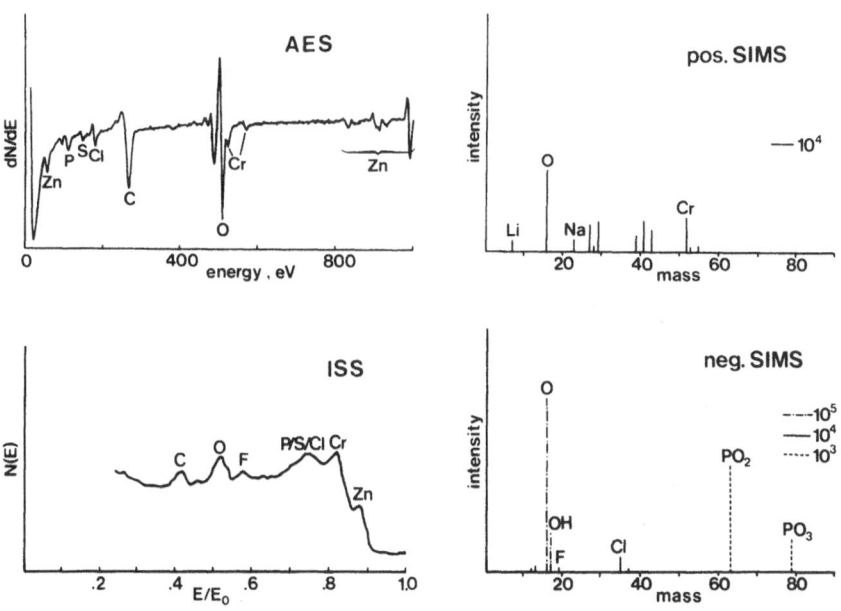

Fig. 6. Typical spectra of the surface of the original spray-phosphatized Zn-coating (Sample 2) obtained by different methods

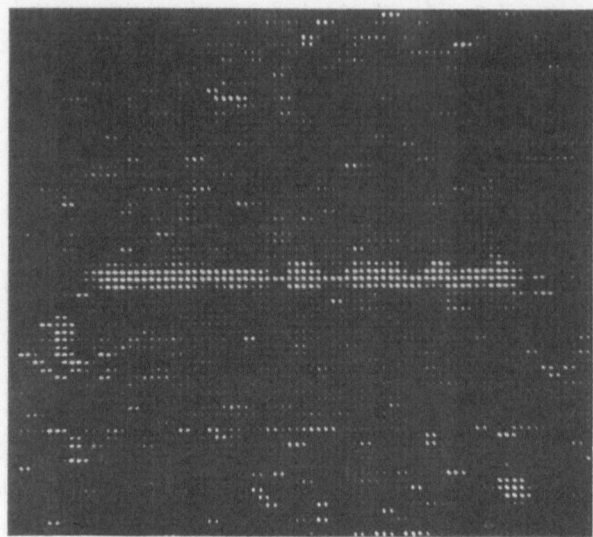

Fig. 7. AES-mapping (80 x 80 points, scanned area q = 0.8 mm²) of P of the surface of the original spray-phosphatized Zn-coating (Sample 2). Along the central straight line, the P-signal is intense due to electron-bombardment induced desorption of the adventitious C-overlayer by a previous line scan

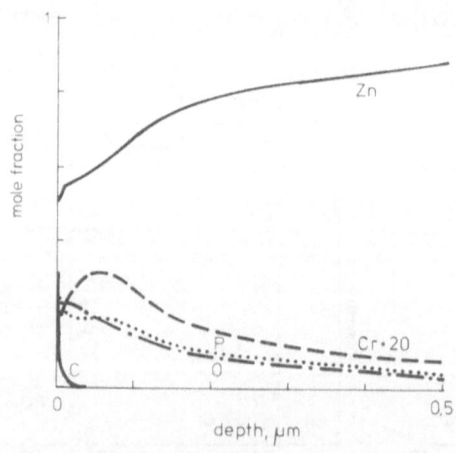

Fig. 8. AES-depth profiles of C, O, P, Cr, and Zn for a spray-phosphatized, Zn-coated steel sheet (Sample 2) showing only the surface-near 0.5 μm. The Cr-intensity is multiplied by a factor of 20

Reasonable quotation of quantitative concentration for the depth profiles within the topography of the surface at hand is justified by AES-line scans in connection with SEM-images. Fig. 9 shows a line scan for P, O, Zn subsequent to the erosion of the superficial C-overlayer.
The signal heights of these elements strongly differ along the scanning trace, essentially caused by surface topography. Since the ratios of the signal heights are nearly constant, the mass fractions calculated with the aid of sensitivity factors are nearly constant. During depth profile analysis, the material erosion by ion sputtering was controlled by investigation of the sample surface with SEM and EDX.

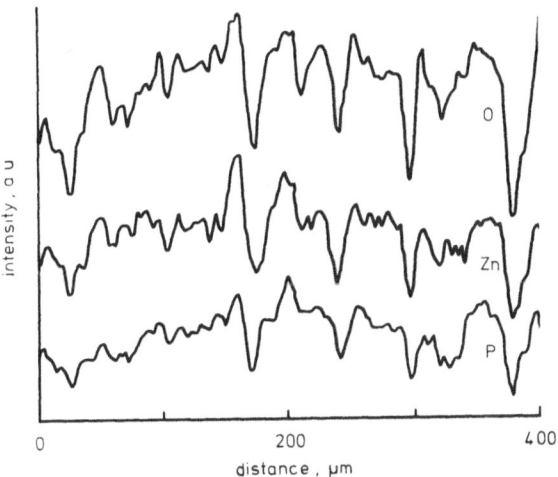

Fig. 9. AES-line scans (160 points per line), of O, P, and Zn for a spray-phosphatized Zn-coated steel sheet (Sample 2) after the desorption of the adventitious C-overlayer, showing the uniform topographic effect

In Fig. 10 the phosphate layer (Fig. 5) is partially removed, so that the underlying Zn-layer becomes sputtered, forming small cones. A typical line scan of those partly bare sample surfaces is shown in Fig. 11. The P- and O-signals are precisely correlated, while the Zn signal shows no correlation due to uncovered regions of metallic Zn.
At further sputtering, the original topography of the phosphate layer (Fig. 5) is embossed onto the underlying Zn-layer (Fig. 12a). Regions which had previously been covered by thick phosphate crystals for a longer period, appear as an elevated relief. This topography is even carried into the underlying Fe material base, as shown in Fig. 12b. The displayed structure is formed by star-like hopeite crystals. Zn and remains of Zn phosphate are still detected on the top of this structure (light regions).

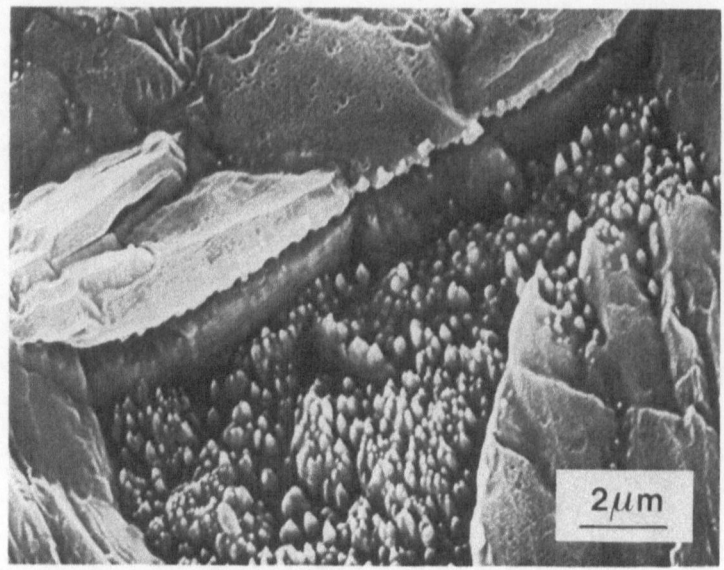

Fig. 10. SEM-image of the spray-phosphatized Zn-coated steel sheet (Sample 2) after ion sputtering. The surface exhibits areas covered by hopeite crystals as well as already discovered areas showing small cones of Zn

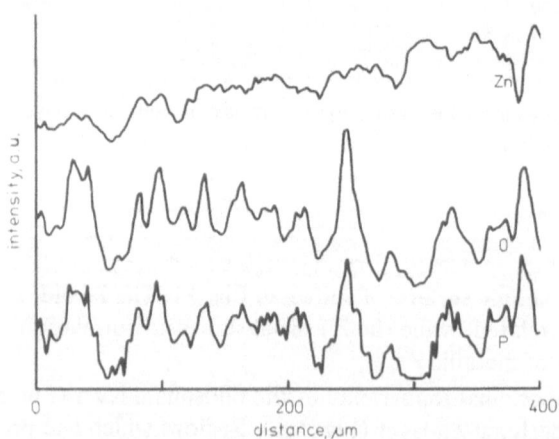

Fig. 11. AES-line scans (160 points per line) of O, P, and Zn for a spray-phosphatized, Zn-coated steel sheet (Sample 2) after ion sputtering (sputter time: 60 min). The subjacent Zn-layer is exposed by partial removal of the hopeite crystals (Fig. 10). Therefore, the correlation of the signals of P, O, and Zn (Fig. 9) is removed for Zn

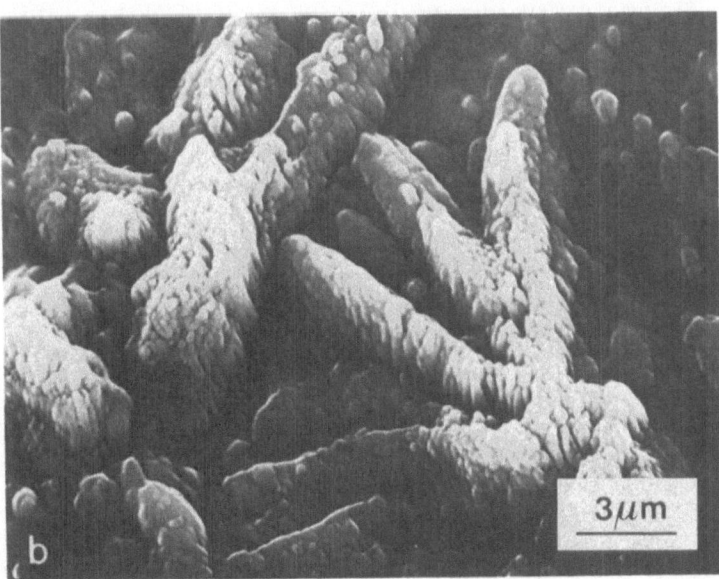

Fig. 12. SEM-images of the spray-phosphatized, Zn-coated steel sheet (Sample 2) after continual ion sputtering. (a) Typical structure of the Zn-layer caused by imprinting the original topography of the hopeite layer into the subjacent Zn-layer; (b) similar structure of the Fe-sheet, showing that the topographical impression reaches the Fe-base bordering the Zn-layer

3.3 Post-Passivated Phosphate Layer

The phosphatized Zn-coatings are normally passivated, in order to arrive at an additionally improved protection against corrosion.
The very thin passivation layers, about 50 nm as calculated by mass balance of Cr, cannot be rendered visible by SEM, although the surface is entirely passivated. This was proved by scanning EDX, AES, and SIMS. According to the topography, the Cr-covering differs correspondingly (cf.[11]).
As known from the production process, the quality of the corrosion-protective layer depends on the chromate concentration of the post-passivating bath solution. The effect of different chromate concentrations upon the Cr-content of the protective layer is shown in the AES depth profiles in Fig. 13. The depth profile of an undigested phosphate layer (Sample 2; spread Cr only) is sketched for comparison (broken line). The Cr-content reaches a maximum at a medium concentration of chromate in the solution; this is also confirmed by depth profiles using SIMS.

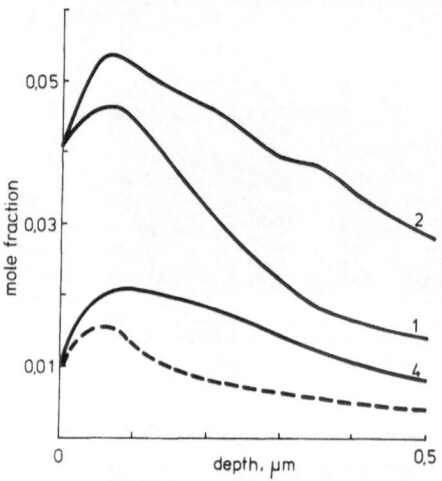

Fig. 13. AES-depth profiles of Cr for an additionally passivated steel sheet (Sample 3). The relative concentration of chromate in the passivating bath is indicated by the multiplier on the solid lines; the broken line displays the profile of Cr for the phosphatized sheet without further digestion (Sample 2)

4. Conclusion

The analysis of the galvanized steel sheets by the methods used indicates a dense poly-crystalline Zn-coating. This coating contains only minor amounts of impurities which do not disturb the further refinement processes. The phosphate layer consists of flake-like hopeite crystals strongly adherent to the Zn-coating. The coverage amounts to 95 % of the surface. The maximum of the Cr content within the protective layer corresponds to an optimum of quality in view of corrosion protection for the technical product. Quantitative analytical statements are limited to layers within the 0.1 μm region, by the properties of the sample (natural roughness). The lateral resolution is limited by time consumption and equipment conditions.

To supplement depth profile analyses with AES the roughening effect was investigated by line scans across a sputter crater.

Results of such line scans may be summarized as follows:

1. After sputtering the superficial C contamination, signal intensities in respect of P, O, and Zn are highly correlated.

2. Within deeper regions of the sputter crater, the correlation of Zn with P and O disappears.

3. As soon as P, O, and Zn do not correlate completely, Fe signals are measured.

Considering only these results of the line scans, a pure Zn-layer between the phosphate layer and the Fe bulk cannot be conclusively identified. This is caused by the roughness within the sputter crater which lies at about 2 to 5 μm. Only when analyzing with a lateral resolution $\leqslant 1$ μm, the details of the structures in the sputtered crater can be resolved.

For further detailed investigations, we will, in addition, apply high lateral resolution scanning Auger microprobe, electron microprobe, glow discharge optical emission spectroscopy, as well as electro-chemical surface erosion methods in the analysis of those samples.

Summary

Investigations of Phosphate Coatings of Galvanized Steel Sheets by a Surface-Analytical Multi-Method Approach

Corrosion protective coatings on galvanized steel sheets have been studied by a combination of SEM, EDX, AES, ISS and SIMS. Analytical statements concerning such rough, poly-crystalline and contaminated surfaces of technical samples are quite difficult to obtain. The use of a surface-analytical multi-method approach overcomes, the intrinsic limitations of the individual method applied, thus resulting in a consistent picture of those technical surfaces. Such results can be used to examine technical faults and to optimize the technical process.

References

1. dpa (deutsche Depeschen-Agentur) concerning SURTEC. Berlin, September 1981.
2. W. Wiederholt, Die chemische Oberflächenbehandlung von Metallen zum Korrosions-schutz, Schriftenreihe Galvanotechnik, R. Weiner (ed.). Saalgau/Württ.: Eugen G. Leuze. 1963.
3. M.F. Ebel and J. Wernisch, Mikrochim. Acta [Wien], Suppl. VII, 1977, 153.
4. M. Gettings and A. J. Kinloch, Surf. Interface Anal. 1, 189 (1979).
5. E. Janssen, Mikrochim. Acta [Wien], Suppl. IX, 1981, 221.
6. W.H. Kool, E.J. Mittemeijer, and D. Schalkoord, Mikrochim. Acta [Wien], Suppl. IX, 1981, 349.
7. H. Vetters, F. Hoffmann, and O. Schaaber, Mikrochim. Acta [Wien], Suppl. IX, 1981, 385.
8. T. Wakano and K. Usuki, Tetsu-to-Hagané 66, 9, 177 (1980).
9. H. Puderbach, H. Gotta, and K.-H. Gottwald, Beitr. elektronenmikroskop. Direktabb. Oberfl. 8, 439 (1975).
10. L.E. Davis, N.C. MacDonald, P.W. Palmberg, G.E. Riach, and R.E. Weber, Handbook of Auger Electron Spectroscopy, 2nd ed. Eden Prairie. Minn.: Perkin-Elmer-Corp. 1976.
11. S. Baumgartl and P. Busch, 15. Kolloq. Elektronenmikroskop. Direktabb. u. Analyse v. Oberflächen (EDO). Bremen, Sept. 1982.

Correspondence and reprints: R. Klockenkämper, Institut für Spektrochemie und ange-wandte Spektroskopie, P.O.B. 778, D-4600 Dortmund, Federal Republic of Germany.

Mikrochimica Acta [Wien], Suppl. 10, 159–176 (1983)
© by Springer-Verlag 1983

Deutsche Forschungs- und Versuchsanstalt für Luft- und Raumfahrt (DFVLR),
Institut für Werkstoff-Forschung, Köln, Federal Republic of Germany

Surface- and Microanalytical Investigations
of the Chemical Constitution of the Grain Boundary Phase
in Dense Silicon Nitride

By

H. J. Dudek, W. Braue, and G. Ziegler

With 15 Figures

(Received February 1st, 1983)

1. Introduction

Ceramic materials based on silicon nitride exhibit a greater high temperature
strength for temperatures above 1000 °C than conventional super alloys
employed at present for turbine engine design. As suitable structural components for high temperature application two groups of Si_3N_4 materials are of
special interest: *1)* dense hot-pressed Si_3N_4 (HPSN) and sintered Si_3N_4
(SSN) showing strength data between 550 and 900 MN/m^2 and *2)* reaction-
bonded Si_3N_4 (RBSN) characterized by a residual porosity from 15 to 30%
and maximum strength of 350 MN/m^2. Further development of dense Si_3N_4
focuses mainly on the optimization of high-temperature properties. Beside
the short-term strength, outstanding properties are the oxidation behavior
and the long-term strength under mechanical and cyclic thermal load which
can be improved by controlling the microstructure and, in particular, by improving the chemical and structural constitution of the mostly amorphous
grain boundary phase. Dense Si_3N_4 can only be achieved via liquid-phase
sintering where in the case of HPSN and SSN MgO, Y_2O_3 or $Y_2O_3 + Al_2O_3$
respectively are usually added as sintering aids. In Fig. 1 the evolution of the
liquid-phase sintering process for MgO-fluxed HPSN is demonstrated schematically as proposed by Ref.[1]. At temperatures below 1400 °C MgO reacts
with the amorphous silica layers on the surfaces of α-Si_3N_4 grains of the
starting powder forming magnesium silicate forsterite Mg_2SiO_4. Above

160   H. J. Dudek et al.:

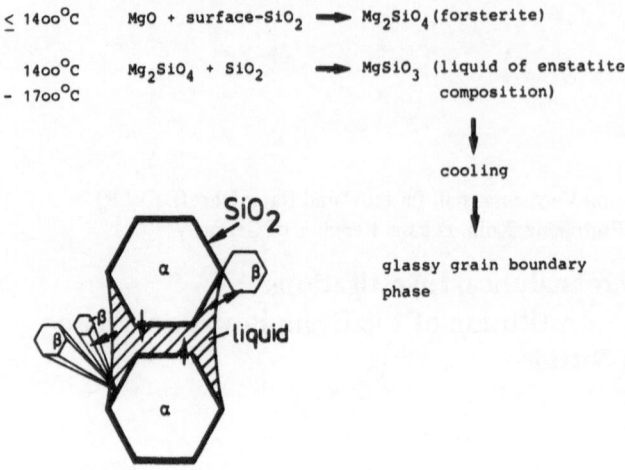

SINTERING OF MgO FLUXED Si₃N₄

$\leq 1400°C$ MgO + surface-SiO₂ ➡ Mg₂SiO₄ (forsterite)

1400°C   Mg₂SiO₄ + SiO₂ ➡ MgSiO₃ (liquid of enstatite
- 1700°C   composition)

cooling

glassy grain boundary
phase

Phase assemblage after cooling

- α - and β-Si₃N₄
- amorphous intergranular grain boundary layer
- amorphous grain boundary phase at triple junctions
- minor crystalline matrix phase (forsterite)

Fig. 1. Schematical illustration of the liquid-phase sintering process for MgO-fluxed HPSN

1400 °C forsterite becomes unstable under excess silica environment resulting in the formation of a liquid of enstatite ($MgSiO_3$) composition which during cooling converts to an oxinitride glass[2]. Simultaneously to this stage of sintering the α/β transformation of Si_3N_4 is activated which means the solution of the α-Si_3N_4 phase in the newly formed liquid, and the precipitation of the likewise hexagonal β-Si_3N_4 modification takes place. In contrast to the equiaxial development of the α-Si_3N_4 phase β-Si_3N_4 exhibits an elongated grain morphology resulting in an intense linking of the microstructure. It is mainly this difference in grain morphology of the Si_3N_4 phases which governs the thermo-mechanical properties of dense Si_3N_4 for temperatures in the range up to 1000 °C. Above 1000 °C the properties controlling key parameter is the behavior of the glassy grain boundary phase which, upon reaching its softening point, results in a severe decrease of strength due to grain boundary sliding and separation under load. As shown by electron microscopic investigations[3,4] besides β-Si_3N_4 and residues of the α-phase, the HPSN phase assemblage consists of a very thin intergranular glassy grain boundary layer, an amorphous phase concentrated at triple grain junctions and a crystalline magnesium and silicon containing phase. In the microstructure of Y_2O_3-fluxed SSN also a crystalline (yttrium and silicon containing) phase can coexist with β-Si_3N_4 [5].*

As the high temperature properties of dense Si_3N_4 are primarily controlled by the characteristics of the grain boundary phase, application of surface- and microanalytical methods [Auger electron (AES)- and X-ray photoelectron spectroscopy (XPS), analytical scanning transmission electron microscopy (STEM)] focuses on the development of more refractory grain boundary phases with improved chemical constitution. The influence of the minor crystalline matrix phases on HPSN and SSN properties has not yet been investigated in detail. In this context characterization of these phases is achieved in this study by backscattered electron imaging and X-ray distribution patterns for both fractured and polished surfaces in HPSN and SSN materials containing various types and amounts of densification aids.

2. Scanning Electron Microscopic Investigations by Means of Backscattered Electrons and X-Ray Microanalysis

Beside the conventional secondary electron (SE) images backscattered electron (BSE) images have been sucessfully employed for microstructural analysis in materials research (for review see[6]). This technique has been applied for identification and imaging the distribution of the minor crystalline matrix phases in HPSN and SSN materials. Ground and polished as well as fractured surfaces were prepared from HPSN and SSN samples.

Starting with ground and polished SSN samples containing Y_2O_3 as densification aid, identification of the minor crystalline matrix phase is easily achieved. Figs. 2a, b show a SE and BSE image respectively. Two phases exhibiting different atomic numbers can be distinguished even in the SE image, but more clearly in the BSE image. In Fig. 3 an area from the BSE image in Fig. 2b is compared with an yttrium X-ray distribution pattern (Fig. 4)** obtained from identical sample position. By comparison of the BSE image with the yttrium distribution pattern it has to be considered that within the microscope column the BSE detector is situated exactly above the sample in contrast to the X-ray detector which is arranged in a lateral position. Moreover, in the BSE image a greater lateral resolution is achieved than in the

---

*   For further discussion the synonym "minor crystalline matrix phase" is referred to these phases in both HPSN and SSN, as β-Si_3N_4 is always the predominantly crystalline phase.

** For printing purposes transformation of the originally coloured X-ray distribution patterns to a grey level scale was necessary. Arrows in the following figures mark corresponding positions of the minor crystalline matrix phases in SE and BSE images and X-ray distribution patterns respectively.

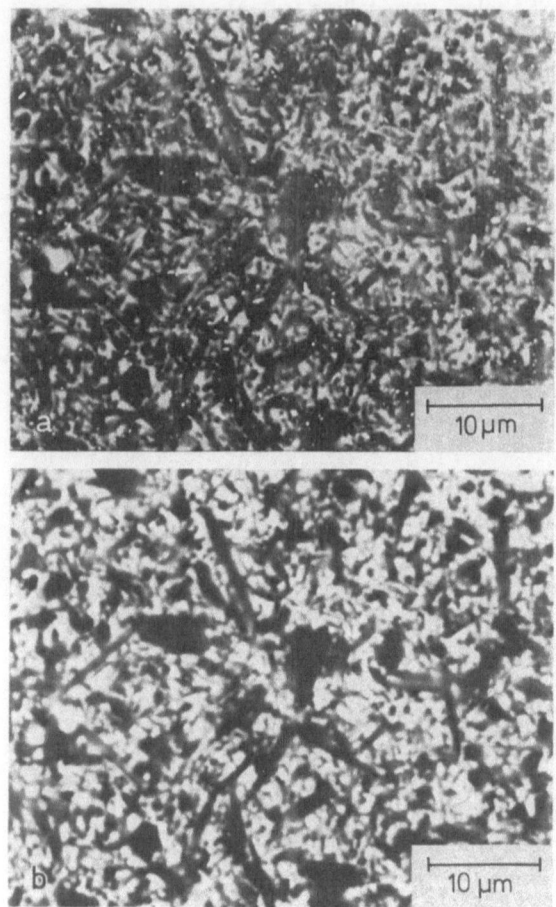

Fig. 2a,b. SE- and BSE images respectively for a polished surface of Y_2O_3-fluxed SSN

X-ray distribution pattern. Keeping in mind these differences a definite coordination between both BSE image and X-ray distribution pattern is possible. The bright areas in the BSE image are attributed to the high yttrium containing minor crystalline matrix phase in SSN. Due to the low difference in mean atomic numbers between β-Si_3N_4 and the minor crystalline matrix phase in MgO-fluxed HPSN, identification of the minor crystalline matrix phase in BSE images renders more difficult than in Y_2O_3-fluxed materials both for polished surfaces and, as shown later, for fractured surfaces. In the SE image of the polished MgO-doped HPSN surface no contrasts between the two matrix phases are visible. In the BSE image (Fig. 5a) there are only minor atomic number contrasts. By comparing the BSE image with the magnesium X-ray distribution pattern in Fig. 5b the grey areas in Fig. 5a can be attri-

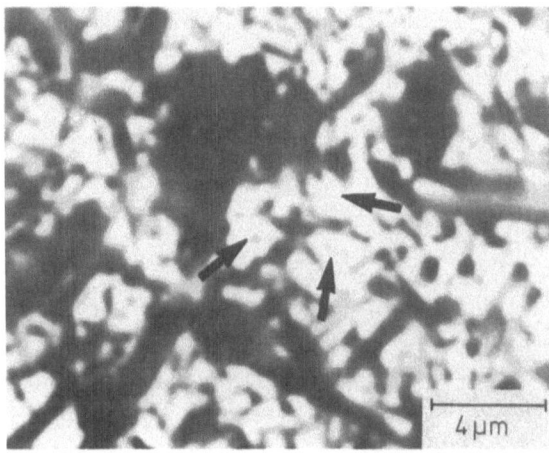

Fig. 3. Enlarged area from BSE image in Fig. 2b (polished surface of Y_2O_3-fluxed SSN)

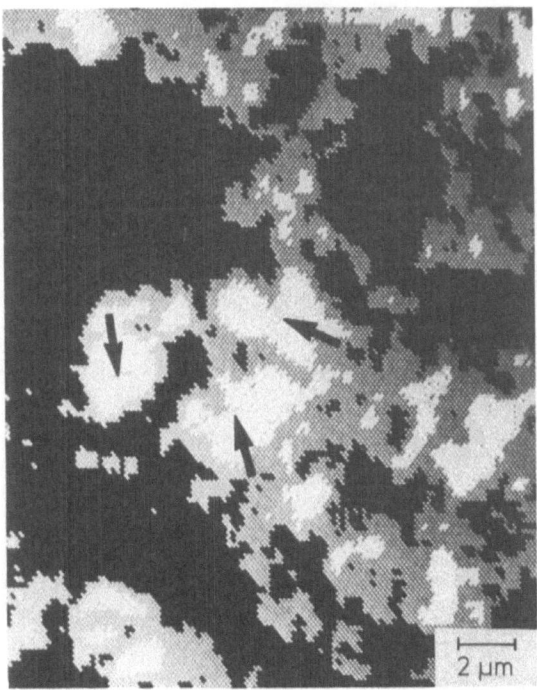

Fig. 4. Yttrium X-ray distribution pattern corresponding to the sample area of the BSE image in Fig. 3

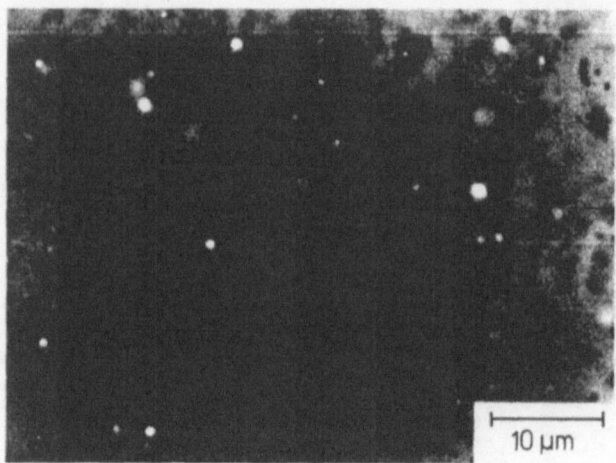

Fig. 5a. BSE image of a polished MgO-fluxed HPSN surface

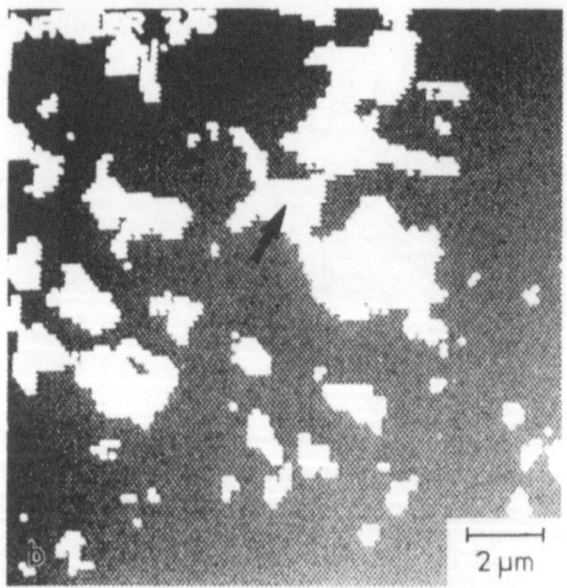

Fig. 5b. Magnesium X-ray distribution pattern corresponding to the sample area of the BSE image in Fig. 5a (polished surface of MgO-fluxed HPSN)

buted to the minor crystalline matrix phase in HPSN (dark areas arise from pores due to preparation).

Microstructural studies of grain morphology by means of fracture surfaces is a routine procedure for dense Si_3N_4 materials. SE and BSE imaging on fracture surfaces may be influenced by the amount of glassy phase coverage which is governed by the fracture mode of the Si_3N_4 material under consideration. Fig. 6a, b show SE and BSE images of the Y_2O_3-fluxed material. In the BSE picture the minor crystalline matrix phase appears bright. The corresponding yttrium X-ray distribution pattern supports coordination of the bright areas with the distribution of the yttrium containing phase within the SSN microstructure (Fig. 7). In contrast to the Y_2O_3-fluxed material, in Figs. 8a, b the SE and BSE images respectively are shown for a fracture surface of 10 wt.% MgO-fluxed HPSN. Focusing the SE image the typical elongated grain morphology of β-Si_3N_4 can only be identified in distinct areas.

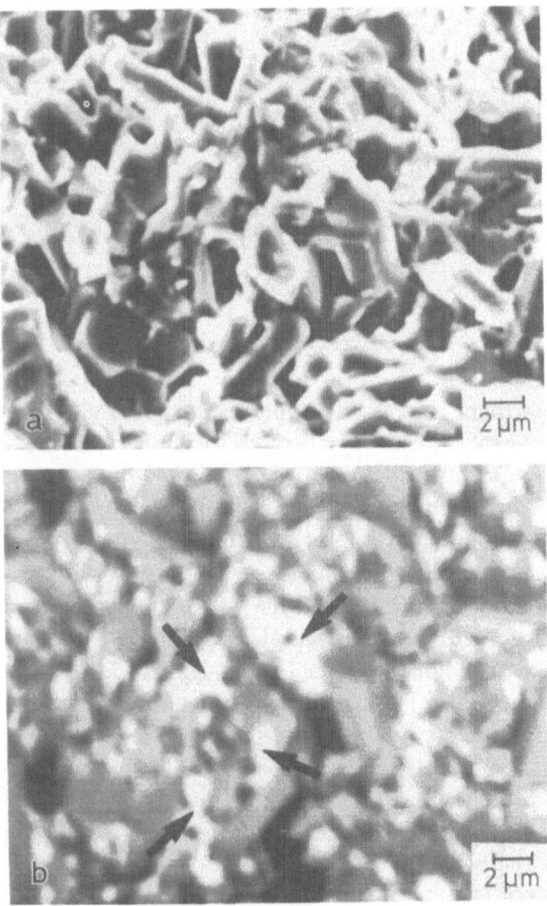

Fig. 6a,b. SE- and BSE images respectively from fracture surface of Y_2O_3-fluxed SSN

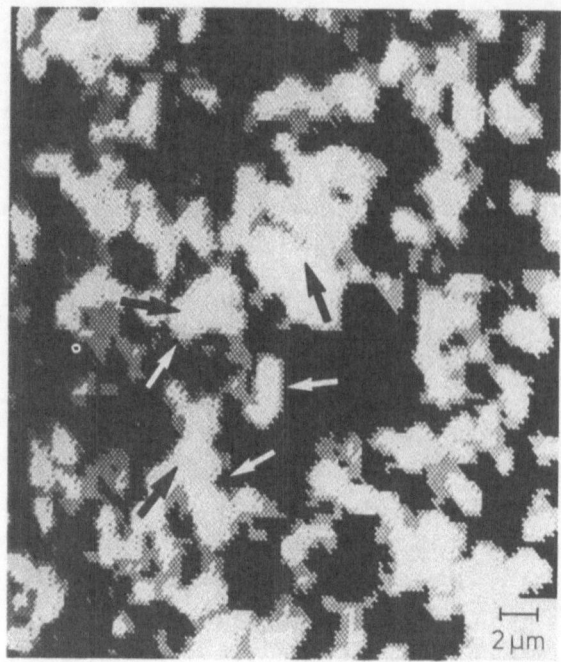

Fig. 7. Yttrium X-ray distribution pattern corresponding to the sample area of the BSE image in Fig. 6b

In the BSE image, however, there is an essential improvement for identification of the grain structure of the β-phase. The marked difference between these two imaging modes concerning fracture surfaces of MgO-fluxed materials may be explained by a higher yield of the SE in the amorphous grain boundary coverage of the predominantly intercrystalline fractured HPSN surfaces than of BSE which penetrate such thin glassy layers only with small interaction. Figs. 9 and 10 show a SE image and the corresponding magnesium X-ray distribution pattern for another sample position of the MgO-fluxed HPSN fracture surface. Although satisfactory imaging of the β-Si_3N_4 phase is given in the SE picture clear identification of the minor crystalline matrix phase itself can only be achieved in the magnesium X-ray distribution pattern. Comparing different amounts of MgO added to HPSN materials it could be demonstrated that the grain size of the minor crystalline matrix phase reaches $1-2$ μm for 10 wt.% MgO added and decreases to $0.1-0.5$μm for samples fluxed only with 1 wt.% MgO.

Concerning the chemical composition of the minor crystalline matrix phases in MgO and Y_2O_3 containing materials, there is support from both phase relationships and data from analytical electron microscopy. Regarding the

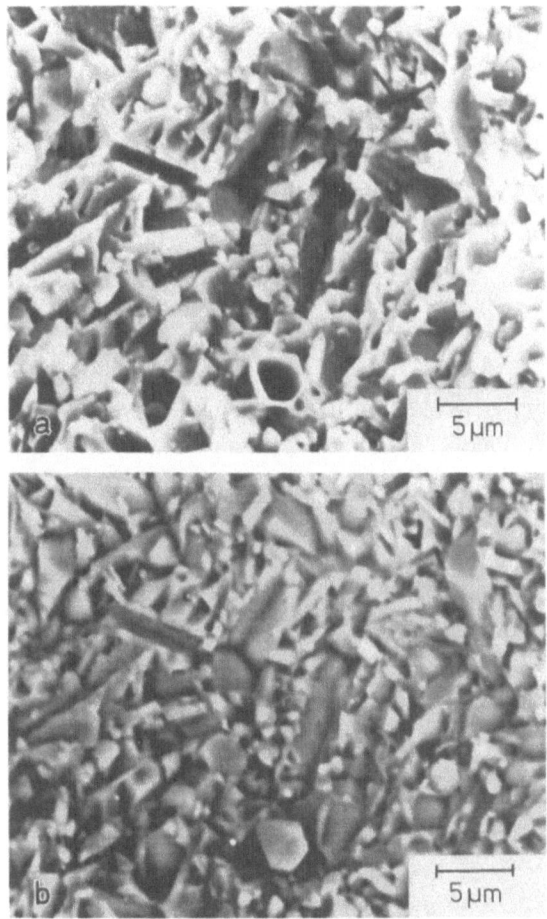

Fig. 8a,b. SE- and BSE images respectively from a fracture surface of 10 wt.% MgO-fluxed HPSN

composition of the minor crystalline matrix phase in Y_2O_3-fluxed materials imaginable phases could be "nitrogen melilithe" $Y_2Si(Si_2O_3N_4)$ or other phases of similar stoichiometry described in the literature for the systems Y-Si-O-N[7, 8] and Y-Al-Si-O-N[9]. Concerning the minor crystalline matrix phase in MgO-fluxed materials X-ray investigations from various authors and microstructural data published in[10] indicate that this phase may predominantly consist of forsterite. The formation of $MgSiN_2$ as another reasonable phase candidate seems unlikely in 10 wt.% MgO-fluxed Si_3N_4. According to ref.[11, 12] its occurence should be restricted to significantly higher MgO contents on the join $MgO-Si_3N_4$. Furthermore, the existence of a stability field

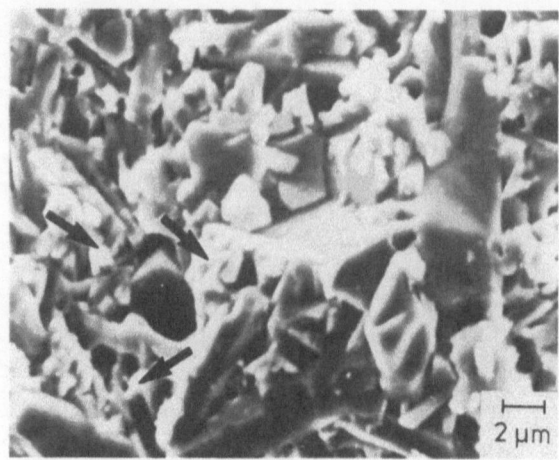

Fig. 9. SE image from a fracture surface of 10 wt.% MgO-fluxed HPSN

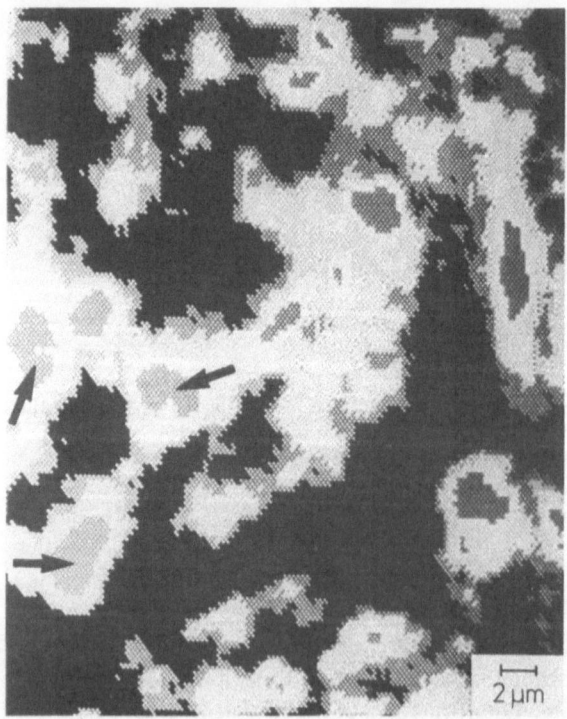

Fig. 10. Magnesium X-ray distribution pattern corresponding to a sample area of the SE image in Fig. 9 (10 wt.% MgO-fluxed HPSN)

for MgSiN$_2$ in the system MgO-Si$_3$N$_4$-SiO$_2$ was generally questioned under real hot-pressing conditions[13]. Minor amounts of enstatite, MgSiO$_3$, resulting from partial crystallization of the glassy grain boundary phase[10], should only be expected after special thermal treatments of the HPSN material.

In a synopsis of the BSE images and X-ray distribution patterns shown, identification of the minor crystalline matrix phases in HPSN and SSN can be achieved in both polished and fractured surfaces. Due to higher atomic number contrasts, imaging of Si$_3$N$_4$ containing densification aids of high atomic number, such as Y$_2$O$_3$, is favoured. As the distribution of the minor crystalline matrix phase in dense Si$_3$N$_4$ and the densification aid are important topics for evaluation of crack propagation mechanisms, application of BSE imaging combined with X-ray distribution patterns is a suitable technique for improving the technology of high-strength sintered materials.

3. Analytical Scanning Transmission Electron Microscopy

Due to its limited layer thickness, characterization of the intergranular grain boundary film can seldomly be achieved by direct chemical profiling and is mainly restricted to imaging by high resolution techniques[14]. Element distribution within the larger glassy pockets concentrated at triple junctions, however, can be monitored using a small spot size analytical scanning transmis-

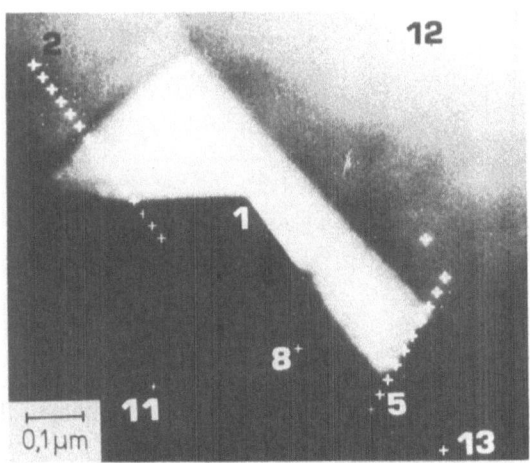

Fig. 11. STEM bright field image from a glassy pocket of 10 wt.% MgO-fluxed HPSN grain boundary phase

sion electron microscope (STEM).* In Fig. 11 a glassy pocket in 10 wt.% MgO-fluxed HPSN thinned by argon ions is shown. For the profile signed "5" the corresponding element distribution was measured with a spot size of 15 nm (Fig. 12). The grain boundary phase specific elements magnesium, calcium and iron can be identified reaching concentrations of 30 wt.% and about 1 wt.% respectively (neglecting amounts of oxygen and nitrogen in the analysis).

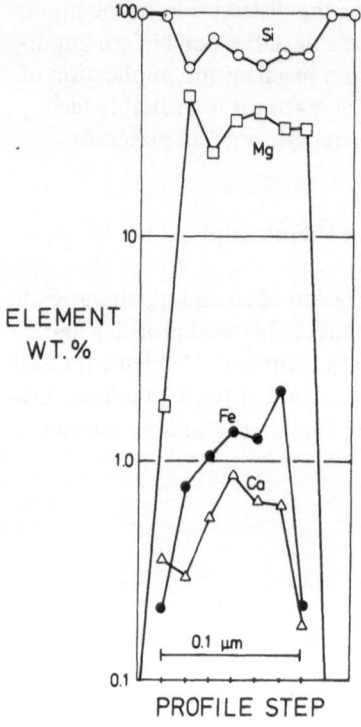

Fig. 12. STEM element distribution corresponding to a glassy pocket profile 5 in Fig. 11

4. Auger Electron Spectroscopy

Due to the predominantly intercrystalline fracture mode of MgO-doped HPSN, Auger electron spectroscopy (AES) is a suitable method for chemical analysis of the thin glassy grain boundary phase coverage in fracture surfaces. Unfortunately AES can only be operated under limited measurement condi-

---

* The authors thank Dr. P. Hagemann, Philips Research Eindhoven, for STEM/EDX measurements and helpful discussions.

tions because of surface charging due to the small electrical conductivity of Si_3N_4 [15,16]. Fig. 13a summarizes AES spectra from air fractured 10 wt.% MgO-fluxed HPSN surface. Using a spot size of about 1 μm nearly identical spectra were obtained at different positions of the fracture surface which may indicate a homogeneous fracture surface coverage compositions within 1 μm dimensions. The element distribution obtained, however, corresponds to an integral composition of both glassy pockets at triple junctions and thin glassy layers on the surface of matrix grains. Fig. 13b shows several AES spectra from oxinitride glasses within the system Ca-Mg-Si-O-N exhibiting different nitrogen contents from 10 to 40 equi. % N which were prepared (courtesy of Prof. Dr. H. Schaeffer, Erlangen) as standard materials matching possible grain boundary phase compositions within the system Si_3N_4-MgO. In these spectra all elements but nitrogen agree with the characteristics of the AES series in Fig. 13a. The high intensity of the nitrogen peak in the AES spectra from HPSN material may result from contributions of sample areas which exhibit no glassy phase coverage. This result corresponds with XPS data (see section 5) which indicate an oxinitride composition of the

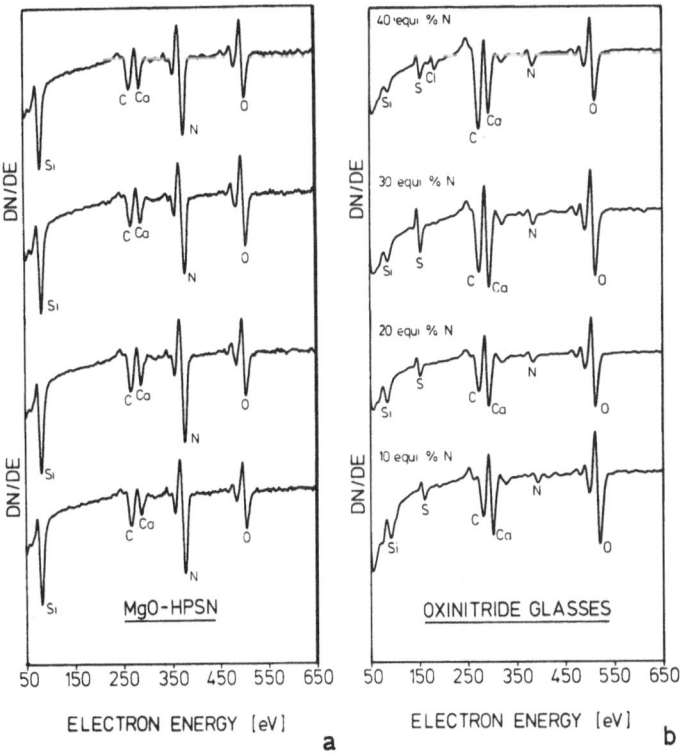

Fig. 13a,b. AES spectra from 10 wt.% MgO-fluxed HPSN and oxinitride glasses in the system Ca-Mg-Si-O-N

grain boundary phase and with experimental data[2] in the system Mg-Si-O-N as well where a significant nitrogen solubility was demonstrated for the glassy interface in HPSN.

For AES investigations of the thin grain boundary film on grain surfaces a highly focused electron beam must be employed due to the limited average grain size for β-Si$_3$N$_4$ of $\leqslant 1$ μm. Fig. 14 shows depth profiling by a series of pulse counting spectra for various argon sputter stages focusing the electron beam on a single β-Si$_3$N$_4$ grain of a 10 wt.% MgO-fluxed HPSN fracture surface. In the unsputtered AES spectrum the grain boundary phase specific

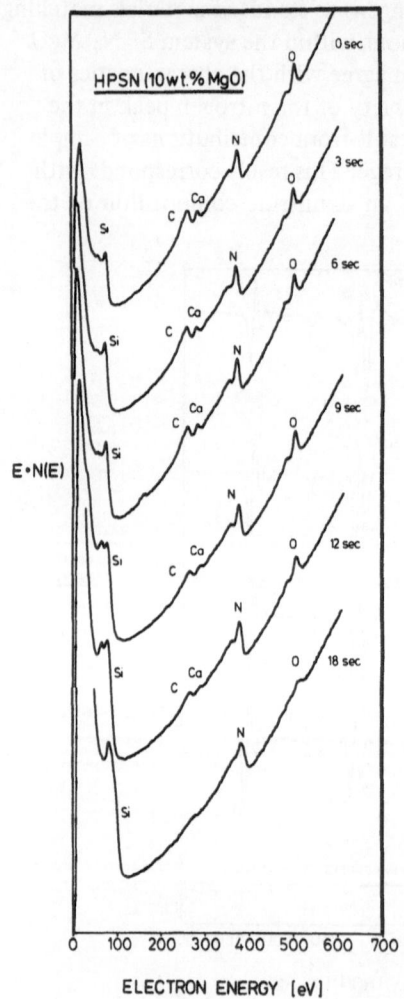

Fig. 14. Auger depth profile of single β-Si$_3$N$_4$ grain sputtering (10 wt.% MgO-fluxed HPSN)

elements silicon, nitrogen, oxygen, calcium, and magnesium (not shown in the electron energy interval presented) can be identified. With increasing sputter time there is a strong signal decrease for calcium, magnesium, and oxygen up to the stage where only the Si_3N_4 components remain. From sputtering parameters an upper estimate of the glassy layer thickness of 5 to 10 nm may be drived which is in the order of the value of $1-3$ nm[17] reported by lattice imaging of the intergranular glassy phase in dense Si_3N_4.

5. X-Ray Photoelectron Spectroscopy

For characterization of the chemical constitution of the grain boundary phase in dense Si_3N_4 X-ray photoelectron spectroscopy (XPS) was applied on HPSN fracture surfaces[18]. Due to the intrinsic lack of lateral resolution for XPS a simultaneous photoelectron emission resulting not only from the glassy phases themselves but also from the minor crystalline matrix phase (forsterite, see section 2) and β-Si_3N_4 must be considered. This is demonstrated in Fig. 15 by comparing the XPS spectrum of 5 wt.% MgO-fluxed HPSN with those from reaction-bonded Si_3N_4 (which exhibits no glassy grain boundary phase), oxinitride glass and single crystal of natural forsterite. As the XPS spectrum of the HPSN grain boundary phase can be formally obtained by weighted superposition of the other three spectra, this effect has to be considered in quantitative XPS analysis.

In Table 1 binding energies and Auger parameters of the grain boundary phase for various HPSN materials are compared with data from various reference samples resulting in a satisfactory agreement of the XPS data from HPSN materials and oxinitride glasses.

6. Conclusions

Depending on the atomic number of the sintering additives BSE imaging may be a powerful tool for identification and evaluation of the distribution of the minor crystalline matrix phases in polished and fractured surfaces of HPSN and SSN materials. Regarding the Y_2O_3-fluxed SSN material, convincing results were obtained for both polished and fractured surfaces. In the case of low atomic number sintering additives as MgO, however, additional information by means of magnesium X-ray distribution patterns is required for discrimination of β-Si_3N_4 and the minor crystalline matrix phase. Furthermore, the improved resolution of the elongated grain morphology of β-Si_3N_4 by BSE imaging of the fracture surface may be helpful for microstructural characterization of this material. The grain size of the minor crystalline matrix phase depends on the amount of the sintering aid added, reaching $1-2$ μm for 10 wt.% MgO-fluxed material and $0.1-0.5$ μm for addition of only 1wt.%

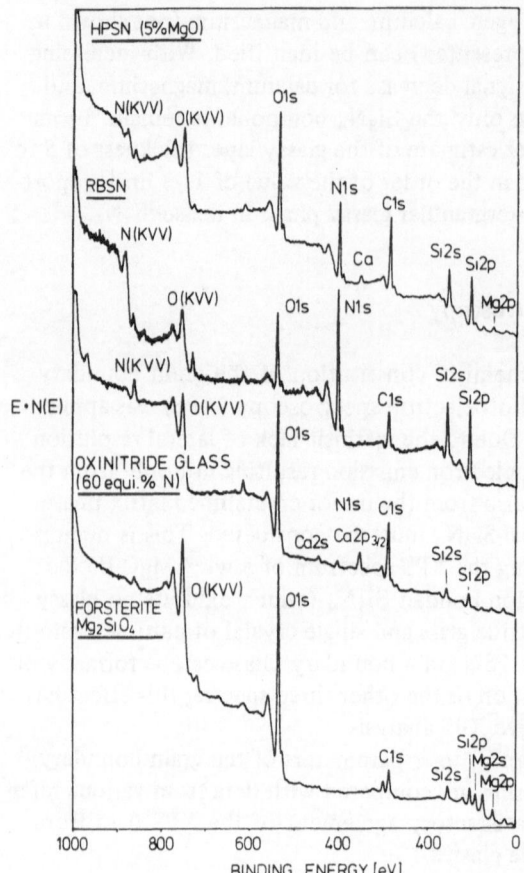

Fig. 15. Comparison of XPS spectra from 5 wt.% MgO-fluxed HPSN, reaction-bonded Si_3N_4, oxinitride glass and natural forsterite

MgO respectively. In the glassy pockets, concentrated in triple junctions of MgO-fluxed HPSN the elements magnesium, calcium and iron were found to be strongly concentrated by using analytical STEM. The chemical constitution of the grain boundary phase was studied by means of AES and depth profiling technique. Possible differences of the intrinsic chemistry between the glassy pockets and the thin intergranular glassy film has to be considered in further research. Regarding the XPS analysis of HPSN fracture surfaces the resulting XPS spectra of the grain boundary phase may be influenced by a signal contribution of both the minor crystalline matrix phase and β-Si_3N_4. From binding energies and Auger parameters an amorphous silicon oxinitride configuration was indicated for the structural state of the HPSN grain boundary layer.

Table 1. Si 2*p* Binding Energies, Auger Si(*KLL*) Kinetic Energies and Modified Auger Parameters α^*_{Si} (eV) for HPSN Grain Boundary Phases and Suitable Reference Materials

	Si 2*p*	Si(*KLL*)	α^*_{Si}
3 wt.% MgO-HPSN, 1780 °C/120 min	101.8	1611.5	1713.3
5 wt.% MgO-HPSN, 1700 °C/ 20 min	102.0	1611.3	1713.3
5 wt.% MgO-HPSN, 1700 °C/ 30 min	101.8	1611.2	1713.0
Si_3N_4 powder (STARCK LC 10)	102.0	1612.0	1714.0
Si_3N_4 powder (GTE 502)	102.0	1612.0	1714.0
Si_3N_4 powder (TOSHIBA)	102.1	1611.8	1713.8
CORNING nitrogen glass	102.6	1610.6	1713.1
CVD silicon nitride films (Ref.[19])	102.1	1612.1	1714.2
Si (Ref.[19])	99.6	1616.3	1715.9
SiO_2 layer on Si (Ref.[19])	103.4	1608.6	1712.0

Summary

Surface- and Microanalytical Investigations of the Chemical Constitution of the Grain Boundary Phase in Dense Silicon Nitride

Identification and distribution of minor crystalline matrix phases in dense hot-pressed and sintered Si_3N_4 were achieved by means of backscattered electron images and X-ray distribution patterns using various types and amounts of densification aids (MgO, Y_2O_3). Both polished and fractured surfaces were investigated. Element concentration profiles within the glassy grain boundary phase pockets at triple junctions of MgO-fluxed hot-pressed material were measured using analytical scanning transmission electron microscopy. Auger electron spectroscopy and the depth profiling technique were employed for evaluation of the chemical constitution of the grain boundary phase. By means of X-ray photoelectron spectroscopy the structural state of the amorphous grain boundary layer was proved to have a silicon oxinitride configuration.

References

1. S. Wild. P. Grieveson, K.H. Jack, and M.J. Latimer, Special Ceramics **5**, 357 (1972).
2. K.H. Jack, J. Mat. Sci. **11**, 1135 (1976).
3. D.R. Clarke and G. Thomas, J. Am. Ceram. Soc. **60**, 491 (1977).
4. L.K.V. Lou, T.E. Mitchell, and A.H. Heuer, J. Am. Ceram. Soc. **61**, 392 (1978).

5. D.R. Clarke and G. Thomas, J. Am. Ceram. Soc. 61, 114 (1978).
6. V.N.E. Robinson, Scanning 3, 15 (1980).
7. F.F. Lange, S.C. Singhal, and R.C. Kuznicki, J. Am. Ceram. Soc. 60, 249 (1977).
8. L.J. Gauckler, H. Hohnke, and T.Y. Tien, J. Am. Ceram. Soc. 63, 35 (1980).
9. R.E. Loehman and D.J. Rowcliffe, J. Am. Ceram. Soc. 63, 144 (1980).
10. D.R. Clarke, N.J. Zaluzec, and R.W. Carpenter, J. Am. Ceram. Soc. 64, 608 (1981).
11. K.H. Jack, Ceramics for High-Performance Application, Proceedings of the Second Army Materials Technology Conference at Hyannis, Nov. 1973 (Brook Hill, Chestnut Hill, Massachusetts, 1974). p. 265.
12. Y. Oyama and O. Kamigaito, Yogyo Kyokai Shi 81, 290 (1973).
13. F.F. Lange, Westinghouse Research and Development Center, Pennsylvania, Office of Naval Research Contract Report No TR-9 (Sept. 1976).
14. D.R. Clarke, Ultramicroscopy 4, 33 (1979).
15. S. Hofmann and L.J. Gauckler, Powder Metallurgy International 6, 90 (1974).
16. S. Hofmann, L.J. Gauckler, and L. Tillmann, Mikrochim. Acta [Wien], Suppl. VI, 1975, 373.
17. P. Greil and J. Weiss, J. Mat. Sci. 17, 1571 (1982).
18. W. Braue, H.J. Dudek, and G. Ziegler, 5th Proc. Intern. Meeting on Modern Ceramics Technologies (CIMTEC), Lignano/Italy, 14–19 June 1982, in press.
19. J.A. Taylor, Applications of Surface Science 7, 168 (1981).

Correspondence and reprints: H. J. Dudek, Deutsche Forschungs- und Versuchsanstalt für Luft- und Raumfahrt (DFVLR), Institut für Werkstoff-Forschung, P.O.B. 90 60 58, D-5000 Köln 90, Federal Republic of Germany.

Mikrochimica Acta [Wien], Suppl. 10, 177–187 (1983)
© by Springer-Verlag 1983

Institut für Allgemeine Physik der Technischen Universität Wien and Metallwerk Plansee,
Austria

Analysis of Reactively Ion Plated Titanium-Nitrid Films

By

S. Laimer, P. Braun, H. Störi, F. Viehböck, P. Rödhammer, and K. Kailer

With 9 Figures

(Received January 19, 1983)

1. Introduction

Hard metal coatings on cutting tools – first applied to cemented carbide
inserts – are presently finding new applications for high-speed-steel (HSS)
tools such as twistdrills, end mills and hob cutters. While the original coating
technique employed was Chemical Vapor Deposition (CVD), Physical Vapor
Deposition (PVD) processes[1] are favored for HSS coatings. The main advantage of PVD lies in the lower process temperatures, well below the annealing
temperature (~ 560 °C) of HSS, as compared to CVD (typically 1000 °C).
The most widely used coating materials for tools employed to date are TiC,
TiN, and Al_2O_3, and multilayer combinations of these refractory compounds. In particular, pseudobinary alloys of type $TiC_x N_y O_{1-x-y}$ are generally chosen for coatings on HSS tools.
In development as well as in quality control of coatings, special attention has
to be paid to the following metallurgical properties of the coatings:

(i) Film morphology. Physical properties depend strongly on microcrystal-
line structure, the isotropic growth being more favorable than a colum-
nar one.

(ii) Crystallografic phases and preferred orientations (textures).

(iii) Chemical composition, in particular depth profiles of ion plated species
and impurities across the film and – very importantly-across the "intermixed" boundary region into the substrate.

These properties determine the technologically important coating characteristics such as adhesion, hardness, thermal conductivity, resistance against oxidation etc.

In the course of an investigation of the "Reactive Ion Plating" process we have examined a large number of TiN coatings with the aid of Scanning-Electron-Microscopy (SEM), Auger-Electron-Spectroscopy (AES), X-Ray-Photoelectron-Spectroscopy (XPS) and X-Ray-Diffraction (XRD). In this paper we present some of our results to demonstrate the potential of these rather sophisticated methods for the optimization of process parameters in coating technology.

2. Reactive Ion Plating

The process of Reactive Ion Plating is illustrated schematically in Fig. 1. The metal to be deposited (in this case titanium) is vaporized in a water-cooled crucible with the aid of an electron gun. The deposition chamber (not drawn) contains an atmosphere of Ar and N_2 ($p \leqslant 10^{-2}$ mbar) which is partially ionized in a gas discharge between the chamber and the negatively biased substrate (S_2 closed). With S_1 closed, the auxiliary anode A serves to increase the degree of ionization (typically in the order of 1%). Metal vapor atoms are ionized by collision with plasma and accelerated toward the substrate where they combine with reactive-gas ions or atoms to form the desired compound. This short introduction to Reactive Ion Plating should suffice to point out the important process parameters: Electron-gun power, substrate potential, auxiliary-anode potential, total gas pressure, partial pressure of reactive gas, and substrate temperature. In their effect on the properties of the film, these parameters are partly interdependent.

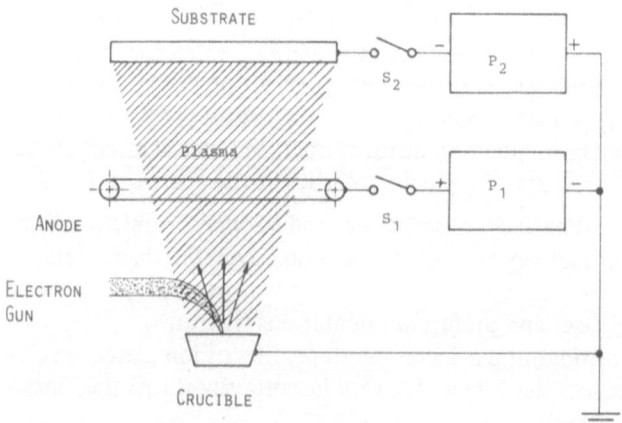

Fig. 1. Schematic illustration of the Reactive Ion Plating process

Table 1. Characterization of TiN Films Deposited with Increasing N_2 Flow

Sample	15	16	17	18
Nitrogen flow ($cm^3 min^{-1}$)	5	25	49	75
Microstructure (SEM)	coarse columnar	fine equiaxed	fine columnar	fine columnar
Phases (XRD)	α-Ti (hex)	α-Ti + TiN_x	fcc TiN_x	fcc TiN_x
Texture (XRD)	⟨001⟩	–	⟨100⟩	⟨100⟩
N (at%) (AES)	10	43	51	51
N (at%) (XPS)	16	49	51	49

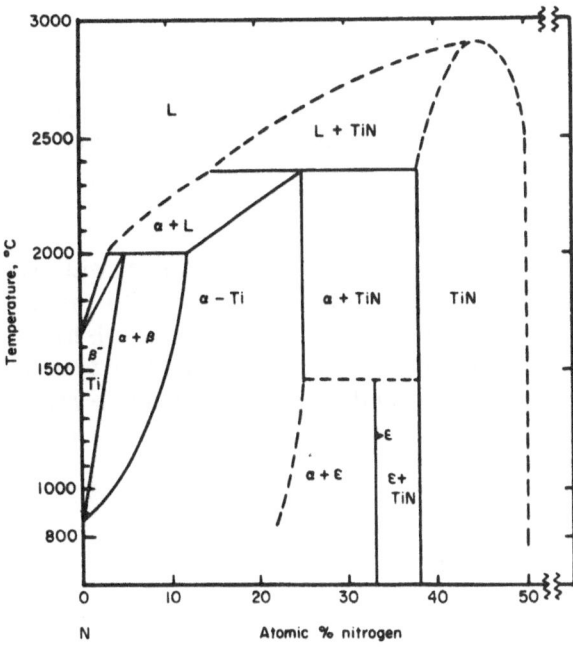

Fig. 2. Phase diagramm Ti-N

In an attempt to optimize the partial pressure of N_2, we have made a series of runs with increasing flow of N_2, keeping all other parameters constant. TiN_x-films of approximately 3 μm thickness were deposited onto molybdenum sheet. The pertaining parameters are given in Table 1, along with a summary of the film properties to be discussed later. When increasing the partial pressure of N_2, we expect — by looking at the phase diagram[2] (Fig. 2) — to pass at first a region of α-Ti with increasing amount of interstitially dissolved N, then two-phase regimes of (α-Ti + Ti_2N) and (Ti_2N + TiN_x), and finally to arrive at the fcc TiN_x which exhibits a broad range of homogeneity. In our study a stoichiometric composition $TiN_{1.0}$ was aimed for.

3. Results of Film Analyses

3.1 SEM

The scanning electron micrographs (Figs. 3a–c) of our samples fractured in liquid nitrogen show the coarse-columnar structure of α-Ti (Fig. 3a), the extremely fine-grained film in the two-phase-region (Fig. 3b), and the fine-columnar TiN-film with fcc structure (Fig. 3c).

3.2 AES

The primary goal of our AES investigations was to determine the atomic ratio N/Ti across the TiN film and to observe the pseudodiffusion zone at the film-substrate interface. Our method of quantitative analysis[3] of the Auger spectra of TiN films will now be discussed in some detail, using the spectrum of our external standard — a single crystal of $TiN_{0.98}$ — for illustration.

Comparing the differentiated Auger spectrum of $TiN_{0.98}$ (Fig. 4) with that of pure Ti (Fig. 5) it is seen that the N *KLL*-peak coincides with the Ti *LMM*-peak at 387 eV. For the purpose of evaluation the two signals have to be separated. This can be achieved by determining the ratio of the peak-to-peak-heights (H_{pp}) of the Ti_{387} signal and the Ti_{417} signal from an Auger spectrum of pure Ti (Fig. 5). Thus we may write for our $TiN_{0.98}$ standard

$$H_{Ti} = H_{417} \tag{1a}$$

$$H_N = H_{387} - H_{417} \cdot \frac{H(Ti_{387})}{H(Ti_{417})} \tag{1b}$$

For a given TiN sample the N/Ti-ratio is then obtained from Eq. (2).

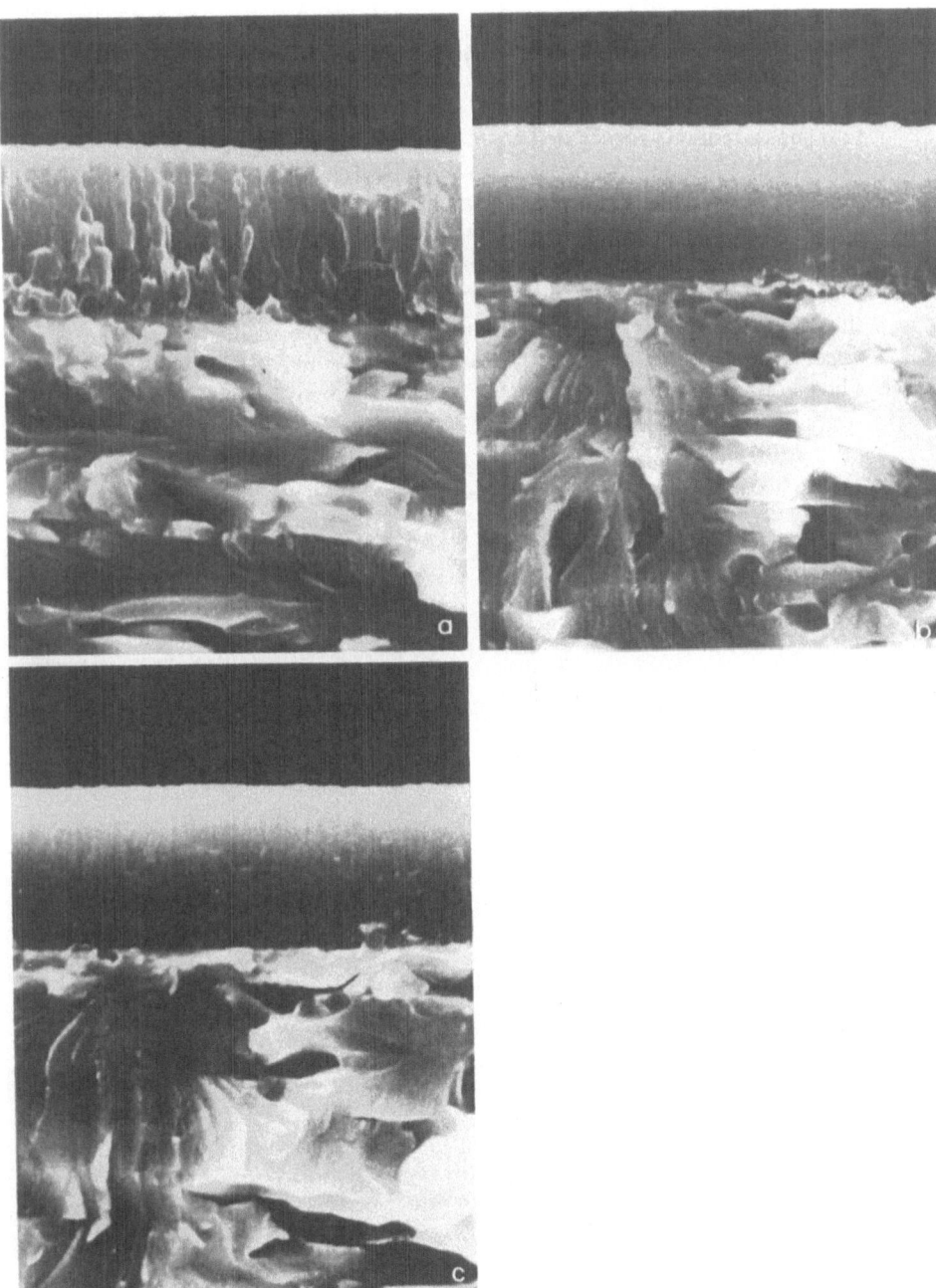

Fig. 3. TiN films on Mo substrate, deposited with increasing N_2 flow: (a) 5 cm^3 min^{-1}; (b) 25 cm^3 min^{-1}; (c) 49 cm^3 min^{-1} SEM (x 10 000)

$$(N/Ti)_{sample} = (N/Ti)_{standard} \cdot \frac{(H_N/H_{Ti})_{sample}}{(H_N/H_{Ti})_{standard}} \qquad (2)$$

Our measurements revealed the presence of O and C in many of the ion plated TiN films. On the other hand our investigations on TiC, TiO_2 and oxidized Ti showed the influence of C and O on the ratio $H(Ti_{387})/H(Ti_{417})$. On the basis of the latter data we were able to correct the H_{pp}'s for the presence in our coatings of C and O impurities in first order.

Fig. 6 shows the results of our analysis for a series of films, including those shown in Figs. 3a–3c. For low nitrogen flow, there exists a region of steep linear increase in the N/Ti ratio; saturation is attained at a N/Ti-ratio of ~ 1. The error bars are estimated to be ± 0.10 in the N-rich regime (N/Ti ⩾ 0.8).

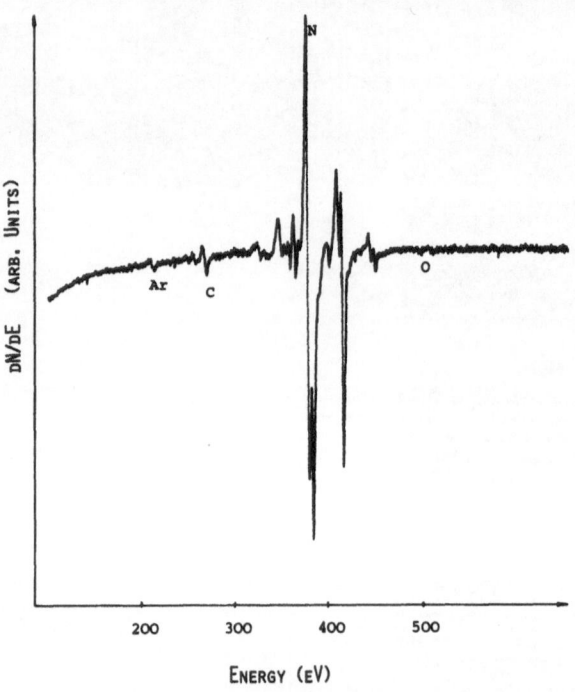

Fig. 4. Auger Electron Spectrum of $TiN_{0.98}$ used as external standard

To study the intermixed boundary region between film and substrate, TiN films of 0.4 μm thickness have been prepared. The H_{pp}'s of six different Auger lines were continuously monitored while sputter-etching through the film using a beam of Krypton ions. A depth profile obtained in this way is

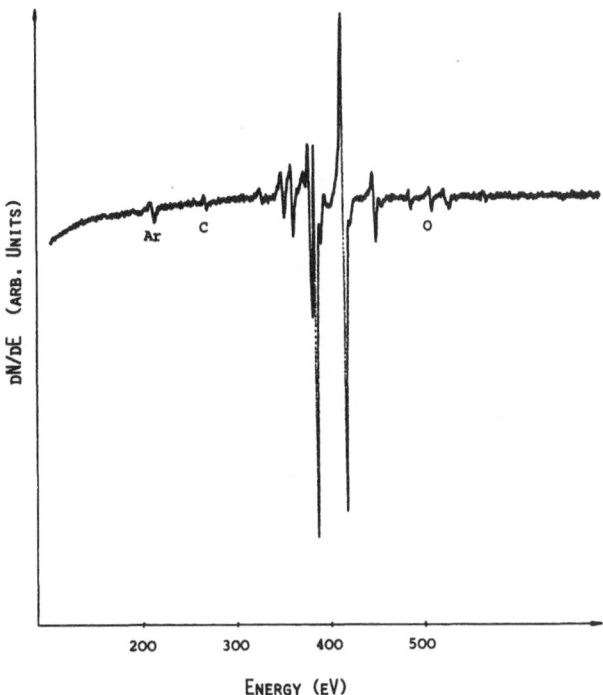

Fig. 5. Auger Electron Spectrum of Ti

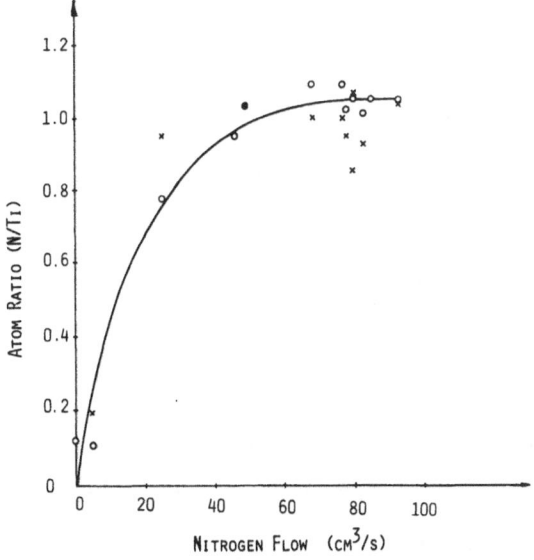

Fig. 6. Dependance of the atomic ratio (N/Ti) on N_2-flow. Results of AES (*o*) and XPS (*x*). Full line is guide-to-the-eye for AES results, only

shown in Fig. 7. The Ti_{417} signal is nearly constant. However H_{pp} of the 387 eV-line (Ti + N) is varying periodically. Concurrently H_{pp} of O_{512} is oscillating with opposite phase. This effect could be traced to an evaporation rate of Ti changing with the periode of rotation of the crucible. Thus we conclude that oxygen – which is always present as an impurity in the plasma – is taken up into the film (as TiO) whenever the nitrogen flow is insufficient for stoichiometric TiN to be formed.

In accordance with previous work[4] we find no preferential sputtering of Ti or N in our films. Due to atomic mixing the resolution of depth has decreased to estimated 500 Å upon reaching the interface to the substrate. Still the observed tails for both the Ti and the Mo signal are considerably wider, indicating a width of 500 – 1000 Å for the pseudodiffusion zone. We note that no interlayer pointing to insufficient sputter cleaning of the surface prior to film deposition is observed.

A further information obtained from the Auger depth profile concerns the possible take-up of Ar (present in the plasma) into the film: no Ar line was detect in any spectrum.

3.3 XPS

Complimentary to AES XPS was employed to gain further information on the chemical composition of the TiN films. A highly resolved XPS spectrum of $TiN_{0.98}$ is given in Fig. 8, showing the Ti-$2p_{3/2}$ and Ti-$2p_{1/2}$ peaks at

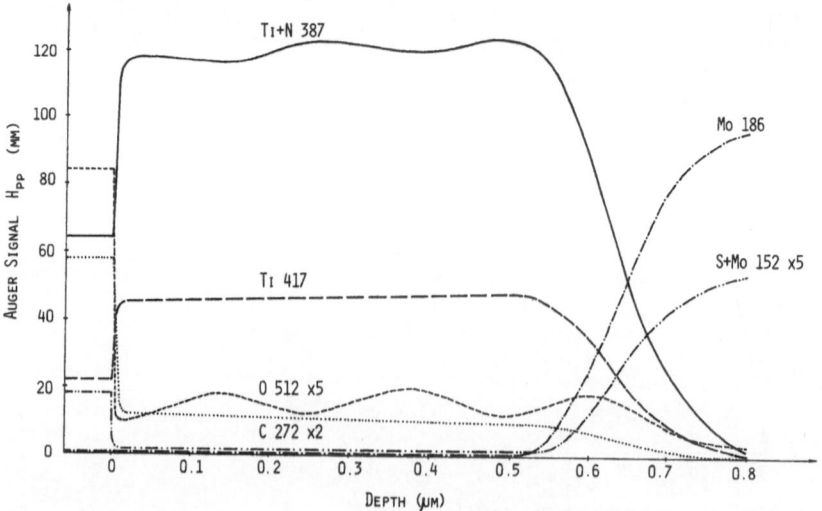

Fig. 7. Depth profile of reactively ion plated TiN film on Mo substrate. Note the differing scale factors

454.4 eV and 460.3 eV, respectively. Since we lack the outfit to sputter-etch the surface while scanning, another Ti-$2p_{3/2}$ peak (at ca. 458 eV) emerges with due to formation of TiO_2 on the surface. Thus for quantitative analysis of the Ti-$2p_{3/2}$ peak intensity, peaks have to be deconvoluted. No such difficulty is encountered with the N $1s$ peak at 397.9 eV.

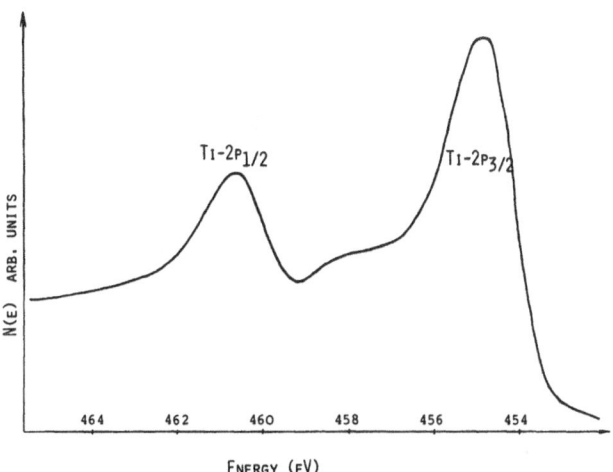

Fig. 8. Detail of the XPS Spectrum of $TiN_{0.98}$ (Ti-$2p_{3/2}$ and Ti-$2p_{1/2}$)

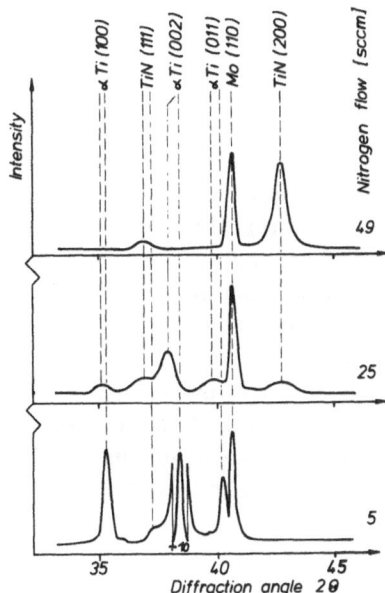

Fig. 9. Detail of X-ray diffraction spectra of TiN film deposited with increasing N_2 flow

XPS spectra have been evaluated in analogy to Auger spectra, using the same external standard. Here, however, the aera F under a given peak was chosen as a measure for the total photoelectron current. Thus the atomic ratio is

$$(N/Ti)_{sample} = (N/Ti)_{standard} \cdot \frac{(F_N/F_{Ti})_{sample}}{(F_N/F_{Ti})_{standard}} \tag{3}$$

The results of our analyses on a large number of films are plotted in Fig. 5. The scatter in the data points is considerably larger than for the AES results, owing to uncertainties in the base line and deconvolution of the Ti peaks.

The chemical shift of the Ti-$2p_{3/2}$ peak observed on films containing – in addition to N – a high percentage of oxygen and carbon indicates that a pseudobinary alloy of $TiC_x N_y O_z$ is formed whenever a mixture of N_2, O_2 and hydrocarbons is present as reactive component in the plasma. This analysis is also confirmed by the results of XRD to be discussed next.

3.4 XRD

The X-ray diffraction spectra taken from a series of TiN_x films prepared with increasing N_2-flow are shown in part in Fig. 9. They reflect the sequence of phases expected from the phase diagram (Fig. 2). A pronounced preferred orientation of film growth is observed: in the hexagonal α-Ti phase, which is formed at 5 cm^3 min^{-1} N_2-flow, the c-axis $\langle 001 \rangle$ lies vertical to the substrate surface, whereas in the cubic TiN_x phase (N_2-flow: 49 cm^3 min^{-1}) preferred growth occurs along the $\langle 100 \rangle$-direction. For intermediate flows (25 cm^3 min^{-1}) a two-phase regime is attained, with the fcc TiN_x-phase emerging besides the α-Ti. Here the structure is very fine-grained (Fig. 3b); hence no pronounced texture is observed.

In all cases the peaks are strongly broaded presumably owing to inhomogeneities in the N/Ti atomic ratio. It should be noted that these films are not in thermodynamic equilibrium, and drastic changes in diffraction patterns are observed upon annealing at temperatures higher than the deposition temperature. The broadening forbids a precise determination of the lattice parameter, which in conjunction with literature values would yield the N/Ti ratio both in the α-Ti and fcc-TiN_x phases. However, the lattice parameter of sample No. 17 ($a_0 = 4.232$ Å) is already very close to the literature value[2] for stoichiometric TiN ($a_0 = 4.235$ Å) and does not change upon increasing the N_2-flow beyond 49 cm^3 min^{-1}. This corroborates the results of AES and XPS given in Fig. 6.

4. Discussion

For a quantitative analysis of the N/Ti atomic ratio in TiN films AES has proved to yield the most reliable results. One mayor uncertainty is due to the use of a single crystal of TiN as an external standard, the influence of morphology on the Auger peak intensities being unknown. The correction for matrix effects due to impurity contents of C and O is essential; still it leaves us with a possible error of about ± 10%.

The results of XPS are in general agreement with AES data. For XPS the larger scatter of data could be strongly reduced by the use of a differentially pumped ion gun to keep the surface free of oxygen.

Summary

Analysis of Reactively Ion Plated Titanium-Nitride Films

The information contained in the results of REM, AES, XPS and XRD has been found to yield a rather complete characterization of TiN films on molybdenum substrates which have been prepared by the Reactive Ion Plating process. These analytical techniques proved to be a highly valuable tool in optimizing our process parameters. They may become even more valuable in the characterization and optimization of TiN coatings of HSS where preliminary results show the formation of the interface film – substrate to be highly sensitive to the parameters of the coating process.

References

1. D.G. Teer, in Proc. Conf. Ion Plating and Allied Techniques, Edinburgh 1977, p. 13.
2. L.E. Toth, Transition Metal Carbides and Nitrides. New York: Academic Press. 1971.
3. S. Laimer, Master Thesis, Technische Universität, Vienna, Austria, 1982.
4. J.-E. Sundgren, Thesis, Linköping Inst. of Technology, Linköping, 1982.

Correspondence and reprints: Univ.-Ass. Dr. H. Störi, Institut für Allgemeine Physik, Technische Universität Wien, Karlsplatz 13, A-1040 Wien, Austria.

Mikrochimica Acta [Wien], Suppl. 10, 189–198 (1983)

Institut für Angewandte und Technische Physik der Technischen Universität Wien, Austria

Quantitative Surface Analysis Without Reference Samples

By

M. F. Ebel, H. Ebel, U. Paschke, G. Zuba, and J. Wernisch

With 3 Figures

(Received January 19, 1983)

Quantitative analysis without reference samples means that the correlation between measured data and sample composition is described by a theoretical expression. Thus, neither reference samples nor pure element standards have to be employed. For X-ray photoelectron spectroscopy (XPS) and X-ray induced Auger electron spectroscopy (XAS) these theoretical correlations can be given. Besides, a good knowledge of the "fundamental parameters" is necessary.

For two reasons quantitative analysis without reference samples is of high importance for XPS and XAS:

i) These two techniques provide informations from a depth smaller than 10 nm. Comparing the results of these surface techniques to the results of chemical analysis it must be considered that chemical analysis investigates bulk material and not only surfaces.

ii) The unknown sample as well as reference samples are always covered by a contamination layer (sometimes also by an oxide layer). These overlayers cause a decrease of line intensities. As the mean free paths of the ejected electrons depend essentially upon the kinetic energy of the electrons quantitative results are strongly influenced when electrons with remarkable differences of their kinetic energies have to be used.

To derive the theoretical correlation for XPS as well as XAS the analytical situations for these two methods are compared to X-ray fluorescence (XRFA) which is also a process based upon photoabsorption. Eq. (1) gives the fluorescence count rate for monochromatic primary excitation[1,2].

$$n_i^k = q \cdot x_\lambda \cdot \kappa_i^k \cdot \frac{\Omega}{4\pi} \cdot \tau_{\lambda i} \cdot \underbrace{\frac{S_i^k - 1}{S_i^k}} \cdot \underbrace{\omega_i^k \cdot p_i^k} \cdot \underbrace{\frac{c_i}{\mu_{ic}}} \cdot \underbrace{\frac{1}{1 + \frac{\mu_{\lambda c}}{\mu_{ic}} \cdot \frac{\cos \epsilon}{\cos \alpha}}} \qquad (1)$$

$$\underbrace{}_{\text{I}} \qquad \underbrace{}_{\text{II}} \qquad \underbrace{}_{\text{III}} \quad \underbrace{}_{\text{IV}} \qquad \underbrace{}_{\text{V}}$$

$$\downarrow \qquad\qquad\qquad \downarrow \qquad\qquad\qquad \downarrow$$
$$\text{Instr.} \qquad\qquad \text{Element} \qquad\qquad \text{Sample}$$

Investigating a sample with m components the concentration of the m elements is found by iteration together with $\sum_{j=1}^{m} c_j = 1$. There exists a linear dependence of the concentrations on the mass absorption coefficients $\mu_{\lambda c} = \sum_{j=1}^{m} c_j \cdot \mu_{\lambda j}$ and $\mu_{ic} = \sum_{j=1}^{m} c_j \mu_{ij}$. For a quantitative analysis without reference samples the knowledge of the "fundamental parameters" included in Eq. (1) is necessary[3].

XPS – Theoretical Considerations

Employing Eq. (1) for quantitative XPS-analysis some of the factors have to be modified. This can be done more easily when instrumental, elemental and sample influences are considered.

I – Instrument

$$q \cdot x_\lambda \cdot \kappa_i^k \cdot \frac{\Omega}{4\pi} \longrightarrow q \cdot x_\lambda \cdot \frac{\Omega}{4\pi} \cdot [A + B \cdot E_i^k + C \cdot (E_i^k)^2]$$

The above given relation shows that the detector efficiency becomes unity. A multiplication with $[A + B \cdot E_i^k + C \cdot (E_i^k)^2]$ becomes necessary to describe the dependence of the transmission of the spectrometer on the energy of the photoelectrons[4].

II – Element

$$\tau_{\lambda i} \cdot \frac{S_i^k - 1}{S_i^k} \longrightarrow 10^{-24} \cdot \frac{L}{A_i} \cdot \sigma_c^{1s} \cdot \left(\frac{\sigma_i^k}{\sigma_i^{1s}}\right) \cdot \left[1 - \frac{\beta_i^k}{2} \cdot \frac{3\cos^2 \vartheta - 1}{2}\right]$$

In XPS-literature[5] the photoabsorption cross-section for a defined X-radiation is given in barns/atom and in multiples of σ_c^{1s}. The expression $[1 - \frac{\beta_i^k}{2} \cdot \frac{3\cos^2 \vartheta - 1}{2}]$ describes the asymmetry in the ejection of photoelectrons[6].

III − Element

$$\omega_i^k p_i^k \longrightarrow 1$$

Characteristic photoelectrons are the result of a single process without relaxation and also the production of photoelectrons by coherent scattered X-rays is neglected.

IV − Sample

$$\frac{c_i}{\mu_{ic}} \longrightarrow \frac{c_i}{1/\rho_c \cdot \Lambda_{ic}^k}$$

The problem becomes much easier when changing over from mass absorption coefficients to mean free paths of the photoelectrons Λ_{ic}^k in the sample c with an energy E_i^k. As Λ_{ic}^k values do not exist for all elements it is advantageous to use the energy dependence of the mean free path of the photoelectrons $\Lambda_{ic}^k = \Lambda_c(\text{ref}) \cdot (E_i^k)^x$ where $\Lambda_c(\text{ref})$ is the mean free path of the photoelectrons in the specimen c at an energy of 1 keV, thus, being a constant for the sample under investigation. In literature the exponent x varies between 0.5 and 0.8[8]. Now we can write

$$\frac{c_i}{\mu_{ic}} \longrightarrow c_i \cdot \rho_c \cdot \Lambda_c(\text{ref}) \cdot (E_i^k)^x$$

V − Sample

$$1 + \frac{\mu_{\lambda c}}{\mu_{ic}} \cdot \frac{\cos \epsilon}{\cos \alpha} \longrightarrow 1 \qquad \text{together with} \qquad \mu_{ic} \gg \mu_{\lambda c}$$

VI − Sample

An expression has to be included which does not exist for XRFA: Due to shading a factor R (identical for all photoelectron lines of one specimen) has to be taken into account. Besides the influence of an overlayer (thickness d) has to be regarded. Exp $(-d/\Lambda_{if}^k \cdot \cos \epsilon)$ expresses this influence on the count rate for a certain take-off direction ϵ of the photoelectrons. It can be rewritten using the energy dependence of the mean free path of the photo-electrons. This means that the factor to be included is given by

$$R \cdot \exp\left[-d/\Lambda_f(\text{ref}) \cdot (E_i^k)^x \cdot \cos \epsilon\right]$$

Having now compared XRFA and XPS the following Eq. (2) gives the correlation between measured photoelectron count rates n_i^k and the concentrations c_i.

$$n_i^k = q \cdot x_\lambda \cdot \kappa_i^k \cdot \frac{\Omega}{4\pi} \cdot [A + B \cdot E_i^k + C \cdot (E_i^k)^2] \cdot 10^{-24} \cdot \frac{L}{A_i} \cdot \sigma_c^{1s} \cdot \left(\frac{\sigma_i^k}{\sigma_c^{1s}}\right) \times$$

$$\times \left[1 - \frac{\beta_i^k}{2} \cdot \frac{3\cos^2\vartheta - 1}{2}\right] \cdot c_i \cdot \rho_c \cdot \Lambda_c(\text{ref}) \cdot (E_i^k)^x \cdot R \cdot \exp[-d/\Lambda_f(\text{ref}) \cdot (E_i^k)^x \cdot \text{co}$$

$$(2)$$

To determine the concentrations of the m elements of a sample from measured data n_i^k and n_j^k ($j = 1, 2, \ldots m$)

$$c_i = 100 \cdot \frac{\dfrac{n_i^k}{N_i^k}}{\displaystyle\sum_{j=1}^{m} \dfrac{n_j^k}{N_j^k}} \tag{3}$$

is used with

$$N_i^k = [A + B \cdot E_i^k + C \cdot (E_i^k)^2] \cdot (E_i^k)^x \cdot \frac{\sigma_i^k}{\sigma_c^{1s}} \cdot \frac{1}{A_i} \left[1 - \frac{\beta_i^k}{2} \cdot \frac{3\cos^2\vartheta - 1}{2}\right] \times$$

$$\times \exp[-d/\Lambda_f(\text{ref}) \cdot (E_i^k)^x \cdot \cos\epsilon]. \tag{4}$$

This evaluation can be done by a pocket calculator (e.g. HP 41 CV). The following "fundamental parameters" have to be known:

$A, B, C, \epsilon, \vartheta$	manufacturer of the instrument
A_i	handbook
$\dfrac{\sigma_i^k}{\sigma_c^{1s}}$	Scofield[5]
β_i^k	Reilman[6]
E_i^k from E_{bi}^k	Siegbahn[9]
$x = 0.7$	Wagner[8]

$d/\Lambda_f(\text{ref})$ for electrons of 1 keV is determined by a separate experiment (variation of take-off angle[10] or subshell twin method[10]).

XAS – Theoretical Considerations

Eq. (1) can also be modified to perform a quantitative XAS analysis without reference samples. Again instrumental, elemental and sample influences have to be compared for XRFA and XAS. It is found that corrections due to instrument and sample are comparable to XPS (see XPS-theoretical consideration) and only elemental influences have to be considered.

II – Element

$$\tau_{\lambda i} \cdot \frac{S_i^k - 1}{S_i^k} \xrightarrow{\hspace{1cm}} 10^{-24} \cdot \frac{L}{A_i} \cdot \sigma_c^{1s} \cdot \left(\frac{\sigma_i^k}{\sigma_c^{1s}}\right)$$

Cross-sections[5] can be used as in XPS. The Auger electron distribution shows no asymmetry.

III – Element

$$\omega_i^k \cdot p_i^k \xrightarrow{\hspace{1cm}} (1 - \omega_i^k) \cdot \eta_i^k$$

Here the fluorescence yield $\omega_i^{k\,[11]}$ gives the Auger yield and $\eta_i^{k\,[12]}$ is its percentage for the measured Auger line.

Thus, the following Eq. (5) gives the correlation between measured XAS data and the concentrations

$$n_i^k = q \cdot x_\lambda \cdot \kappa_i^k \cdot \frac{\Omega}{4\pi} \cdot [A + B \cdot E_i^k + C \cdot (E_i^k)^2]\, 10^{-24}\, \frac{L}{A_i} \cdot \sigma_c^{1s} \cdot \left(\frac{\sigma_i^k}{\sigma_c^{1s}}\right) \times \tag{5}$$

$$\times (1 - \omega_i^k) \cdot \eta_i^k \cdot c_i \cdot \rho_c \cdot \Lambda_c(\text{ref}) \cdot (E_i^k)^x \cdot R \cdot \exp[-d/\Lambda_f(\text{ref}) \cdot (E_i^k)^x \cdot \cos\epsilon].$$

The determination of the concentrations of the m components of a sample is performed (comparable to XPS) from measured data n_i^k and n_j^k ($j = 1, 2, ...m$)

$$c_i = 100 \cdot \frac{\dfrac{n_i^k}{N_i^k}}{\displaystyle\sum_{j=1}^{m} \dfrac{n_i^k}{N_j^k}} \tag{6}$$

with

$$N_i^k = [A + B \cdot E_i^k + C \cdot (E_i^k)^2] \cdot \frac{1}{A_i} \cdot \left(\frac{\sigma_i^k}{\sigma_c^{1s}}\right) \cdot (1 - \omega_i^k) \cdot \eta_i^k \cdot (E_i^k)^x \times$$

$$\times \exp[-d/\Lambda_f(\text{ref}) \cdot (E_i^k)^x \cdot \cos\epsilon]. \tag{7}$$

The necessary fundamental parameters are

A, B, C, ϵ manufacturer of the instrument

A_i handbook

$\dfrac{\sigma_i^k}{\sigma_c^{1s}}$ Scofield[5]

$1 - \omega_i^k$ Bambynek[11]

η_i^k Handbook of Auger Electr. Spec.[12]

E_i^k from E_{bi}^k Siegbahn[9]

$x = 0.7$ Wagner[8]

Again d / Λ_f (ref) has to be determined by a separate experiment.

Experimental Examples

In Fig. 1 XP-spectra of a ternary Au-Ag-Cu alloy are depicted and evaluation is shown. The area beneath the photoelectron line after background subtraction (half width x maximum) represents the photoelectron count rate. Investigations were performed on a great number of ternary and binary Au-Ag-Cu alloys and the results of quantitative XPS-analysis without reference samples show a good agreement with the chemical analysis of these samples (Fig. 2a). This is also to be seen when plotting the distribution of the XPS-concentrations versus $\Delta c = c$ (XPS) $- c$ (chem) as shown in Fig. 2b.

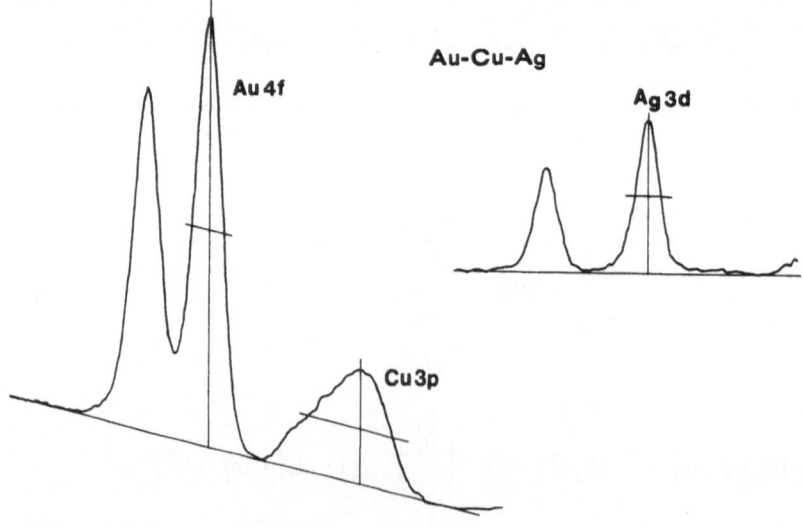

Fig. 1. Example for evaluation of XP-spectra (ternary Au-Cu-Ag alloy)

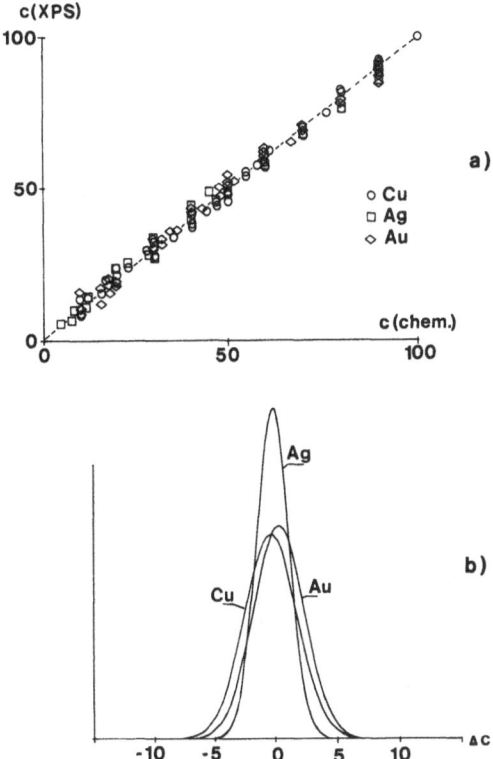

Fig. 2. (a) XPS-concentration versus chemical concentration for ternary and binary alloys of the system Au-Cu-Ag: (b) distribution of the XPS-concentrations in dependence of $\Delta c = c\,(\text{XPS}) - c\,(\text{chem})$ of the alloys of Fig. 2a

As an example for XAS-analysis the results of investigations performed on an Ag-Cu alloy are depicted (Fig. 3). The measured spectra are differentiated by computer. The Auger electron count rate is represented by the peak-to-peak height of the differentiated Auger line multiplied by the distance of the two extremes. As XP-lines are also measured the XAS results can be compared with XPS results.

Conclusions

From a great number of experiments – not only performed on alloys – we can say that the results provided by quantitative XPS-analysis without reference samples deviate less than ± 2 wt.% from the bulk concentration. As far

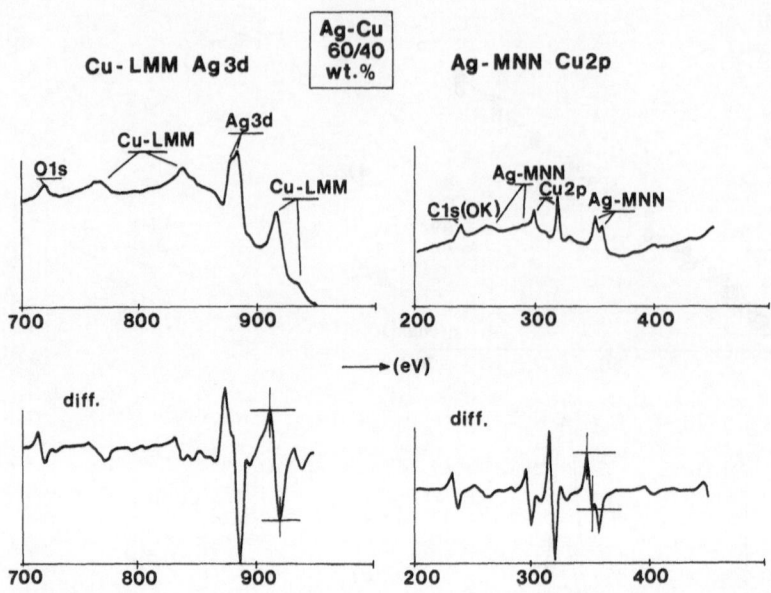

Fig. 3. Measured XA-spectra and differentiated spectra which depict the evaluation of XA-spectra (Ag-Cu alloy)

as XAS results are concerned the accuracy is not yet of this quality. Further consideration will be given to η_i^k and the mode of evaluation of the measured spectra in the near future.

Nomenclature

n_i^k	count rate of k-radiation of element i (fluorescence) or photoelectron count rate from subshell k of element i (XPS)
q	cross-section of detected X-ray or photoelectron beam
x_λ	number of X-ray quanta per second and cm^2 normal to the direction of the incident beam of monochromatic quanta of wavelength λ
κ_i^k	detector efficiency; for XPS mostly unity
Ω	solid angle of detection
$\tau_{\lambda i}$	photoabsorption coefficient of radiation λ in element i
S_i^k	absorption edge jump
ω_i^k	fluorescence yield
p_i^k	transition probability

$\mu_{\lambda c}$	mass absorption coefficient of compound c for radiation of wavelength λ
μ_{ic}	mass absorption coefficient of compound c for radiation of element i
α	angle between direction of incident X-radiation and the normal to the sample surface
ϵ	angle between take-off direction of (fluorescence radiation) photoelectrons and the normal to the sample surface
$[A+B \cdot E_i^k + C \cdot (E_i^k)^2]$	spectrometer function, describes the dependence of the spectrometer transmission on the energy of the photoelectrons (A, B and C are constants which depend upon the instrument)
E_i^k	photoelectron energy from subshell k of element i excited by a defined X-radiation (in keV)
c_i	concentration of element i in wt.%
σ_i^k	photoabsorption cross-section for subshell k of element i for a given X-radiation
σ_c^{1s}	photoabsorption cross-section for subshell $1s$ of element c and a given X-radiation
L	Loschmidt's number
A_i	Atomic weight of element i
β_i^k	asymmetry parameter
ϑ	angle between incident X-radiation and the direction of detected photoelectrons
ρ_c	density of compound c
Λ_{ic}^k	mean free path of k-photoelectrons from element i in compound c
$\Lambda_c(\text{ref})$	mean free path of photoelectrons in the compound c at an energy of 1 keV
Λ_{if}^k	mean free path of k-photoelectrons from element i in the overlayer f of thickness d
d	thickness of overlayer (oxide, contamination)
R	percentage of observed photoelectrons due to shading
E_{bi}^k	binding energy of k-photoelectrons from element i: $h\nu = E_i^k + E_{bi}^k$ (energy of X-ray quanta of monochromatic radiation = $h\nu$)
η_i^k	percentage of Auger yield

Acknowledgement

We are indebted to the "Fonds zur Förderung der wissenschaftlichen Forschung in Österreich" (Project Nr. 4272) for financial support.

Summary

Quantitative Surface Analysis Without Reference Samples

X-ray photoelectron spectroscopy and X-ray induced Auger electron spectroscopy can be applied in quantitative analysis without reference samples. The theory is presented in comparison to X-ray fluorescence analysis. The simplicity of evaluation of measured data is shown and practical examples are given.

References

1. J. Sherman, Spectrochim. Acta 7, 283 (1955);
   J. Sherman, Spectrochim. Acta 11, 466 (1959).
2. T. Shiraiwa and N. Fujino, Jpn. J. Appl. Phys. 5, 886 (1966);
   T. Shiraiwa and N. Fujino, Bull. Chem. Soc. Jpn. 40, 2289 (1967).
3. J.W. Criss and L.S. Birks, Analyt. Chemistry 40, 1077 (1968).
4. M. Vulli and K. Starke, J. Microsc. Spectrosc. Electron. 3, 45 (1978).
5. J.H. Scofield, J. Electron Spectrosc. Relat. Phenom. 8, 129 (1976).
6. R.F. Reilman, A. Msezane, and S.T. Manson, J. Electron Spectrosc. Relat. Phenom. 8, 389 (1976).
7. D.R. Penn, J. Electron Spectrosc. Relat. Phenom. 9, 29 (1976).
8. C.D. Wagner, L.E. Davis, and W.M. Riggs, Surf. Interface Anal. 2, 53 (1980);
   J. Szajman, J. Liesegang, J.G. Jenkin, and R.C.G. Leckey, J. Electron Spectrosc. Relat. Phenom. 23, 97 (1980); Yu Hsiang-Pu (Peking Centre of Physical and Chemical Analysis): private communication.
9. K. Siegbahn, C. Nordling, A. Fahlman, R. Nordberg, K. Hamrin, J. Hedman, G. Johansson, T. Bergman, S. Karlsson, T. T. Lindgren, and B. Lindgren: ESCA Atomic, Molecular and Solid State Structure Studied by Means of Electron Spectroscopy. Uppsala: Almquist & Wiksell. 1967.
10. M.F. Ebel, H. Ebel, and K. Hirokawa, Spectrochim. Acta 6, 461 (1982).
11. W. Bambynek, B. Craseman, R.W. Fink, H.U. Freund, H. Mark, C.D. Swift, R.E. Price, and P. Venugopala Rao, Rev. Modern Physics 4, 716 (1972).
12. Handbook of Auger Electron Spectroscopy, ed. Physical Electronics Industries Inc. 1976.

Correspondence and reprints: Univ.-Doz. Dr. Maria F. Ebel, Institut für Angewandte und Technische Physik, Technische Universität Wien, Karlsplatz 13, A-1040 Wien, Austria.

Mikrochimica Acta [Wien], Suppl. 10, 199–210 (1983)
© by Springer-Verlag 1983

Metaalinstituut TNO, Apeldoorn, The Netherlands; Mannesmann, Forschungsinstitut, Duisburg, Federal Republic of Germany, and The University of Western Ontario, London, Canada

Improved Methods of Quantitative Electron Probe Microanalysis of Carbon-$\phi(\rho z)$ Compared to Other Methods

By

A. P. von Rosenstiel, P. Schwaab, and J. D. Brown

With 3 Figures

(Received January 19, 1983)

The quantitative determination of carbon in metals and carbides by electron probe microanalysis is difficult because the low energy of the carbon $K\alpha$ line leads to low efficiency of X-ray generation (the X-ray yield is only about 1%) and detection (the long wavelength requires large d-spacing crystals and special thin window detectors). In addition to this, other more fundamental problems remain. Because absorption coefficients are large, greater accuracy is required of the models for the absorption correction because this correction even for electron energies as low as 4 keV is still quite large. Complicating matters is the fact that mass absorption coefficients which are a necessary parameter in the absorption correction are difficult to measure accurately and hence are not well known. A further factor which could lead to doubts about the success of any quantitative analysis scheme for carbon, is that the C $K\alpha$ line results from electron transitions involving valence or conduction electrons, and the number and energy of these is structure and bonding dependent. The consequences are that the C $K\alpha$ line changes shape and width in different compounds and hence the usual measurement of k-ratios from peak heights may have to be supplanted by peak integral measurements. By necessity this is a more time consuming process which the analyst would like to avoid.

Despite other complicating factors[1, 14] such as carbon-free sample preparation, elimination of contamination effects, possible line overlap and proper background determination, reliable measurements of carbon in steels have been performed in daily routine using comparable standards and appropriate calibration graphs[1-4].

A number of studies have already been carried out on low atomic number quantitative electron probe microanalysis. Duncumb and Melford[5] analysed carbides using an absorption correction based on Monte Carlo calculations. Kohlhaas and Scheiding[6,7] used a ZAF approach which included the simplified Philibert absorption correction[8] and published a table of recommended mass absorption coefficients which resulted in improved agreement between known and measured compositions for their analysis conditions. Weisweiler in a series of papers[9-12] examined the C $K\alpha$ line shape from different forms of carbon and some carbides and produced a series of correction factors to take into account line shape effects. He also reviewed mass absorption coefficients for the low atomic number characteristic lines[13] and made measurements of some mass absorption coefficients in an electron probe microanalyser. In a search for a good carbon standard, Weisweiler examined a number of different homogeneous carbon containing materials and recommended vitreous carbon[10]. Ruste[14] also examined carbide analysis, carefully evaluating background contributions from L lines of metals. He based his correction procedure on a ZAF method in which the full Philibert absorption correction equation is used, but with somewhat modified expressions for excitation potential, J, electron absorption, σ, the matrix parameter, h, and surface ionization $\phi(0)$. Once again, recommended mass absorption coefficient values were published which improved analytical results using his equations and measurement conditions. More recently, Love[15] used the same ZAF approach with the full Philibert absorption correction, but recommended different parameters in the correction equations.

Experimental Measurements

13 well characterized and commercially available carbides ranging from B_4C to WC were chosen for this study. These carbides in grain sizes of 10 to 50 μm were carefully analyzed by wet chemical analysis and X-ray diffraction, see Table 1.

These carbide powders were mixed in a ratio of 1:50 with aluminium Powder (60 to 200 μm grain size) and cold pressed at 8 tons/cm^2 into cylinders of 20 mm ϕ and 8 mm thickness in order to achieve a carbon free sample preparation. Density achieved was 96% of the compact metal.

These metallographic samples were prepared in the usual way, e.g.: polished with diamond paste, carefully cleaned and finally polished with Al_2O_3.

Analysis was performed on two microprobes with different take off angle (30° and 52.5°, viz, Siemens Elmisonde and ARL-EMX) to investigate the effect of the take-off angle on the correction procedure. Furthermore, to

Table 1. Chemical Composition of Carbide Samples

Sample	% C	% N	% B	% Si	% Ti	% V	% Cr	% Zr	% Nb	% Mo	% Ta	% W	% Al	% Mn	% Fe	% Cu	% Ni	% O
B_4C	19.00	0.49	78.50	0.00		0.00	0.01			0.01			0.00	0.00	0.13	0.00	0.10	0.13
α-SiC	29.90	0.091		69.60		0.02	0.00			0.02			0.08	0.04	0.40	0.00	0.22	0.04
TiC	20.50	<0.01		0.02	79.10	0.00	0.00			0.01			0.00	0.00	0.04	0.00	0.08	0.05
VC	18.50			0.03		80.46	0.09			0.02			0.00	0.00	0.04	<0.01	<0.01	0.30
$Cr_{23}C_6$	5.60	0.28		0.14		0.00	91.50			0.02			0.67	0.00	0.26	0.00	0.08	1.10
Cr_3C_2	13.20	0.057		0.04		0.00	86.30			0.02			0.00	0.00	0.00	0.00	0.10	0.14
ZrC	11.70	0.083		0.06		0.00	0.01	87.50		0.01			0.00	0.00	0.06	0.00	0.07	0.20
NbC	11.10	0.23		0.00		0.00	0.00		88.00	0.00			0.00	0.00	0.05	0.00	0.01	0.20
Mo_2C	5.86			0.61		0.00	0.05			93.00			0.02	<0.01	0.05	<0.01	0.04	0.27
TaC	6.40	0.055		0.00		0.00	0.01			0.00	91.60		0.00	0.00	0.03	0.00	0.01	0.06
WC	6.10	0.021		0.00		0.00	0.00			0.034		93.60	0.06	0.00	0.03	0.00	0.01	0.02
(W,Ti) C, 70/30 %	9.80	0.044		0.00	26.50	0.14	0.00			0.06		62.80	0.00	0.00	0.00	0.00	0.00	0.04
(W,Ti) C, 50/50 %	13.00	0.11		0.00	46.60	0.15	0.01			0.07		39.40	0.00	0.00	0.00	0.00	0.01	0.08

study the energy dependence of the matrix correction, EPMA measurements were performed at electron energies of 4, 6, 8, 10 and 12 keV and sample currents of 300 to 100 nA. The actual electron energies were directly determined on the specimen itself by measurement of the short wavelength cut off of the X-ray continuum with an accuracy of about 20 eV. Pure metal standards were used for the metallic components and a well defined stoichiometric Fe_3C standard for the measurements of carbon. A well controlled gas jet anti-contaminator (pure oxygen) was used throughout all measurements. Peak counting was applied for at least 10 individual areas, viz. particles on both specimen and standard. The counting time was adjusted for an accumulation of at least 50.000 counts for carbon and 25.000 counts for the metal per individual area, viz. particle. Peak area measurements were performed by graphical integration of the C $K\alpha$ peak of the carbide relative to Fe_3C. Background determination for the metal was made by conventional "off peak" measurements on both carbide and pure metal standard. Background determination for carbon was done using the unchanged "on peak" position on the respective pure element standards (including pure Fe for the C background on Fe_3C).

Special attention was given to the proper wavelength position of the respective carbides, optimized pulse height analyzer settings to avoid spectral overlap[14] and towards high stability and reproducibility of the instruments.

Calculation Methods

The measured k-ratios have been converted to composition using a number of different calculation schemes to compare their accuracy in quantitative analysis. The first approach is a ZAF calculation with the simplified absorption correction of Philibert[8] and atomic number correction of either Reed and Springer[16] or Philibert and Tixier[17]. These correction procedures failed drastically, primarily because the simplified Philibert method is really not applicable in such high absorption systems. No further reference will be made to this method.

The second approach is the modified ZAF approach proposed by Ruste[14] in which the full Philibert absorption is used. In it, he has substituted the $\phi(0)$ expression of Reuter[18] for $R(0)$ and modified values for h and σ. Philibert and Tixier's atomic number correction was also used. Heinrich's mass absorption coefficients[19] were used. The mass absorption coefficients for the carbon $K\alpha$ radiation are given in Table 2.

The third approach uses the $\phi(\rho z)$ equation of Parobek and Brown[20]. In this method, assuming that the distributions of X-ray production as a function of depth, ρz, in the sample ($\phi(\rho z)$ curves) are known as a function of matrix electron energy and absorption edge energy, then the k-ratio, k_A for the element A in an unknown specimen U relative to pure element standard S can be written

Table 2. Mass Absorption Coefficients for the C $K\alpha$ Line

Absorber	Ruste[14]	Weisweiler[11]
C	2373.	2280.
Si	36980.	30500.
Ti	8093.	6900.
V	9236.	7500.
Cr	10482.	7700.
Fe	13300.	10800.
Zr	31130.	45000.
Nb	24203.	15800.
Mo	15500.	12500.
Ta	20000.	8500.
W	21580.	10500.

$$k_A = \frac{W_A \; \phi_U(\rho z) \, e^{-\mu_A^U \rho z \, \csc.} \, d\rho z.}{\phi_S(\rho z) \, e^{-\mu_A^S \rho z \, \csc.} \, d\rho z.} \tag{1}$$

where W_A is the weight fraction of A in the unknown and μ is the mass absorption coefficient.

Note that $\phi(\rho z)$ will be different both in shape and area for the unknown and pure element standard and that the areas under the $\phi(\rho z)$ curves give the total intensities generated per unit concentration and hence the atomic number correction. Parobek and Brown approximated the $\phi(\rho z)$ curves by the equation

$$\phi(\rho z) = D.k.n.R^{n-1} \cdot e^{-(kR)^n} \tag{2}$$

where $R = \rho z - \rho z_0$, and the parameters D, k, n and ρz_0 depend on Z and A of the matrix and the electron and absorption edge energy. These 4 empirical parameters were obtained by fitting by a least squares procedure to a large number of experimentally measured $\phi(\rho z)$ curves.

The calculation of composition from k-values follows the usual iterative method of approximating the composition from the measured k-values, calculating k-values from that composition then comparing with the measured. New composition estimates are made and so on until measured and calculated k-values agree. The only significant difference in the $\phi(\rho z)$ approach is that numerical integration of Eq. (1) is used instead of the evaluation of ZAF factors.

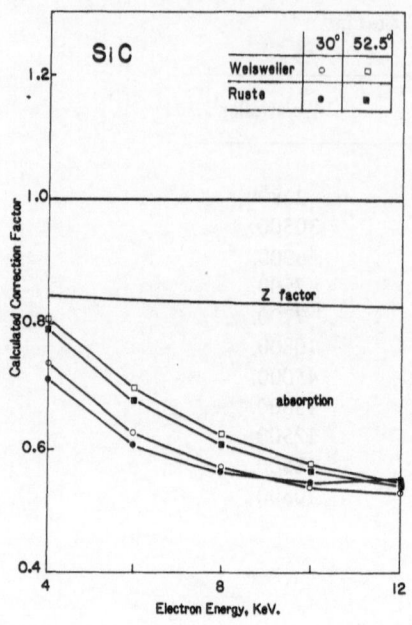

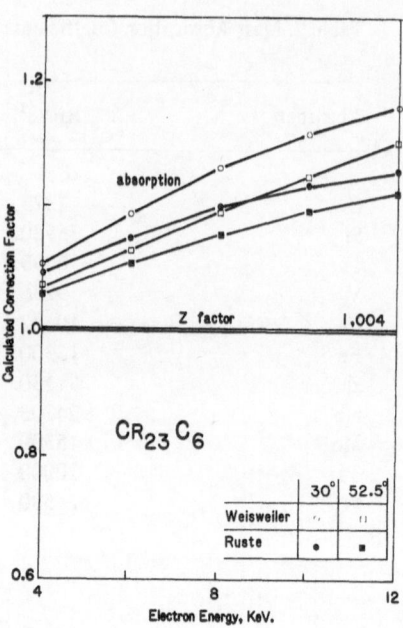

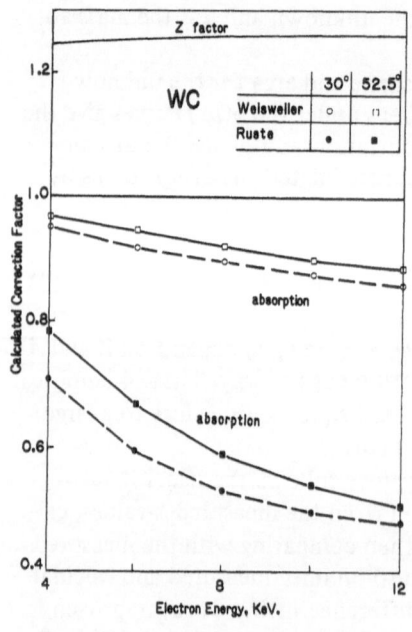

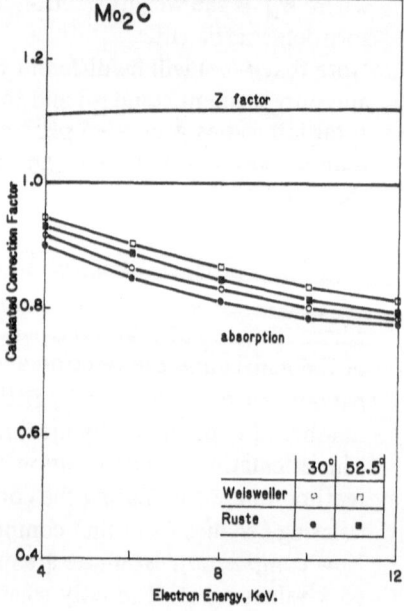

Fig. 1. Calculated correction factors

The fourth approach is another $\phi(\rho z)$ approach but using a Gaussian equation for $\phi(\rho z)$[21, 22], i.e.,

$$\phi(\rho z) = \gamma_0 \left(1 - \frac{\gamma_0 - \phi(0)}{\gamma_0} \, e^{-\beta \rho z}\right) e^{-(\alpha \rho z)^2}. \tag{3}$$

α, β, and γ_0 are again parameters which depend on A and Z of the matrix and electron and absorption edge energy. The advantage of the Gaussian expression is that the parameters can now be derived from electron scattering laws. Thus not only measured $\phi(\rho z)$ curves but also theoretical advances in describing electron scattering can be used to improve the parameters.

Discussion of the Results

Four systems; SiC, $Cr_{23}C_6$, Mo_2C, and WC are examined in detail as they are representative of most of the situations which arise with respect to the magnitude and sign of the corrections applied. First in Fig. 1 are shown the absorption and atomic number correction factors obtained from the Gaussian equation with two sets of mass absorption coefficients shown in Table 2. Starting from the known compositions, the correction factors are calculated relative to the Fe_3C standard. Although slight changes in detail might occur if other correction procedures were used in deriving these Figures, the broad conclusions would remain the same.

The first important general effect to notice is that the atomic number (Z) factor is essentially independent of electron energy. The variation is at most 2 or 3%. Of course there can be no difference with X-ray take-off angle since this factor depends on the primary beam parameters alone. On the other hand, the absorption (A) factor depends strongly on the electron energy and the X-ray take-off angle.

In SiC, the A and Z factor are both less than 1 and differ significantly from 1. Thus a very large correction is required in this system. In $Cr_{23}C_6$, the Z factor is almost exactly 1.0 (as compared to Fe_3C) so that any correction applied must be due to absorption. In Mo_2C, the two corrections are roughly equal and opposite so that the resulting overall correction factor is small, but increases slightly with electron energy. In WC, the Z correction is large, while the A correction depends strongly on the chosen set of mass absorption coefficients.

Fig. 2 shows calculated k-ratios from the Gaussian in comparison with measured k-ratios using peak intensities. In all systems, the measured variation with energy parallels the calculated variation, especially when Weisweiler's mass absorption coefficients are used. Note that the measurements for WC and Mo_2C at 4 keV are suspect. Two major problems remain. The first concerns the absolute values. Remembering that the Z factor is almost constant

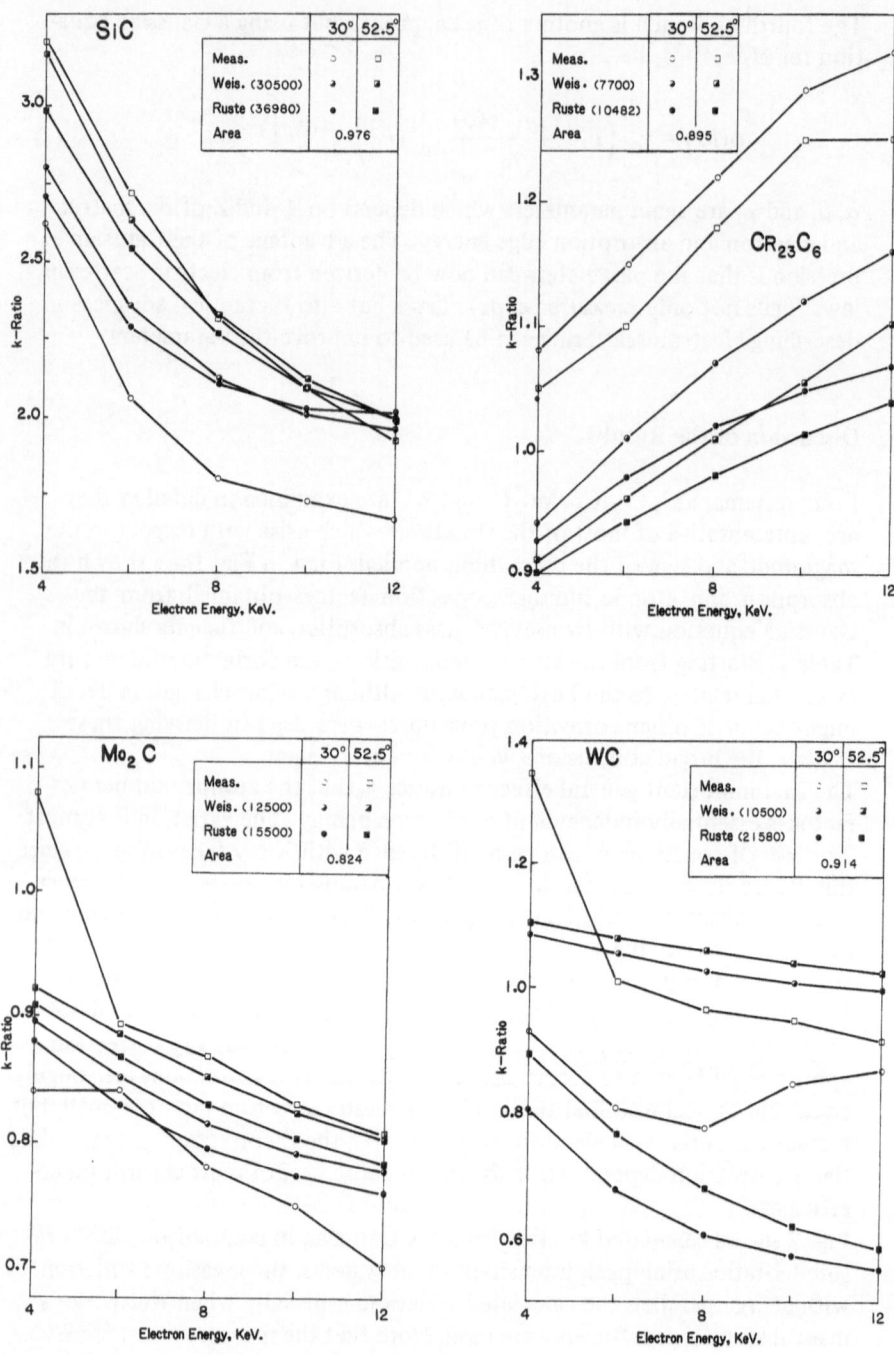

Fig. 2. Comparison of calculated and measured *k*-ratio

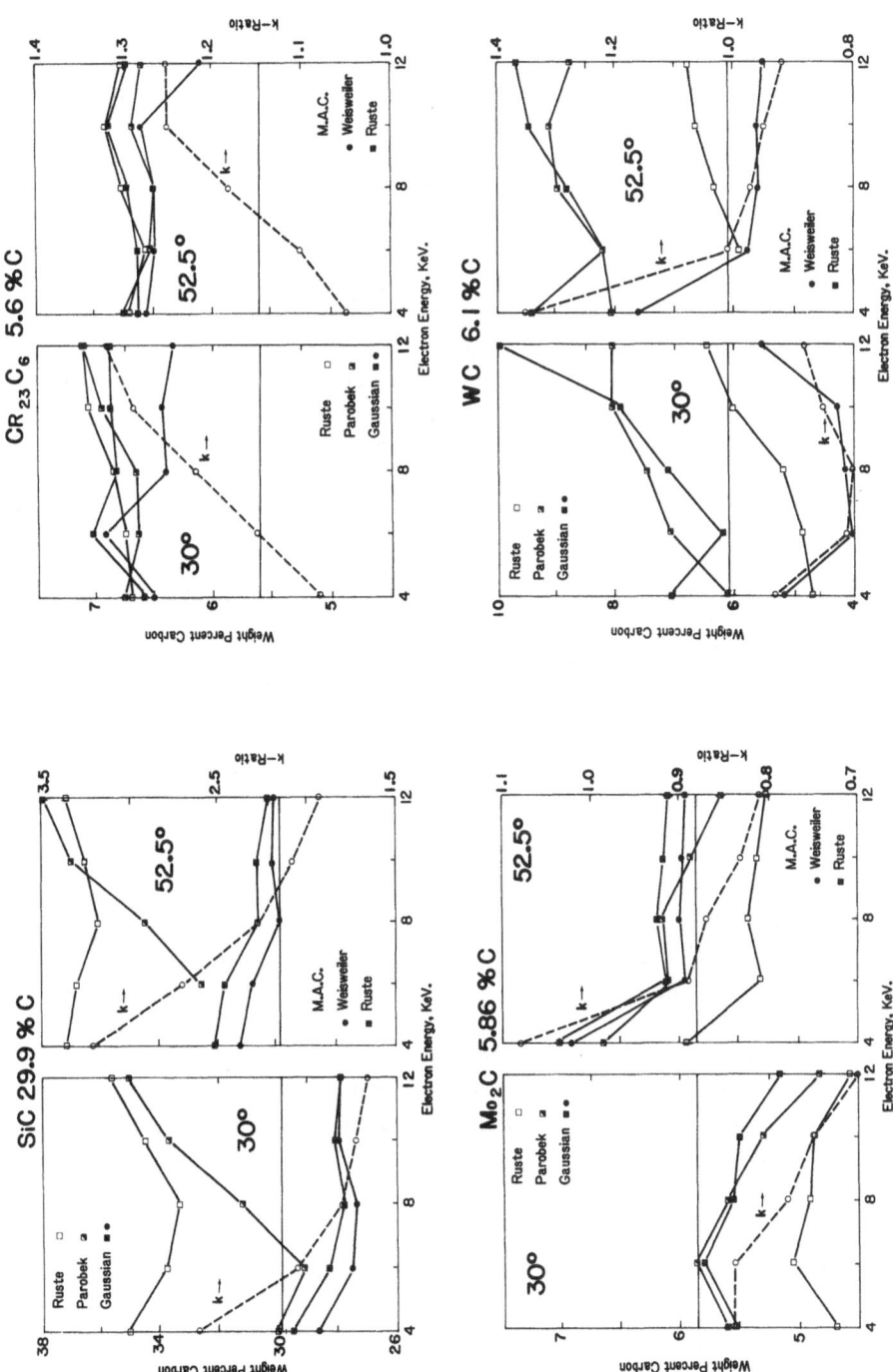

Fig. 3. Compositions obtained with different matrix corrections

with energy, this would seem to be at fault. However, $Cr_{23}C_6$ has the same average atomic number as Fe_3C, yet the calculated values lie too low. Thus the Z factor cannot account for this difference in this case. If k-values measured by integrated areas are used, then the peak k-values should be multiplied by 0.895 bringing values into much closer agreement. These area factors for each system are given in Fig. 2. While the $Cr_{23}C_6$ case offers almost unambiguous proof that integrated peak intensities must be used, agreement in the Mo_2C and WC systems would become poorer. Part of this may be a consequence of adjusting Gaussian parameters while using peak k-values. This is being investigated further.

The second problem is the differences between 30° and 52.5° results. Measured differences almost always exceed calculated differences. In the Gaussian equation the differences result mainly from the magnitude of constants in α and β. Optimization of these constants is also under investigation.

The practice of quantitative analysis does not proceed from composition to k-values but rather in the other direction. A comparison of the calculation procedures in quantitative analysis is shown in Fig. 3. Compositions calculated from k-values derived from peak intensity measurement using Ruste's, the Parobek and Brown and Gaussian $\phi(\rho z)$ methods. The measured k-values are also plotted to give some idea of the variation with energy which occurs.

For SiC, the Ruste procedure gives results which are 15 to 20% high, but have little variation with energy. Parobek's equation has an obvious energy dependence. The Gaussian however gives good results with either set of mass absorption coefficients. For $Cr_{23}C_6$, all methods give comparable results, with all calculated compositions too high. Integrated area measurements in this case would lead to excellent agreement for all methods. For Mo_2C, the Ruste results are a bit low but independent of energy. The $\phi(\rho z)$ methods give better agreement but tend to be slightly low at 30° and high at 52.5° take-off angle. The WC results are the most complex. First, the Ruste values are quite reasonable but still show a significant energy dependence. The $\phi(\rho z)$ methods both have a strong energy dependence when Ruste's mass absorption are used. However with Weisweiler's smaller value of μ for tungsten, the Gaussian results are independent of energy. The compositions are quite low at 30° however. Using area rather than peak measurements would not improve the results for WC and Mo_2C for any of the calculation methods.

Conclusions

The Gaussian equation for $\phi(\rho z)$ curves shows great promise in the quantitative analysis of carbides. The $Cr_{23}C_6$ data suggests that peak area measurements of intensity ratios will be necessary for quantitative analysis unless a factor relating peak to area ratios can be established and verified experimentally. Since in most of the carbide systems, the compositions calculated from k-ratios do not vary with electron energy, factors which lead to a different Z dependence of the atomic number correction will need to be carefully examined.

Acknowledgement

The authors wish to thank the European Community for Coal and Steel for the financial support of this investigation and for permission for publication of these preliminary results.
Thanks are also due to their colleagues at both TNO and Mannesmann for their help and experimental skill during these investigations.

Summary

Improved Methods of Quantitative Electron Probe Microanalysis of Carbon-ϕ (ρz) Compared to Other Methods

Measurements of C $K\alpha$ k-ratios relative to a Fe_3C standard have been made on a series of well characterized binary carbides ranging from boron carbide to tungsten carbide in two different electron microprobes having take-off angles of 30° and 52.5° respectively and at electron energies in the range from 4 to 12 keV. Where possible, k-ratios for the metallic element relative to the pure standard were measured in the same energy range. K-ratios were determined for peak and integrated line profile measurements. Distinct and systematic deviation were observed for the integrated line intensity as compared to the simple peak measurement. Conversion of k-ratios into weight % was obtained by two conventional ZAF programs (Siemens, MAGIC) a specially modified ZAF programme for light element analysis (Ruste) and two $\phi(\rho z)$ programs (Brown-Parobek and the Gaussian Brown-Packwood equation). Heinrich's, Ruste's and Weisweiler's mass absorption coefficients were used.
The ZAF programs often gave too high C values, especially at low kV and strong deviations with increasing acceleration voltages. Ruste's method also showed overcorrection in most cases but a smaller energy dependance. Similar results were obtained with the Brown-Parobek equation. Promising results were obtained with the Gaussian $\phi(\rho z)$ method. However, integrated peak measurements for the carbon $K\alpha$-line or appropriate correction factors may be necessary for general analysis of carbon.

References

1. S. Baumgartl, A. Büchner, K. Dreyer, P. Schwaab, H. Stender, and M. Vetters, Arch. Eisenhüttenwes. 50, 85 (1979).
2. J.S. Duerr, R.E. Ogilvie, Analyt. Chemistry 44, 2361 (1972).
3. E. Eichen, J. Tabock, and K.R. Kinsman, Metallography 5, 151 (1972).
4. G. Runnsjö, Skandinavian Metallurgy 9, 199 (1980).
5. P. Duncumb and D. Melford, in 4th Int. Conf. on X-Ray Optics and Microanalysis, R. Castaing, P. Deschamps, and J. Philibert (eds.). Paris: Herman. 1967. p. 240.
6. E. Kohlhaas and F. Scheiding in 5th Int. Conf. on X-Ray Optics and Microanalysis, G. Möllenstedt and K.H. Gaukler (eds.). Berlin-Heidelberg-New York: Springer-Verlag. 1969. p. 193.
7. E. Kohlhaas and F. Scheiding, Arch. Eisenhüttenwes. 41, 97 (1970).
8. J. Philibert, in 3rd Int. Conf. on X-Ray Optics and Microanalysis, H.H. Pattee, V.E. Cosslett, and A. Engström (eds.). New York: Wiley. 1963. p. 379.
9. W. Weisweiler, Mikrochim. Acta [Wien] 1970, 17.
10. W. Weisweiler, Mikrochim. Acta [Wien] 1975 I, 611.
11. W. Weisweiler, Mikrochim. Acta [Wien] 1975 II, 168.
12. W. Weisweiler, Mikrochim. Acta [Wien] 1972, 145.
13. W. Weisweiler, Mikrochim. Acta [Wien] 1970, 744.
14. J. Ruste, J. Microsc. Spectrosc. Electron. 4, 123 (1979).
15. G. Love and V.D. Scott, Scanning 4, 111 (1981).
16. G. Springer, Neues Jahrb. Mineral. 106, 241 (1967).
17. J. Philibert and R. Tixier, in Quantitative Electron Probe Microanalysis, NBS. Spec. Publ. 298, 1968. p. 13.
18. W. Reuter, in 6th Int. Conf. on X-Ray Optics and Microanalysis, G. Shinoda, K. Kohra, and T. Ichinokawa (eds.). Tokyo: U. Tokyo Press. 1972. p. 121.
19. K.F.J. Heinrich, in The Electron Microprobe. New York: Wiley. 1966.
20. L. Parobek and J.D. Brown, X-Ray Spectrom. 7, 26 (1978).
21. R.H. Packwood and J.D. Brown, X-Ray Spectrom. 10, 138 (1981).
22. J.D. Brown and R.H. Packwood, X-Ray Spectrom. 11, 187 (1982).

Correspondence and reprints: A. P. von Rosenstiel, Metaalinstituut TNO, Apeldoorn, The Netherlands.

Mikrochimica Acta [Wien], Suppl. 10, 211–216 (1983)

[1]Philips GmbH Forschungslaboratorium Hamburg, Federal Republic of Germany, and
[2]Metaalinstituut TNO, Apeldoorn, The Netherlands

Quantitative Electron Probe Microanalysis of Sputtered FeC Dry Lubrication Films

By

P. Willich[1], A. P. von Rosenstiel[2], and N. Drost[2]

With 1 Figure

(Received January 19, 1983)

FeC-layers were prepared on various substrate materials (Si, SiO_2, glass, steel) by radio frequency sputtering. Typical applications may be in the fields of dry lubrication and wear resistance coatings. Chemical analysis is essential for the discussion of sputtering parameters and friction coefficients. This requires also the determination of low concentrations of argon and oxygen, which are incorporated during the sputtering process.

Carbon concentrations of these films range from about 5 to 65 wt%. This enables some interesting studies of quantitative EPMA of carbon in Fe-C systems. Variations of $CK\alpha$ lineshape is discussed in connection with the application of different matrix correction procedures.

Experimental

Measurements were performed by a minicomputer controlled CAMEBAX electron microprobe (X-ray take-off angle 40°) equipped with three crystal spectrometers. The depth range of the emitted X-rays was reduced to the level of film thicknesses ranging from 0.5 to 1 µm by electron energies of 10 keV for $FeK\alpha$ and 5 keV for the $K\alpha$-lines of argon, oxygen and carbon. Counting time was adjusted to 30 sec for the peak position and to 10 sec for both background positions, respectively. For a beam current of 100 nA the analytical precision for a concentration of 1 wt% was about 5% for argon (PET crystal) and 7% for oxygen (ODPb crystal).

Iron was determined by a pure element standard. For the other elements the following compound standards were used: Fe_3C with controlled stoichiometric composition

(6.69 wt% C). Si-Ar thin film ($d \simeq 1.5$ μm, 10.4 wt% Ar): This composition was analyzed by EPMA using an extrapolation procedure for the determination of the 'pure argon' counting rate[1]. Fe-O thin film ($d \simeq 1$ μm, 4.8 wt% O): The composition was obtained by EPMA using a standard of coated $Y_3 Fe_5 O_{12}$ combined with an empirical correction for the influence of the carbon coating.

Results and Discussion

Broadening of the $CK\alpha$ Line

Within the resolution of the used ODPb crystal no variations in the position of the peak maxima of $CK\alpha$ measured on Fe_3C and the sputtered films could be found. Table 1 shows the K-ratios ($I_{specimen}/I_{standard}$) of all elements calculated from conventional peak height measurements. In addition K-ratios from carbon peak areas were obtained by graphical integration (planimeter). The variations of the $CK\alpha$ lineshape viz. FWHM with increasing concentration of carbon is given in Fig. 1, where delta $K = K$ (peak area)-K (peak height) is plotted versus K (peak height). Positive values of delta K indicate broadened lines of $CK\alpha$ on sputtered films as compared to the Fe_3C stand-

Table 1. K-Ratios of FeC-Layers Based on Measurements of Peak Heights and Peak Areas (for $CK\alpha$)

specimen	K-ratio = $I_{specimen}$ / $I_{standard}$				
	$FeK\alpha^1$	$CK\alpha^1$	$CK\alpha^2$	$ArK\alpha^1$	$OK\alpha^1$
Fe_3C	0.923	1.000	1.000		
1	0.886	0.924	0.680	0.167	0.139
2	0.852	1.223	1.090	0.012	0.140
3	0.837	1.351	1.260	0.016	0.194
4	0.801	1.583	1.570	0.016	0.509
5	0.783	1.803	1.870	0.034	0.378
6	0.668	2.638	2.990	0.019	0.409
7	0.601	3.021	3.510	0.071	0.806
8	0.580	3.233	3.790	0.072	0.828
9	0.557	3.625	4.320	0.326	0.034
10	0.438	5.145	6.370	0.008	0.720
11	0.319	6.439	8.110	0.329	0.036
12	0.255	6.965	8.820	0.562	0.006

[1] peak height   [2] peak area

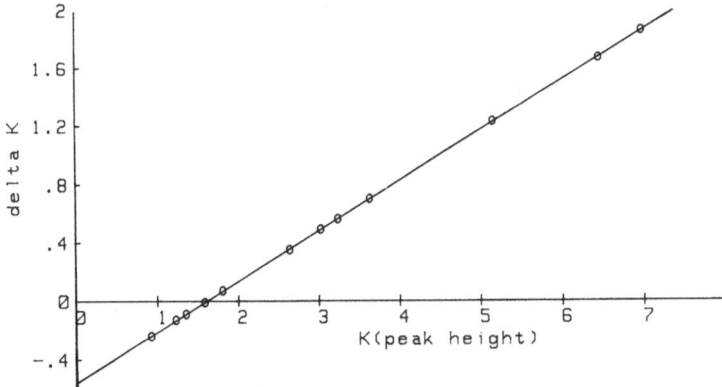

Fig. 1. Systematic variation of the line width of $CK\alpha$ from measurements of sputtered FeC-layers.

delta $K = K$ (peak area) $- K$ (peak height)

ard. This may be explained by morphological variations from an almost crystalline to a more amorphous state of the FeC-layers. The once established linear function of Fig. 1 enables an empirical correction of the K-ratios of arbitrary sputtered FeC materials measured by the convenient peak height method.

Comparison of Matrix Correction Procedures

Conventional ZAF data reduction was performed by an on-line program called 'COR2EX'[2], which may be regarded as a minicomputer version of the NBS COR2 program. Calculated mass absorption coefficients used for $FeK\alpha$ and $ArK\alpha$ fit the values of Heinrich[3]. For $CK\alpha$ the mass absorption coefficients proposed by Weisweiler[4] (for C and Fe) and Henke[5] (for Ar and O) were used. For $OK\alpha$ the data given by Love[6] (for Fe and O) and Henke (for Ar and C) were chosen. Results given in Table 2 indicate systematic errors up to 20% for the analysis of carbon without correction for the increase of $CK\alpha$ line width. All-element totals of analyses based on corrected K-ratios range from 96 to 102 wt%. For an electron energy of 5 keV an accuracy of about 5% may be estimated for the determination of carbon in this particular case, although ZAF correction was never designed with light element analysis in mind.

The matrix correction proposed by Ruste[7] is regarded to be more suitable to the analysis of light elements[8]. When applied to FeC films, values for carbon (Table 3) and also for the other elements are almost identical to those obtained by the COR2EX program. However, due to the low electron

Table 2. Systematic Analytical Errors Caused by the Variation of the $CK\alpha$ Line Width. Compositions Calculated by Conventional ZAF Matrix Correction

specimen	wt% (COR2EX)						
	Fe	Ar	O	C^1	tot	C^2	tot
Fe_3C	94.1			6.7	100.8	6.7	100.8
1	90.8	1.5	0.7	6.3	99.3	4.6	97.5
2	88.3	0.1	0.7	8.2	97.3	7.3	96.5
3	87.1	0.1	1.0	9.1	97.3	8.5	96.7
4	84.2	0.2	2.6	10.8	97.8	10.7	97.8
5	82.8	0.3	2.0	12.3	97.4	12.8	97.8
6	72.8	0.2	2.3	18.2	93.5	20.7	96.0
7	66.5	0.7	4.6	21.2	93.0	24.6	96.6
8	64.6	0.7	4.8	22.7	92.8	26.7	96.9
9	62.1	3.1	0.4	26.4	92.0	31.4	96.8
10	50.6	0.1	4.7	36.1	91.5	44.3	100.0
11	37.8	3.2	0.3	47.9	89.2	58.6	100.0
12	30.7	5.5	<0.1	54.0	90.2	65.4	101.8

[1] peak height   [2] peak area

Table 3. Concentrations of Carbon from ZAF Calculations Compared to the Results Obtained by the Procedures of Ruste and Packwood/Brown

specimen	COR2EX		Ruste		Packwood/Brown	
	wt% C	tot	wt% C	tot	wt% C	tot
1	4.6	97.5	4.6	97.3	4.6	98.1
2	7.3	96.5	7.3	96.1	7.3	97.4
3	8.5	96.7	8.5	96.5	8.4	97.6
4	10.7	97.8	10.7	97.7	10.5	98.4
5	12.8	97.8	12.8	97.9	12.5	98.5
6	20.7	96.0	20.6	96.0	19.9	96.4
7	24.6	96.6	24.4	97.0	23.4	96.6
8	26.7	96.9	26.5	97.3	25.3	96.6
9	31.4	96.8	31.2	97.2	29.7	95.8
10	44.3	100.0	43.8	100.7	41.2	98.0
11	58.6	100.0	57.9	100.4	54.2	96.1
12	65.4	101.8	64.7	102.1	60.5	97.7

energy of 5 keV absorption corrections seem to be reduced to a level where differences in the correction formula as well as uncertainties of the applied mass absorption coefficients become less critical (absorption correction factors for $CK\alpha$ range from 1 to 1.2).

Data reduction by the method of Packwood and Brown is based upon a Gaussian expression to describe $\phi(\rho z)$ curves[9]. Optimized values of the Gaussian parameters have been recently published[10]. Concentrations of iron, argon and oxygen correspond to those obtained by COR2EX and the model of Ruste. Table 3 shows an almost random and narrow distribution of 95.8 to 98.4 wt% for the all-element totals calculated by the method of Packwood and Brown. This indicates a somewhat improved accuracy of the carbon concentrations obtained by this new model of microprobe data reduction.

Summary

Quantitative Electron Probe Microanalysis of Sputtered FeC Dry Lubrication Films

Microprobe analysis was applied to deposited FeC-layers with thicknesses of 0.5 to 1 μm. This included the determination of trapped argon and oxygen with concentrations ranging from 0.1 to 5 wt%. Measurements were performed by electron energies of 5 keV (Ar, O, C) and 10 keV (Fe). The FWHM of the $CK\alpha$ line showed a systematic increase with increasing carbon concentration. This variation may be represented by a linear function allowing straightforward correction of the measured K-ratios. An accuracy of about 5% for carbon concentrations ranging from 4 to 65 wt% was obtained by conventional ZAF and the method of Ruste. Randomly distributed all-element totals of 95.8 to 98.5 wt%, calculated by the model of Packwood and Brown, indicate a somewhat improved accuracy of the carbon concentrations obtained by this method.

References

1. P. Willich and D. Obertop, X-Ray Spectrom. **11**, 32 (1982).
2. J. Henoc, C. Conty, and M. Tong, Proceedings of the 14th MAS Conference. San Antonio, 1979, p. 281.
3. K.F.J. Heinrich, The Electron Microprobe. New York: Wiley. 1966.
4. W. Weisweiler, Mikrochim. Acta [Wien] 1970. 744.
5. B.L. Henke and E.S. Ebisu, Advances in X-Ray Analysis **17**, 150 (1974).
6. G. Love, M.G.C. Cox, and V.D. Scott, J. Phys. D: Appl. Phys. **7**, 2131 (1974).
7. J. Ruste, J. Microsc. Spectrosc. Electron. **4**, 123 (1979).

8. G. Love and V.D. Scott, Scanning 4, 111 (1981).
9. R.H. Packwood and J.D. Brown, X-Ray Spectrom. 10, 138 (1981).
10. J.D. Brown and R.H. Packwood, X-Ray Spectrom. 11, 187 (1982).

Correspondence and reprints: Dr. P. Willich, Philips GmbH Forschungslaboratorium Hamburg, Vogt-Kölln-Strasse 30, D-2000 Hamburg 54, Federal Republic of Germany.

Mikrochimica Acta [Wien], Suppl. 10, 217–229 (1983)
© by Springer-Verlag 1983

Kernforschungszentrum Karlsruhe, Institut für Material- und Festkörperforschung,
Karlsruhe, Federal Republic of Germany

Experimental and ZAF Correction Aspects of Carbon Analysis in Steels: Application to the Carburization of Irradiated Uranium-Plutonium Carbide Fuel Pin Claddings

By

H. Kleykamp

With 11 Figures

(Received January 19, 1983)

The quantitative carbon analysis in steels by electron probe microanalysis (EPMA) has been on debate for more than one decade when the ZAF correction procedure was extended to the light element analysis with material data becoming available at that time. The conditions for a reliable analysis are a hydrocarbon low vacuum in the electron optical column of the instrument, a qualified sample preparation, selected standards similar in composition to the target and a well-developed ZAF correction computer programme.

Pumping Systems and Anticontamination Devices

A liquid nitrogen cooling finger next to the sample is a successful remedy to suppress carbon contamination on the sample surface. If this device cannot be mounted due to spatial reasons or to γ leakage in shielded instruments for the analysis of radioactive specimens, turbomolecular pumps are better suited than oil diffusion pumps. In Fig. 1, the carbon contamination count rate vs. counting time is given for the point analysis on a steel surface using different types of pumps in the same instrument (Jeol JRXA 50). The incubation period is followed by a linear count rate increase $tg\,\alpha$ in all cases. The tested turbomolecular pump was superior to the oil diffusion pump; this behaviour is quantified in Fig. 1 by the slope $tg\,\alpha = 0.07$ counts/s² for the

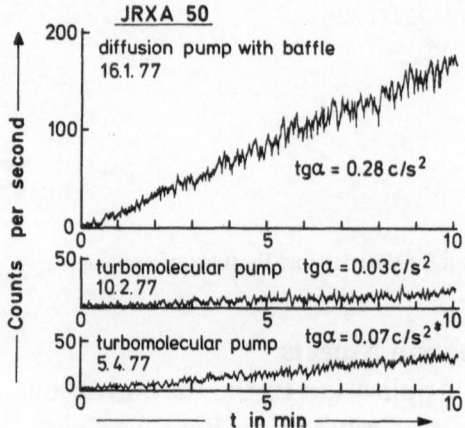

Fig. 1. Carbon contamination count rate on a steel surface using different pumping systems. $E_0 = 15$ kV, $I_{beam} = 100$ nA, level 0.75 V, window 2.5 V. * = steady-state operation

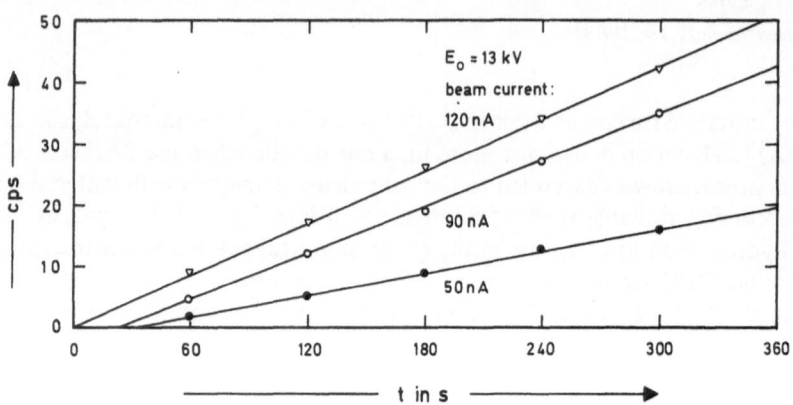

Fig. 2. Carbon contamination count rate on a metal surface demonstrating the incubation period at low beam currents

turbomolecular pump during steady-state operation compared to the baffle cooled oil diffusion pump by $tg\ \alpha = 0.28$ counts/s^2. If a turbomolecular pump is not installed in an instrument, quantitative carbon analyses in steels can be made with oil diffusion pumps using low beam currents. The carbon $K\alpha$ count rate as a function of the counting time is plotted in Fig. 2 for different beam currents after 60 second intervals. It can be deduced by back-extrapolation of the curves to zero that the time up to the carbon contamination onset was e.g., about 30 s for 90 nA beam current for the instrument used in this experiment. This procedure gives evidence on the best test conditions of carbon analysis in electron probe microanalyzers without any adulteration by contamination of cracked hydrocarbons.

Sample Preparation

The carburization of the stainless steel clads of uranium-plutonium carbide fuel pins during reactor operation is of great interest for the advanced fuel development. Numerous analyses by EPMA on irradiated pins revealed a carburisation on the inner steel surface in the tenth percent range[1, 2, 3] whereas results of other investigators on similar samples and irradiation conditions are a factor of ten higher[4, 5, 6]. The reasons for this discrepancy may be due to different embedding techniques and ZAF correction methods. Preliminary tests by EPMA have shown that the carbon point analysis is only successful by embedding the samples in a synthetic resin free matrix e.g., Wood's metal, and by polishing with diamond free paste e.g., alumina. The carbon concentration profile in the steel cladding of an irradiated carbide fuel pin is shown in Fig. 3. The clad was embedded into two different media and was polished with two different abrasive materials. The chlorine profile simultaneously recorded indicates a surface contamination of the steel by the synthetic resin "Araldite" of which epichlorohydrine H_2C——$CH \cdot CH_2Cl$ ‾O‾

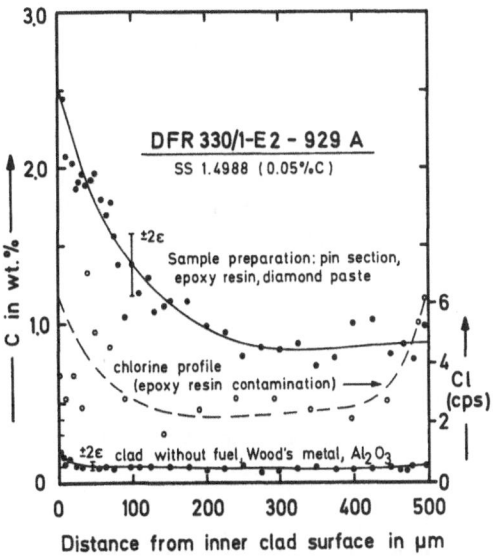

Fig. 3. Carbon concentration profiles in the steel clad of a sodium bonded uranium-plutonium carbide fuel pin section due to different sample preparation methods

is an ingredient. As a consequence, a reliable carbon analysis by EPMA is guaranteed only by embedding of the samples in carbon free metallic materials.

X-Ray Spectrum and Detection Limit of Carbon in Steels

The carbon $K\alpha$ line is superimposed by lines of the steel components. The composition of the favoured austenitic steels as cladding materials in fast breeder reactor fuel pins are given in Table 1. The quantitative carbon analysis is aggravated by the location of higher order L X-ray lines of iron, chromium, nickel and manganese (see Fig. 4) which can be suppressed only in part by discrimination. In addition, the carbon $K\alpha$ line of stainless steel is displaced to longer wavelength $\lambda = 4.46_4$ nm compared to diamond ($\lambda = 4.45$ nm[7]) or natural graphite ($\lambda = 4.45$ nm[8]). The carbon peak of the stainless steel 1.4970 has been found to be symmetrical, whereas that of the stainless steel 1.4988 has in addition a shoulder on the short wavelength side. With respect to the interference with higher order lines of the steel components, it has been found that the background can be measured in the best way at 3.8 nm and 5.5 nm.

Carbon point analyses have been made in different austenitic steels with known carbon contents which have been analyzed chemically by coulometric titration. A linear relationship has been found between the count rate and the concentration of carbon in five different steel samples (Fig. 5). The background corrected counts at zero carbon concentration is supposed to be mainly due to the second order Mn $L1$ line which cannot be resolved from the C $K\alpha$ line in spite of discrimination. Using the well known formula for the detection limit, $2\sqrt{2B} \cdot c_0/(P-B)$, where P are the peak counts, B are the background counts and c_0 is the concentration of the respective element in the sample, a detection limit of 0.007% C has been calculated under the experimental conditions given within Fig. 5 (Cameca MS 46). The detection limit of carbon in irradiated steels is up to a factor of two higher due to the increased background radiation.

Table 1. Composition of Austenitic Steels (Standard Values)

W.-Nr.	C	N	Fe	Cr	Ni	Si	Mn	Mo	V	Nb	Ti
1.4970	0.09	0.01	66.1	15.0	15.3	0.4	1.5	1.2	–	–	0.47
1.4988	0.05	0.08	65.3	16.3	13.6	0.6	1.3	1.3	0.76	0.85	–
AISI 316*	0.05	–	64	18	13	0.5	1.8	2.5	–	–	–

* equivalent to 1.4401

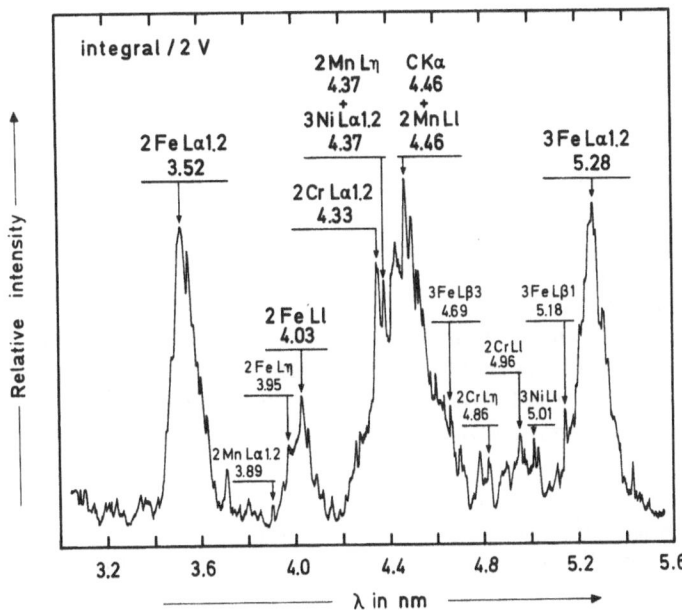

Fig. 4. X-ray spectrum of stainless steel 1.4988 between 3.1 and 5.5 nm

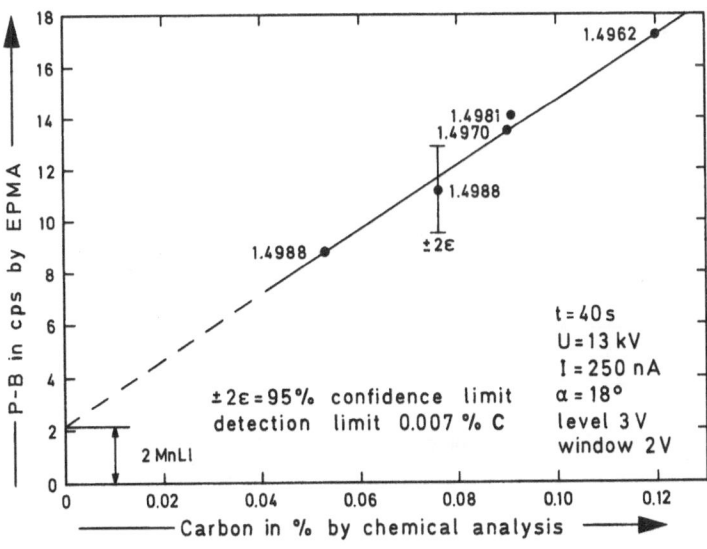

Fig. 5. Carbon count rate measurements by electron probe microanalysis in austenitic steel standards with known composition

ZAF Correction for Carbon in Steels

The general problems of the carbon analysis in steels by EPMA have been recently reviewed by Baumgartl et al.[9]. It was pointed out that the composition of the standards should be similar to that of the sample for quantitative analysis. Nevertheless, there is a number of papers in which diamond or graphite standards are used for the ZAF correction of carbon in steels[4, 5, 6]. This procedure results in misleading interpretations of the overcorrected carbon concentrations.

The atomic number correction is the most uncertain one with respect to the physical models and the material data used though the reliability of the absorption correction by extension of the Philibert-Heinrich formalism to light element analysis and the right choice of mass attenuation coefficients μ for C $K\alpha$ radiation is also permanently under discussion. The present state of quantitative EPMA is discussed in periodical intervals[10, 11, 12, 13]. The uncertainty in the atomic number correction made by use of Bethe's law and by alternative choices of the ionization potential J (Fig. 6) in the stopping power term prohibits the use of pure element standards for quantitative anal-

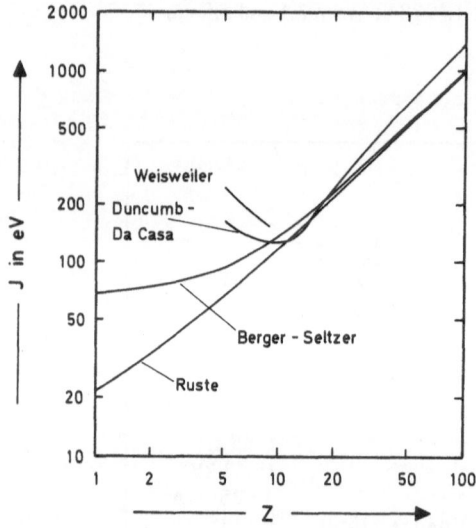

Fig. 6. Ionization potential as a function of the atomic number

ysis of light elements. Furthermore, the different modes in summing up the concentration averaged stopping power $\bar{S}$ and backscatter coefficient $\bar{R}$ have an influence on the ZAF corrected concentration. Recently, results on the quality of different ZAF correction programmes have been published[14, 15] to find out the influence of the individual parameters in the physical models and in the material data sets on the reliability of different correction procedures. However, these calculations are unsuccessful attempts as long as the summations of $\bar{R}$, $\bar{S}$, $\bar{h}$, $\bar{Z}$, etc., in the different models are unknown for the user because these quantities influence the intended statement on the most favourable material data e.g., the ionization potential.

We have developed an own ZAF correction programme which is written in APL, an interactive, time-sharing programme operated by the central KfK computer IBM 370/3033. This programme is versatile, as to the selection of the Thomas[16] or Philibert-Tixier mode[17] in the atomic number correction, of the ionization potential data sets of Berger-Seltzer[18], Duncumb-da Casa[19], Ruste[20], Weisweiler[21] or others, and of the fluorescence yield sets of Colby[22] or Bambynek et al.[23], etc. At the present stage of development of the programme, different alternatives in the summation of concentration averaged quantities e.g., $\bar{S}$ and $\bar{R}$ have been tested.

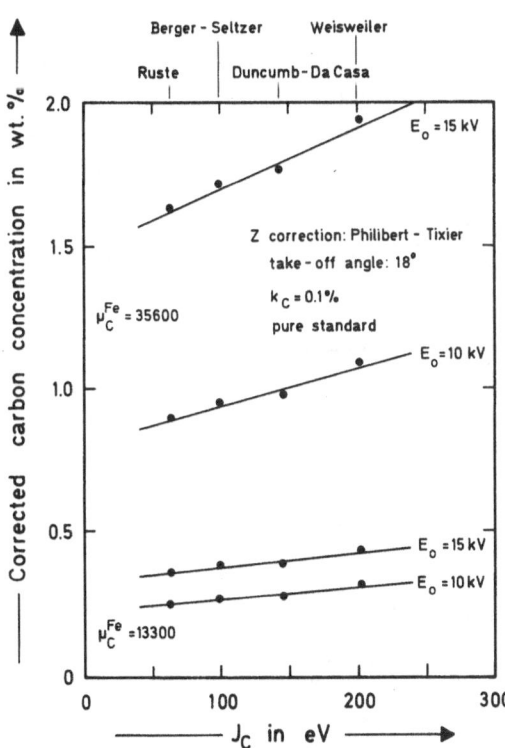

Fig. 7. Influence of the carbon ionization potential J_C and carbon mass attenuation coefficient in iron μ_C^{Fe} on the ZAF correction of carbon in a Fe-C alloy with $k_C = 0.001$

This programme has been used for elucidation of the influence of the above mentioned parameters on the ZAF correction of a hypothetical Fe-C alloy with a given carbon intensity ratio 0.001 referred to a pure element standard. First of all, the parameter study has been made to trace the influence of different carbon ionization potentials (Fig. 6) and mass attenuation coefficients of C $K\alpha$ radiation in iron on the ZAF corrected carbon concentration. The results are given in Fig. 7. There is a difference in the corrected carbon concentration up to 20% relative by replacing the ionization potentials of Ruste[20] preferred in our ZAF programme by those of Weisweiler[21]. We have changed for a time the mass attenuation coefficient of carbon in iron from $\mu_C^{Fe} = 13300$ cm²/g[24] which we use normally in our ZAF programme, to 35600 cm²/g. We have found that there are different sets of mass attenuation coefficients for light element $k\alpha$ radiation in private and commercial ZAF correction programmes which range to e.g., 35600 cm²/g for carbon $K\alpha$ radiation in iron. This change gives a sense of the influence of the obviously too high value on the corrected carbon concentration and can possibly explain the obtained extremely high values up to 4% carbon in carburized steels[4].

Another parameter calculation has been made by varying the mathematical treatment in the stopping power term S and the summation of the concentration averaged backscatter coefficient $\bar{R}$ in the atomic number correction. The results are given in Fig. 8. It has been found that the S term in the simplified atomic number correction of Thomas[16] gives nearly the same results as the complete integration of the ionization cross-section in the Philibert-Tixier model[17]. The difference amounts to only 3% relative for carbon in the considered Fe-C alloy though the carbon overvoltage ratio was extremely

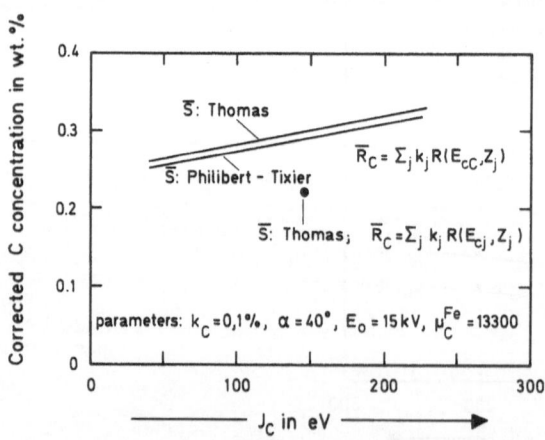

Fig. 8. Influence of the stopping power term S, the backscatter coefficient R and the carbon ionization potential J_C on the ZAF correction of carbon in a Fe-C alloy with $k_C = 0.001$

high, and is negligible in metallic systems with moderate overvoltage ratios e.g., in Al-Fe or Al-Pd that have been tested. However, different summations in the concentration averaged backscatter coefficient $\bar{R}$ result in different ZAF corrected concentration values. Philibert[25] and recently Brümmer[26] pointed out that it is not permitted to sum up over the overvoltage ratios E_{cj}/E_{0j} of all elements $j = 1, ..., i$ in the Z correction procedure for the element i, in our example: $\bar{R}_C = k_C \cdot R\,(E_{cC}/E_{0C},Z_C) + k_{Fe} \cdot R\,(E_{cFe}/E_{0\,Fe},Z_{Fe})$; instead, the summation must be carried out by: $\bar{R}_C = k_C \cdot R\,(E_{cC}/E_{0C},Z_C) + k_{Fe} \cdot R\,(E_{cC}/E_{0C},Z_{Fe})$. The application of the inexact $\bar{R}$ results in a ZAF corrected carbon concentration which is about 27% lower than the exact $\bar{R}$. Unfortunately, there are some correction programmes in use in which the R values of the pure elements weighed by the intensity ratios are taken for calculation of $\bar{R}$.

As a consequence from this parameter study in the ZAF correction of carbon in a hypothetic Fe-C alloy, it seems not expedient to use pure element standards for light element analyses, but mixed standards which are similar in composition to the materials to be analyzed should be preferred.

Carburization of the Stainless Steel Claddings by Uranium-Plutonium Carbide Fuels During Irradiation

In mixed carbide fuels containing impurity oxygen, transport of carbon takes place via the medium carbon monoxide down the temperature gradient which causes a carburization of the stainless steel cladding and an influence of its mechanical properties. The amount is dependent on the clad temperature, the fuel temperature gradient, the carbon-to-metal ratio, the relative density and the oxygen content of the fuel[2]. An example for the inner clad carburization of the stainless steel 1.4988 within the irradiation experiment FR 2 Vg.6D is given in Fig. 9. The carbon intensity ratios have been obtained by point analysis using a 1.4988 steel standard with similar composition. The succeeding ZAF correction has been made for all components in the cladding with changes in the ZAF corrected carbon concentration only up to 10% relative to the measured intensity ratio. There is a steep decrease from 0.20% C at the inner cladding surface and at the highest temperatures to nearly the initial carbon content of 0.05% C in the centre of the cladding. The carburization behaviour of different steels has been analyzed in the irradiation experiment DFR 330/1 (Fig. 10). The carburization of the steels beyond the solubility limit of carbon in the austenitic matrix, 0.02% at 600 °C[27], takes place mainly by formation of $(Cr,Fe)_{23}C_6$ precipitates in the grain boundaries which are small compared to the excited X-ray volume and are invisible in the unetched microstructures. The sensibility of carbide formation in the stainless steel 1.4970 is higher than that in the stainless steel

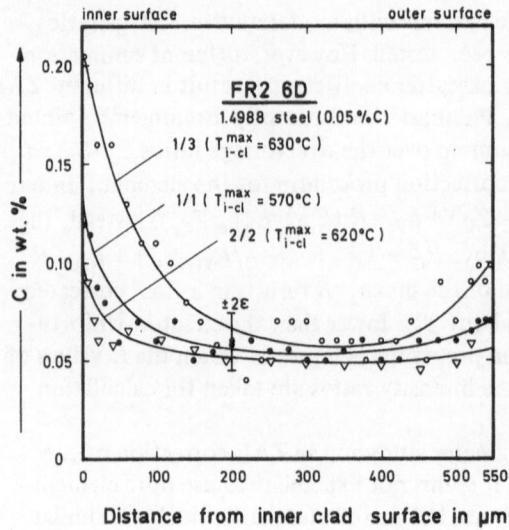

inner surface   outer surface

Fig. 9. Carburization of a 1.4988 stainless steel cladded helium bonded uranium-plutonium carbide fuel pin after 8% burnup at different inner cladding temperatures $T_{i\text{-}cl}$

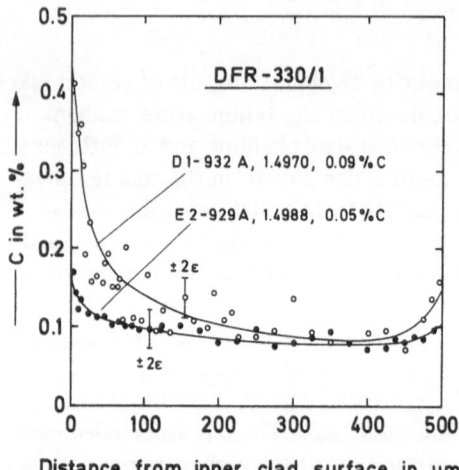

Distance from inner clad surface in µm

Fig. 10. Carburization of 1.4970 and 1.4988 stainless steel cladded sodium bonded uranium-plutonium carbide fuel pins in the maximum heat rating regions after 5% burnup (for inner cladding temperatures see Fig. 11)

1.4988. This behaviour is illustrated in Fig. 11 along the inner clad surface of mixed carbide pins yielding a maximum carburization of 0.3% C for 1.4988 and 0.45% C for 1.4970 at the highest inner clad temperatures of 630 °C.

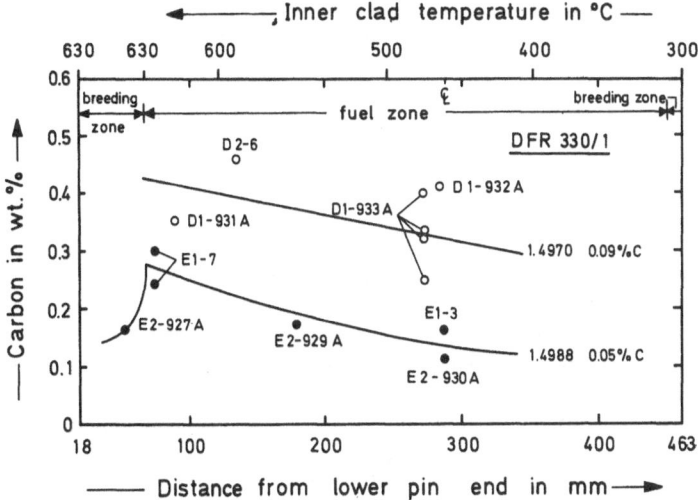

Fig. 11. Inner clad carburization of stainless steels 1.4970 and 1.4988 of sodium bonded uranium-plutonium carbide fuel pins as a function of the inner clad temperature

Acknowledgement

The author is indebted to Dr. I. Müller-Lyda who has written the computer programme in APL for the ZAF correction procedure in quantitative electron-probe microanalysis.

Summary

Experimental and ZAF Correction Aspects of Carbon Analysis in Steels: Application to the Carburization of Irradiated Uranium-Plutonium Carbide Fuel Pin Claddings

The influence of different pumping systems (oil diffusion and turbomolecular) on the carbon contamination rate of the target during light element electron probe microanalysis is explained. It is shown that organic resins used for embedding can result in incorrect carbon count rates of the target. A ZAF correction parameter study has been made on a hypothetical Fe-C alloy with a carbon intensity ratio 0.001 referred to a pure element standard by varying the atomic number correction mode and the concentration averaged summation of the backscatter coefficient as well as the ionization potential and the mass attenuation coefficient of carbon. It is concluded that a reliable carbon analysis can be attained only by use of standards similar to the target. The carburization of irradiated uranium-plutonium carbide fuel pins has been measured as a function of the cladding temperature by this comparative ZAF correction method. Different carburization behaviour has been found for the austenitic steels 1.4970 and 1.4988.

References

1. H. Kleykamp, Reaktortagung Düsseldorf 1976, proc. symp., p. 542.
2. H. Kleykamp, in Advanced LMFBR Fuels, Tucson/Ariz. 1977, proc. symp., p. 166.
3. H. Kleykamp, Reaktortagung Hannover 1978, proc. symp., p. 660.
4. M. Champigny, D. Gauvain, L. Meny, and J. Ruste, J. Microsc. Spectrosc. Electron. **4**, 137 (1979).
5. R. Cytermann, M. Perrot, and J. Rouault, in Int. Conf. on Post-irradiation Examination, Lake District/UK 1980, proc. symp., p. 161.
6. R. Cytermann, M. Perrot, and N. Vignesoult, J. Microsc. Spectrosc. Electron. **6**, 611 (1981).
7. J. Ruste, M. Bouchacourt, and F. Thevenot, J. Less-Common Met. **59**, 131 (1978).
8. W. Weisweiler, Mikrochim. Acta [Wien], Suppl. IV, **1970**, 17.
9. S. Baumgartl, A.R. Büchner, K. Dreyer, P. Schwaab, H. Stender, and H. Vetters, Arch. Eisenhüttenwes. **50**, 85 (1979).
10. K.F.J. Heinrich, Mikrochim. Acta [Wien], Suppl. IV, **1970**, 253.
11. K.F.J. Heinrich, H. Yakowitz, Mikrochim. Acta [Wien] **1970**, 123.
12. S.J.B. Reed, in 8th Int. Conf. on X-Ray Optics and Microanal., Boston 1977, proc. symp., p. 25A.
13. G. Love and V.D. Scott, in 9th Int. Conf. on X-Ray Optics and Microanal., Den Haag 1980, proc. symp., vol. 3, p. 146.
14. A. Boekestein, A.M. Stadhouders, A.L.H. Stols, G.M. Roomans, and E. Peters, in 10th Int. Conf. on Electron Microsc., Hamburg 1982, proc. symp., vol. 1, p. 667.
15. H. Hantsche and G. Kempf, Beitr. elektronenmikrosk. Direktabb. Oberfl. **15**, 99 (1982).
16. P.M. Thomas, Report AERE-R 4593 (1964).
17. J. Philibert and R. Tixier, J. Phys. D **1**, 685 (1968).
18. M.J. Berger and S. M. Seltzer, in Publ. 1133, Ntl. Acad. of Sciences, Washington/D.C. 1964, p. 205.
19. P. Duncumb, P.K. Shields-Mason, and C. da Casa, in 5th Int. Congr. on X-Ray Optics and Microanal., Tübingen 1968, proc. symp., p. 146.
20. J. Ruste, in 8th Int. Conf. on X-Ray Optics and Microanal., Boston 1977, proc. symp., p. 35A.
21. W. Weisweiler, Arch. Eisenhüttenwes. **49**, 555 (1978).
22. J.W. Colby, Adv. in X-Ray Anal. **11**, 287 (1968).
23. W. Bambynek, B. Crasemann, R.W. Fink, H.U. Freund, H. Mark, C.D. Swift, R.E. Price, and P.V. Rao, Rev. Mod. Phys. **44**, 716 (1972).
24. B.L. Henke and E.S. Ebisu, in Adv. in X-Ray Anal. **17**, 150 (1974).
25. J. Philibert, in 5th Int. Congr. on X-Ray Optics and Microanal., Tübingen 1968, proc. symp., p. 114.

26. O. Brümmer, Mikroanalyse mit Elektronen- und Ionensonden, 2nd ed. Leipzig: 1980. p. 98.
27. W.C. Leslie, The Physical Metallurgy of Steels. New York: 1981.

Correspondence and reprints: Dr. H. Kleykamp, Postfach 3640, D-7500 Karlsruhe, Federal Republic of Germany.

Mikrochimica Acta [Wien], Suppl. 10, 231–239 (1983)

Gemeinschaftslabor für Elektronenmikroskopie der RWTH Aachen, Federal Republic of Germany

Quantitative Electron Probe Microanalysis of Borides in Aluminium

By

P. Karduck, H. J. Schürhoff*, and W.-G. Burchard

With 5 Figures

(Received January 19, 1983)

1. Introduction

It is generally accepted that the quality of cast metals depends essentially on grain size and grain shape. Usually a fine grained cast structure is most beneficial. A small grain size can be achieved by means of suitable grain refinement methods. In case of aluminium and its alloys appropriate grain refining additives are available. In industrial praxis titanium or titanium along with boron are added in form of master alloys, which contain Al_3Ti- and TiB_2- particles and are proved to be effective grain refiners.

Alloying both Ti *and* B increases the efficiency of the grain refinement considerably comparing it with the exclusive addition of Ti and reduces the necessary amount of the master alloy.

Although the grain refinement of aluminium with Ti and B is practiced in industrial application the nature of the nucleating particles is not well understood. The investigation of those nuclei is problematic owing to the difficult analysis of the element boron by electron probe microanalysis (EPMA) and the additional complication of a quantitative determination of the boron content in very small particles. This paper is to report a method for quantitative analysis of borides in an aluminium – matrix by means of EPMA. It gives results on the composition of the nuclei effective for grain refinement by the addition of Ti and B. Based upon theories of Cibula[1] and Crossley and

---

* Now Honsel-Werke AG, D-5778 Meschede.

Mondolfo[2] there exists a controversial discussion wether the nuclei have compositions like $Al_3 Ti^2$ or TiB_2 [1], or are even of the type AlB_2 or $(Ti, Al) B_2$ [3].

2. Difficulty of the EPMA of Boron

To determine the nature of borides, which are effective in heterogeneous nucleation, their boron content as well as the concentration of other elements should be known. Although the well known methods for the correction of quantitative EPMA based upon Castaing and Descamps[4,5] have been applied successfully for electron energies above 15 keV and X-ray wavelength $\lambda < 1$ nm, quantitative analysis of the so called light elements is not feasible with adequate accuracy up to now.
Furthermore the analysis of light elements is obstructed by the very low emissivity of their characteristic X-rays and relative wide X-ray-lines. Owing

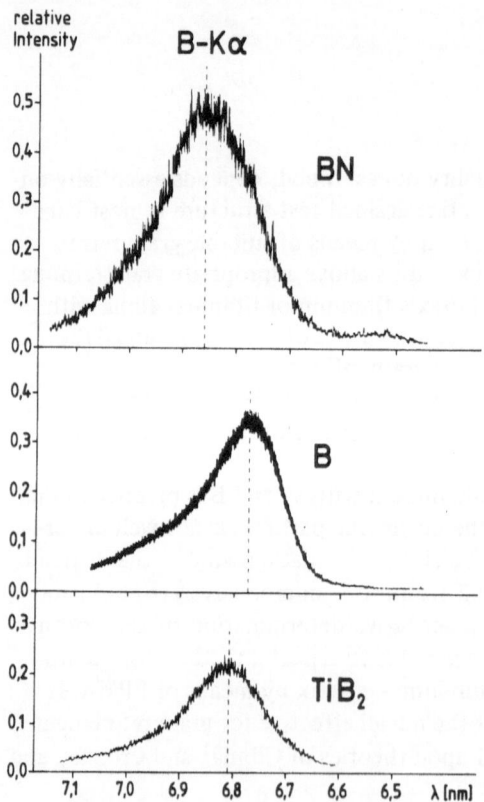

Fig. 1. The K-spectra of B-$K\alpha$ for TiB_2, pure B and BN

to different chemical bonding the X-ray lines of light elements can shift strongly in wavelength and can change their width[6] (Fig. 1). This fact can become a serious source of error for the quantitative analysis, when specimen and standard have very different compositions. These problems can be alleviated with a TiB_2-standard specimen, which should have almost the same composition as the particles to be investigated. This method should exclude matrix effects as well as line shifts and line broadening.

3. Measuring Conditions

Because of the above mentioned difficulties in measuring X-rays of light elements it is very important to optimize the measuring conditions. The following chapter will give an overview on how to obtain optimal parameters for the analysis of the B-$K\alpha$-radiation.

3.1 Accelerating Voltage

As a function of accelerating voltage the net intensity of the B-$K\alpha$-radiation exhibits maxima between 5 and 10 kV (Fig. 2).
The voltage at the maximum intensity changes for different chemical compounds due to the varying matrix effects. On the other hand it is necessary to have at least fair imaging conditions for finding the particles to be analysed in the scanning electron microscope. These prerequisites can be satis-

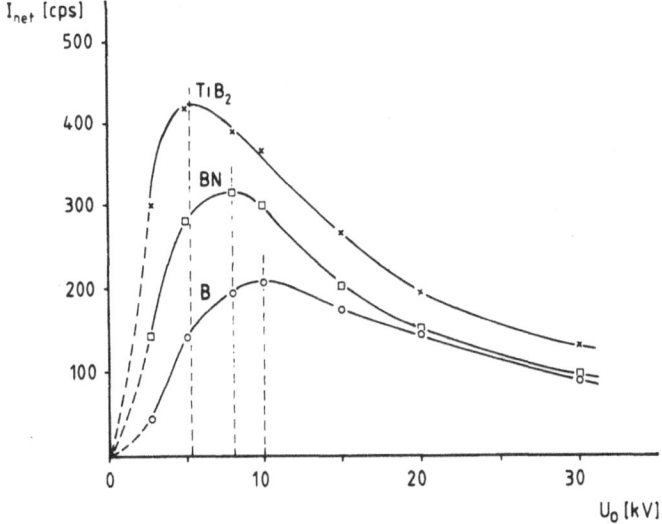

Fig. 2. Net intensity of B-$K\alpha$ for various accelerating voltages and materials

fied by a voltage of 10 kV. For a comparison with another excitation condition some measurements were carried out with 15 kV.

3.2 Anode Potential of the Proportional Counter and Pulse Height Distribution

The suitable anode potential in the proportional counter can be determined by measuring the appropriate pulse height distribution (Fig. 3). At low anode potentials this distribution is very close to the electronic noise and it becomes broader with increasing anode potential. Thus in case of very weak radiation the potential should be so high, that the distribution can easily be separated from the electronic noise. Following Fig. 4 a high anode potential has to be applied to achieve this separation[7]. Because of the relative broad pulse height distribution a suitable "energy window" for discrimination was set to 2 volts. With these conditions a peak to background ratio of 57 could be achieved for B-$K\alpha$-radiation of TiB$_2$.

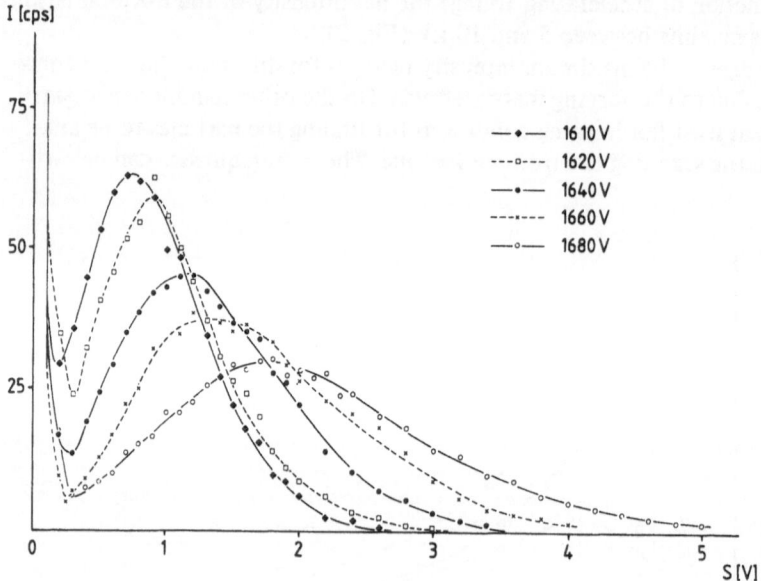

Fig. 3. Pulse height distribution of B-$K\alpha$ for different anode potentials of the proportional counter

3.3 Background Measurement

The background intensity was determined by linear interpolation. This simple method is sufficient because the background is almost constant in the λ-range of the B-$K\alpha$-line.

4. Experimental

A JEOL JXA 50 A was applied as electron microanalyser. A sintered TiB_2-specimen containing 70.54 % Ti and 29.04 % B was used as a standard for the B- and Ti-analysis. Since a sintered TiB_2 specimen can have local variations in composition[8] the standard specimen was moved during the measurement continuously under the constant electron beam to provide a good average intensity representing the mean chemical composition of the specimen.

The investigation of the grain refining particles was carried out on aluminium of 99.9 % purity containing different amounts of Ti and B. For the microanalysis the specimens were used at first in polished and then in etched condition. To display the dendrites of the Al α-solid solution and the nuclei in the centre of the grains the specimens were etched in a solution of 90 ml H_2O, 10 ml H_2SO_4, 95 ml H_2O and 5 ml HF[9] (Fig. 4).

5. Results and Discussion

Hypoperitectic Addition of Ti

For industrial grain refinement with Ti and B the Ti-addition is below the peritectic point of 0.15 % Ti in the Al-Ti phase diagram. Representative for this region one sample with 0.16 % Ti and 0.03 % B and another containing 0.1 % Ti and 0.02 % B were investigated.
In both samples clusters of small TiB_2 particles were found in the centre of the grains (Fig. 4). With respect to their size and shape clusters are similar to the TiB_2-clusters already present in the master alloy[10]. Concerning the real nucleation process some authors contend that the TiB_2-particles present in the melt are substituted by (Al, Ti) B_2 due to the exchange of Ti and Al atoms by diffusion processes[3, 11, 12]. However, according to Moriceau[14] the (Al, Ti) B_2 should have better wetting properties for the liquid α-Al.
To verify what kind of particles are present in the centres of the grains the intensities of the B- and Ti-$K\alpha$-lines respectively were measured by EPMA and compared to the corresponding intensities determined in the TiB_2-standard. This procedure is difficult because in most cases the diameter of the particles is smaller than the diameter of the excitation volume for X-ray production.

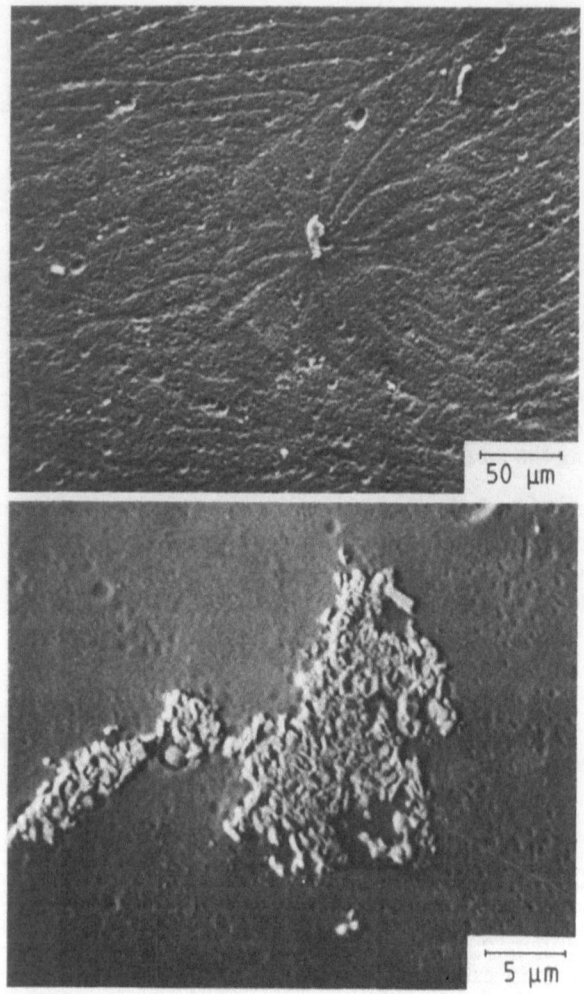

Fig. 4. TiB$_2$-clusters as nuclei in grain centres

The intensities of the particles are always lower than the standard intensities since the α-solid solution between the particles containing an unknown amount of dissolved Ti contributes slightly to the intensity. To obtain information on the real composition of the particles the following ratios between net intensities were compared:

$$q = \left(\frac{I_{Ti}}{I_B} \right)_{specimen} : \left(\frac{I_{Ti}}{I_B} \right)_{standard}$$

q should have the value 1 in case that no Al is substituted in the nuclei and the particles consist of TiB_2. Otherwise q must be less than 1. To obtain a criterion for the magnitude of the α-phase-contribution to the intensity two different excitation energies of 10 and 15 kV were applied. In addition polished and slightly etched samples were investigated. In total 90 clusters were measured. The results are summarized in Table 1:

Table 1

U_0 [kV]	specimen condition	q
10	polished	1.083 ± 0.13
10	etched	0.998 ± 0.11
15	polished	$0.96 \ \pm 0.07$
15	etched	—

With an acceleration voltage of 10 kV the results on the polished as well as the slightly etched specimens are satisfactory. The maximum deviation is 8 %.

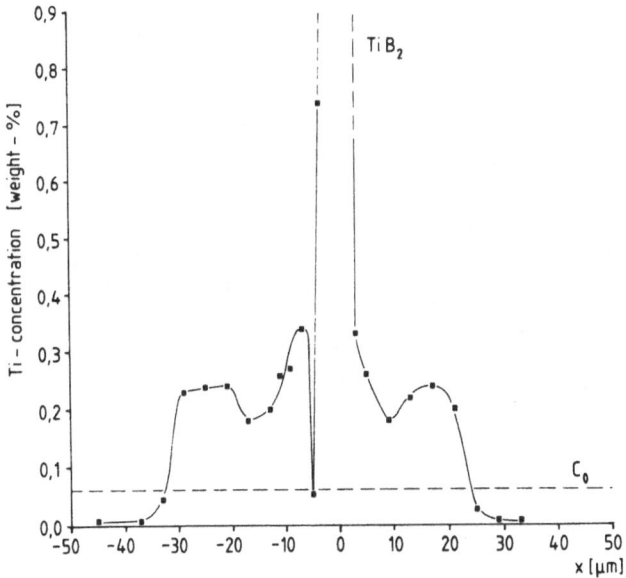

Fig. 5. Ti-concentration profile along a line through two dendrites and the appropriate TiB_2-nucleus

The slightly etched state shows the best result presumably because of the diminished matrix effects of the α-Al. This result shows, that no Al can be substituted inside the particles since $q > 1$.
At 15 kV only the polished specimen provides good results. At 15 kV is $q < 1$, but this can also be caused by the differences in excitation volume between Ti- and B-$K\alpha$-radiations.
From all these results we conclude that the investigated samples with hypo-peritectic Ti-content have TiB_2-clusters to account for heterogeneous nucleation. The efficiency of these clusters can be enhanced by a Ti-enrichment of the α-solid solution in the surroundings of the nucleus. Fig. 5 shows a Ti-concentration profile measured along a trace through one nucleus and two adjacent dendrites which are emerged at the nucleus. Following this profile the Ti-concentration increases in the surroundings of the nucleus up to more than five times the average concentration C_0[13]. All these results reveal the TiB_2-particles as effective nuclei and a clear absence of (Al, Ti) B_2-phases. Of course, it can not be ruled out that very small amounts of Al are dissolved in the TiB_2-particles.

Summary

Quantitative Electron Probe Microanalysis of Borides in Aluminium

A procedure for the quantitative analysis of borides in aluminium was introduced. For this purpose the optimal apparative boundary conditions for the EPMA of boron were worked out. With these conditions a satisfactory peak to background ratio of 57 could be achieved for B-$K\alpha$-radiation. By application of this method the following conclusion should be drawn concerning the kind of nuclei during grain refinement of aluminium with titanium and boron:
For grain refinement of aluminium with titanium and boron in the hypo-peritectic region of the binary system Al-Ti TiB_2-particles in clusters provide the high efficiency of refinement. This entails that the TiB_2-particles already present in the master alloy remain inert in the melt.
Hence, the good efficiency of refinement in this region cannot be attributed to the presence of particles like Al_3Ti, AlB_2 or (Al, Ti) B_2.

References

1. A. Cibula, J. Inst. Metals **76**, 321 (1949/50).
2. F.A. Crossley, L.F. Mondolfo, Trans. AIME **191**, 1143 (1951).
3. L. Bäckerud, Jernkont Ann. **155**, 422 (1971).
4. R. Castaing in Advances Electron. Physics, L. L. Marton and C. Marton (eds.), Vol. 13. New York: Academic Press. 1960. p. 317.

5. R. Castaing and J. Descamps, J. Phys. Radium **16**, 304 (1955).
6. W. L. Baun in Electron Probe Microanalysis, A. J. Tousimis and L. Marton (eds.). New York: Academic Press. 1969. p. 155.
7. J. Ruste, J. Microsc. Soectrosc. Electron. **4**, 123 (1979).
8. M. Hansen, Constitution of Binary Alloys. London: McGraw-Hill. 1958.
9. M. Schippers, Vereinigte Aluminium Werke AG Bonn, priv. communication.
10. H.J. Schürhoff, Dissertation RWTH Aachen, 1982.
11. G.W. Delamore and R.W. Smith, Met. Trans. **2**, 1733 (1971).
12. J.A. Marcantonio and L.F. Mondolfo, Met. Trans. **2**, 465 (1971).
13. H. J. Schürhoff, W. Gruhl, and W.-G. Burchard, to be published in Gießereiforschung.
14. J. Moriceau, Mém. sci. rev. met. **67**, 787 (1970).

Correspondence and reprints: Dr. rer. nat. P. Karduck, GfE der RWTH Aachen, Intzestrasse 5, D-5100 Aachen, Federal Republic of Germany.

Mikrochimica Acta [Wien], Suppl. 10, 241–250 (1983)
© by Springer-Verlag 1983

Metallgesellschaft AG, Metall-Laboratorium, Frankfurt a.m., Federal Republic of Germany

Microprobe Measurements to Determine Phase Boundaries and Diffusion Paths in Ternary Phase Diagrams Taking a Cu-Ni-Al System as an Example

By

G. Rudolph

With 5 Figures

(Received January 19, 1983)

1. Introduction and Objective

The alloy CuNi5Al5, with iron and manganese as additives, is a coin material with high resistance to corrosion and tarnish. This material is also distinguished by its golden colour[1], and is particularly suitable for coins with a high nominal value.

In order to protect high-value coins from counterfeiting clad materials have been used in minting systems since 1964/65. For example, Vereinigte Deutsche Metallwerke AG developed the coin material MAGNIMAT[(R)] which is used in many countries, and consists of magnetizable nickel core and silver white non-magnetic CuNi25 surface material. During the first tests carried out to produce a gold coloured MAGNIMAT by plating nickel with CuNi5Al5Fe1.2Mn0.8, it soon became evident that although plating would guarantee a perfectly satisfactory compound, diffusion annealing at all annealing temperatures would cause brittle intermetallic layers to be formed.

At high temperatures, this finding seems to contradict the available literature[2] which suggests that solubility for aluminium in α-(Cu,Ni) is at least 4.5 mass-% at 900 °C. We thus deduced that there will be no departure from the α-field, and no formation of intermetallic intermediate layers if CuNi5Al5Fe is annealed against nickel at high temperatures ≥ 900 °C. This, however, proved untrue.

The diffusion tests and microprobe measurements carried out in connection with this problem are described in this paper. The primary objective was to measure the diffusion paths in the ternary CuNiAl-phase diagram between the boundary compositions of CuNi5Al5 and nickel as a function of temperature. Secondly, it was intended to assess the position of the α-(Cu,Ni)-phase boundary in the ternary phase diagram published in[2].

2. Tests

The tests were carried out with homogenized samples of the alloy CuNi5Al5 and nickel. Solid nickel sections were provided with conical perforations, and conical rivets made of CuNi5Al5 were driven into these perforations. Since CuNi5Al5 has a higher coefficient of expansion than nickel, it was ensured that the CuNi5Al5 rivets would constantly be in close contact with the nickel during diffusion annealing. The samples were degreased, degassed in quartz ampoules and flushed several times in a stream of dried, high-purity argon, before they were sealed in for annealing. The duration of heat treatment at temperatures of 600, 700 and 800 °C was 16 hours in each case and 8 hours at 900 °C. Following this, the samples were quenched in water.

The diffusion profiles were registered by a CAMECA microprobe of the CAMEBAX MICRO type. Concentration profiles were recorded analogously on metallographic sections on traces perpendicular to the diffusion front. Quantitative measurements were carried out at numerous points on this trace, and converted into mass and atom % values after ZAF correction.

3. Results

3.1 Metallography

Fig. 1 shows the constitution in the diffusion zone. The grey intermetallic θ-phase has formed at the interface at all temperatures and diffusion zones are visible on the nickel side. Also, a small heterogeneous region can be seen on the nickel side. This region is irregular at 900 °C and slightly visible at 800 °C and 700 °C. On the CuNi5Al5 side, precipitation of the θ-phase of type Ni_3Al is clearly visible at 600 °C and still slightly noticeable at 700 °C.

3.2 Microprobe

Fig. 2 shows the 900 °C diffusion profile with the intermetallic phase which has a concentration gradient. The Al-profile shows uphill diffusion on the nickel side in the diffusion zone.

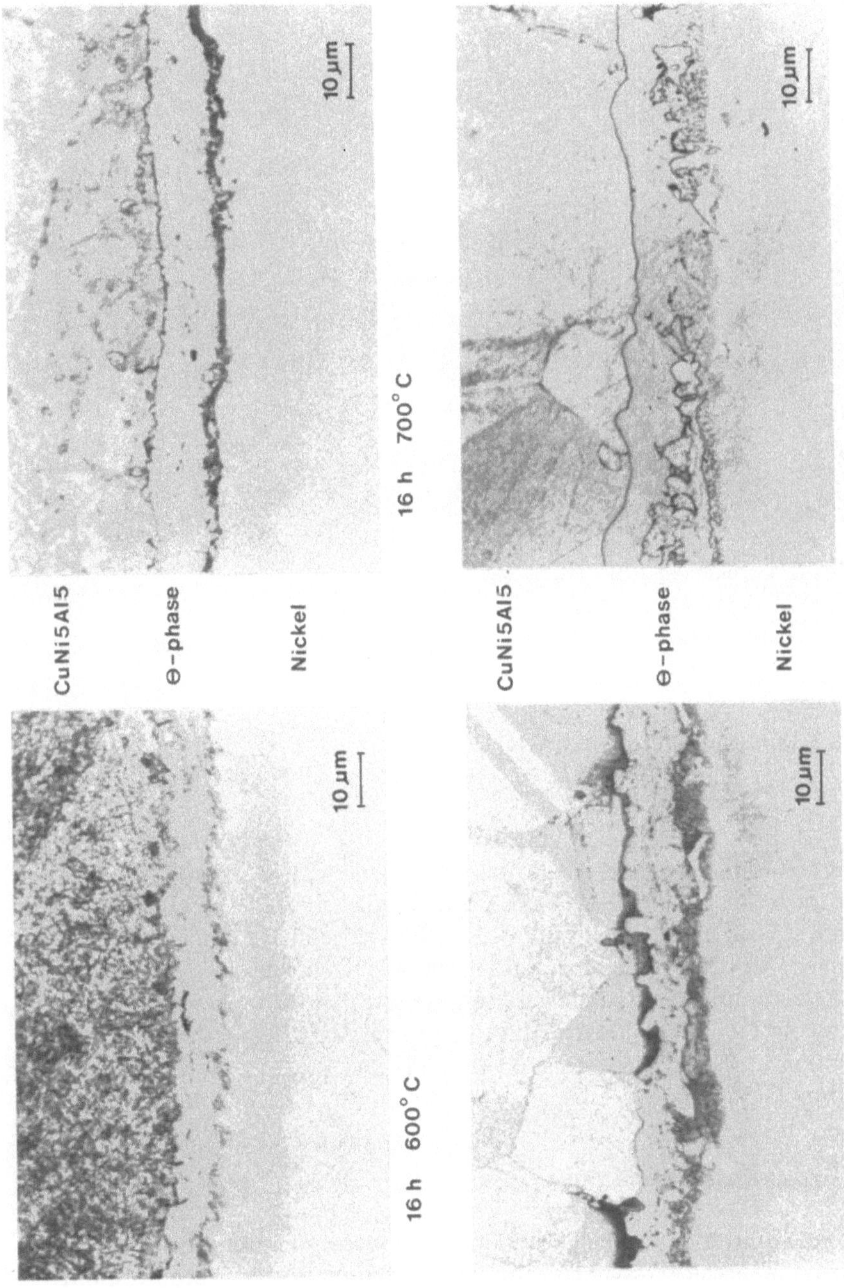

CuNi5Al5

Θ-phase

Nickel

16 h   700°C

10 μm

16 h   600°C

10 μm

CuNi5Al5

Θ-phase

Nickel

8 h   900°C

10 μm

16 h   800°C

10 μm

Fig. 1. Diffusion in the couple Nickel/CuNi5Al5

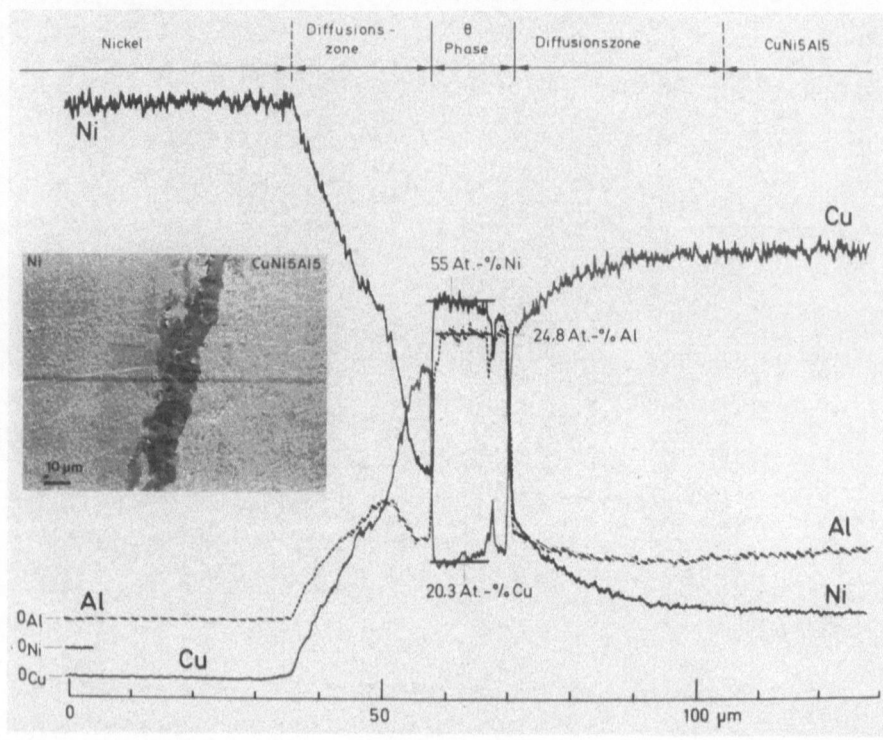

Fig. 2. Diffusion profile Ni/CuNi5Al5, 8h 900 °C

The diffusion profiles at 800 °C are comparable with those at 900 °C except for the lacking uphill diffusion in the Al-profile. At 700 °C (Fig. 3), and even more at 600 °C, the super-saturated CuNi5Al5 solid solution is no longer stable. The locally occurring peaks in the Al and Ni concentration profiles and the simultaneously falling Cu values show that the θ-phase has precipitated. At $T = 700$ °C, there seems to be no further precipitation in the solid solution in the diffusion zone on the CuNi5Al5 side below a value of approx. 3.5 % Al.

3.3 Quantitative Evaluation

The concentrations at the phase boundaries were determined from the concentration path curves by means of extrapolation. For this purpose, a vertical line was drawn at the phase boundaries at the point corresponding to half the concentration value. The degree of accuracy is $\leqslant 1$ μm.

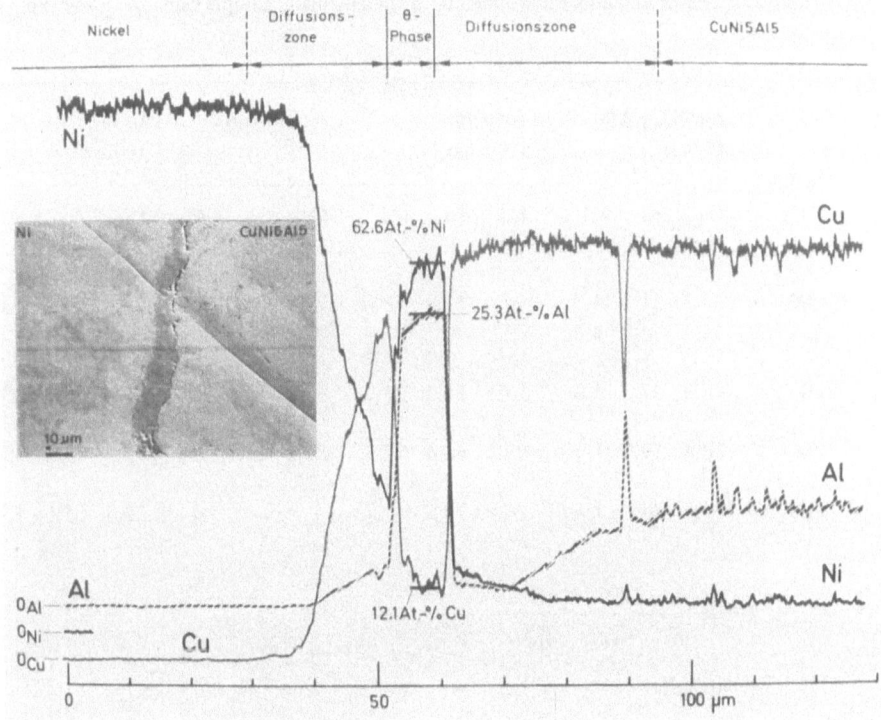

Fig. 3. Diffusion profile Ni/CuNi5Al5, 16h 700 °C

The concentrations at the phase boundaries thus ascertained are summarized in Table 1 and were marked in the ternary CuNiAl-phase diagrams of the respective test temperature (Figs. 4 and 5), along with the quantitative ternary value combinations measured along the concentration path curves[2].

At 900 °C, the diffusion path in the (Ni,Cu)-solid solution approaches the $(\alpha+\theta)$-phase boundary fairly steeply, following which it practically runs along this boundary (Fig. 4), but around 1 % Al inside the published $(\alpha+\theta)$ two-phase field*. Two values were determined for the phase boundary α-Ni/$(\alpha+\theta)$ (Table 1) and are marked in the diagram. The margin of error is estimated to be around ± 0.5 mass %. With regard to Al, these phase boundary concentrations are slightly higher than the published values[2]. In the α-(Gu,Ni)-solid solution, a bending diffusion path is observed. The phase boundary to the $(\alpha+\theta)$-field (Table 1) indicates a lower solubility of the

---

* All data are given in mass-% unless otherwise mentioned.

Table 1. Phase-Boundaries Determined with the Diffusion Couple CuNi5A15 / Nickel (mass-%)

Temp. (°C)	α-(Ni,Cu,Al)/ (α+θ)			(α+θ)/θ (Ni,Cu)₃Al			α-(Cu,Ni,Al)/ (α+θ)			(α+θ)/θ (Ni,Cu)₃Al		
	Cu	Ni	Al	Cu	Ni	Al	Cu	Ni	Al	Cu	Ni	Al
900	64.9	31.1	4.0	24.6	62.4	13.0	74.8	21.0	4.2	29.8	57.2	13.0
	51.2	44.0	4.8									
800	57.3	37.7	5.0	21.5	65.3	13.2	83.8	14.0	2.2	27.4	59.3	13.3
700	78.4	19.7	1.9	22.3	65.4	12.3	87.2	11.4	1.4	15.2	71.3	13.5
600	72.4	26.0	1.6	16.0	72.3	11.7	–	–	–	25.3	61.6	13.1

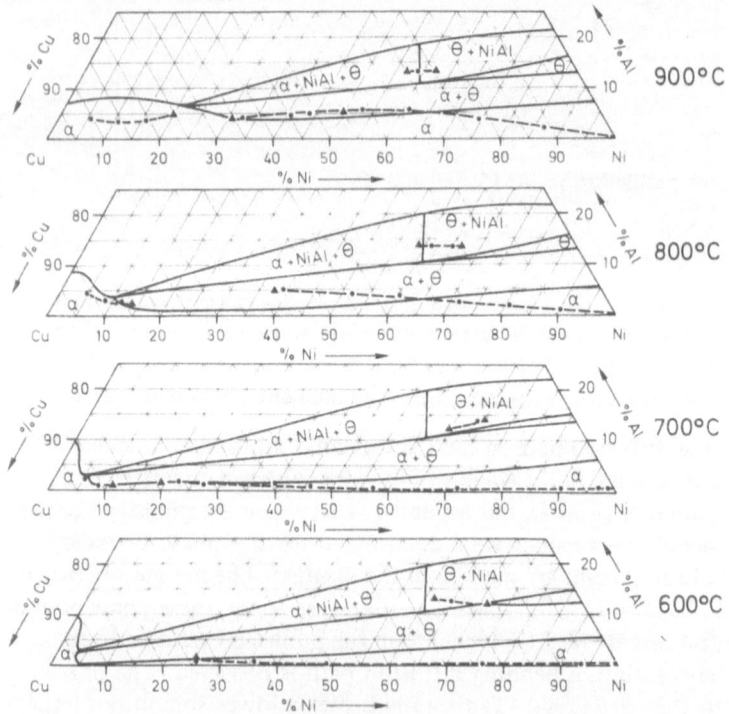

Fig. 4. Sections of the ternary CuNiAl phase diagram[2]. Diffusion paths in the α-(Cu,Ni)-solid solution (●) and phase boundaries (▲) determined by microprobe analysis (mass-%)

copper solid solution for nickel plus aluminium than published[2]. The
θ-phase which is in equilibrium with the α-nickel and α-copper is of the type
(Ni,Cu)$_3$Al with 25 at.% Al (13.0 mass-% Al), and has a high solubility for
copper, Table 1.
At 800 °C, there is a clear discrepancy between the measured values and the
position of the ($\alpha+\theta$) miscibility gap published in[2]. On the nickel side, the
diffusion path extends far into the ($\alpha+\theta$)-field. The phase boundary is
reached at 5 % Al. In[2], however, the solubility at 800 °C for aluminium in
CuNi40 is given as around 2 % Al. In the copper corner of the phase diagram,
at 14 % Ni, 2.2 % Al we also find a slightly higher solubility for aluminium
and nickel in copper. Once again, the θ-phase is exactly of the (Ni,Cu)$_3$Al-
type with 25 at.% Al and exhibits high solubility for copper up to 27.4 %,
Table 1.
At 700 °C and 600 °C, it was observed that the diffusion paths, in accordance with the published data, run practically parallel with the NiCu axis well
inside the α-(Ni,Cu) solid solution, and only reach the phase boundary to the
($\alpha+\theta$) miscibility gap at relatively high copper-levels, Table 1. The value determined at 700 °C in the copper corner for the α-(Cu,Ni)/($\alpha+\theta$)-miscibility
gap (11.5 % Ni; 1.4 % Al, rest copper) is in conformity with[2]. The θ-phase
of the type (Ni,Cu)$_3$Al has higher Al values at 700 °C (12.3 % Al) than at
600 °C (11.7 % Al), Table 1.

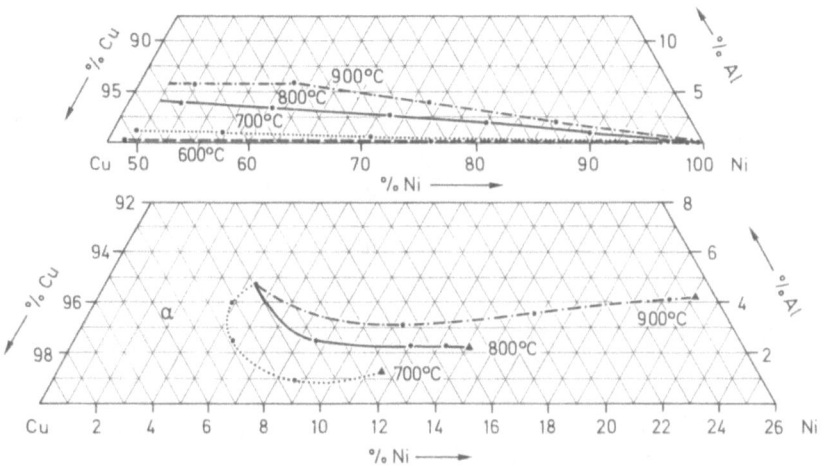

Fig. 5. Temperature dependence of the diffusion paths in the α-(Ni,Cu)- and α-(Cu,Ni)-
solid solution of the CuNiAl phase diagram (mass-%)

Fig. 5 shows the diffusion paths within the α-nickel- (top of Fig. 5) and the
α-copper-solid solution (bottom of Fig. 5): According to this, the course of
the diffusion path is dependent on temperature. As temperature increases,

a rise of the curves to higher Al values can be observed. The tendency of the curves to bend towards copper is less marked in the α-copper solid solution at higher temperatures.

4. Discussion

In accordance with literature, for example covering CuSnZn[3], AlMgZn[4], or CuNiZn[5] systems, the analysis has shown that the diffusion paths in the ternary system do not run alongside the connecting line between the two ternary initial concentrations, but that they assume bending, S-forms.

It may be stated as a general rule, without taking into account chemical potential differences for example, that diffusion paths bend away from the more rapid component[6]. This also holds true for the CuNiAl system under analysis here. The diffusion data in Table 2 show, that the diffusion coefficients of Al in Cu are higher than those for Ni in Cu. This means that concentration evens out more rapidly for Al than Ni in the copper corner.

Table 2. Diffusion Data

	Composition (at.%)	Temperature (°C)	D (cm^{2}s^{-1})	D_0 (cm^{2}s^{-1})	Q $\frac{kcal}{g \cdot atom}$	Reference
Al in Cu	3 % Al	700	0.5×10^{-10}	–	–	7
"	3 % Al	900	15×10^{-10}	–	–	7
Ni in Cu	40 % Ni	700	2×10^{-11}	–	–	8
"	10 % Ni	1000	4×10^{-10}	–	–	8
Al in Ni	0.08	800–920	–	1.1	59.5	8
Cu in Ni	Trace	800–1050	–	0.724	61.0	8

In the nickel corner of the phase diagram, virtually straight diffusion paths can be observed. The diffusion data in Table 2 show that with approximately equal activation energies, the preexponential factor D_0 for Al in Ni is larger than it is for Cu in Ni. This means that the diffusion path at 800 and 900 °C should run above the connecting line from Ni to CuNi5A15. This is confirmed by our measurements. At lower temperatures, copper seems to diffuse more rapidly than aluminium with the result that the diffusion path virtually runs along the Ni-Cu axis. We did not have any diffusion coefficients on hand for the purpose of estimation.

As regards the position of the $(\alpha+\theta)$-phase boundary, the boundary lines determined metallographically by Alexander[2] for various temperatures were confirmed in our microprobe analysis up to 700 °C-values. However, at 800 °C and to a lesser extend at 900 °C, the solubility of α-CuNi40 for aluminium amounts to 5 mass-%. This is especially for 800 °C significantly higher than stated in[2].

The θ-phase of type $(Ni,Cu)_3Al$ which is in equilibrium with α-CuNi contains Cu up to 30 % and has higher Al concentration values than indicated by[2]. Excepting the θ-phase formed at 600 °C, Al concentrations are always in the region of 13 mass-%, that is 25 at.% Al. The solubility value for copper in the θ-phase is not constant. At 800–900 °C, it is about 27–30 % Cu, and it drops to around 22–25 % Cu at 600–700 °C, Table 1.

Acknowledgement

The assistance of O. Häfner in performing the microprobe measurements aided significantly in this work. We are grateful to Prof. Hehenkamp, Göttingen, for valuable comments.

Summary

Microprobe Measurements to Determine Phase Boundaries and Diffusion Paths in Ternary Phase Diagrams Taking a Cu-Ni-Al System as an Example

With the aid of quantitative microprobe tests, diffusion phenomena and phase formation in the ternary CuNiAl system at 600–900 °C were investigated taking as an example the diffusion couple CuNi5Al5-nickel. The diffusion paths in the ternary system are dependent on temperature and assume an S-form in the copper corner of the phase diagram. In the copper corner, the curves swing away from the more rapid component aluminium towards the copper. Due to this non-linear course of the curves, the intermetallic θ-phase of the type $(Ni,Cu)_3Al$ can be observed as a layer at all temperatures in the boundary zone. At 800 °C and to a lesser extend at 900 °C the solubility of α-CuNi40 for aluminium, at around 5 mass-%, is higher than given by[2]. As far as it is possible with the diffusion couple under analysis, the microprobe measurements taken otherwise conform at 700 and 600 °C the position of the phase boundary α-(Cu,Ni)/$(\alpha+\theta)$-miscibility gap indicated in[2].

References

1. U. Heubner, M. Rockel, and G. Rudolph, Z. Metallkunde **73**, 522 (1982).
2. W.O. Alexander, J. Inst. Metals **63**, 163 (1938).
3. F.N. Rhines, R.A. Meussner, and R.T. Dettoff, Trans. AIME **212**, 860 (1958).
4. J.B. Clark and F.N. Rhines, Trans. ASM **51**, 199 (1959).
5. M.A. Dayananda and C.W. Kim, Scripta Metallurgica **16**, 815 (1982).
6. J.S. Kirkaldy and L.C. Brown, Can. Met. Quarterly **2**, 89 (1963).
7. F.N. Rhines and R.F. Mehl, Trans. AIME **128**, 185 (1938).
8. Y. Adda and J. Philibert, La diffusion dans les solids. Paris: Presses Universitaires de France. 1966.

Correspondence and reprints: Dr. phil. Günther Rudolph, Metallgesellschaft AG, Metall-Laboratorium, P. O. Box 3724, D-6000 Frankfurt a. M. 1, Federal Republic of Germany.

Mikrochimica Acta [Wien], Suppl. 10, 251–260 (1983)

[1] Department of Material Science and [2] Department of Mathematical Cybernetics of the Technical University of Dresden, German Democratic Republic, [3] Institute for Chemical Technology of Inorganic Materials of the Technical University of Vienna, Austria

Investigation of (Mo,W)C Mixed Carbides by Electron Probe Microanalysis and Kossel Technique

By

Hans-Jürgen Ullrich[1], Siegfried Rolle[1], Andreas Uhlig[2], Peter Ettmayer[3], and Benno Lux[3]

With 8 Figures

(Received February 1st, 1983)

Tungsten as an element has such a unique combination of outstanding properties that it is very difficult to find a substitute with similar properties. In alloys the task is somewhat simpler: tungsten in high speed steels has been successfully replaced at least partially by molybdenum in the past decades. Attempts to replace tungsten by molybdenum in the hardmetal industry, which takes up more than 50% of the total tungsten production, have been discouraged for a long time by the belief that molybdenum in hardmetal is harmful because of its tendency to form brittle subcarbide (Mo_2C)-needles or brittle complex compounds. It was only recently that Rudy and coworkers[1] showed that tungsten atoms in WC can be replaced by molybdenum atoms without changing the crystal structure. Apparently a partial substitution of W in WC by Mo results in only gradual and minor changes in physical and mechanical properties. Hardmetals sintered with these mixed (Mo,W)C-carbides show nearly the same properties as straight WC-based hardmetals. The microstructure of Mo-substituted (Mo,W)C-Co hardmetals is distinctly different from that of unalloyed conventional WC-Co hardmetals. In (Mo,W)C-Co hardmetals the carbide grains have a zoned structure, which is easily observable in the etched microsection shown in Fig. 1. In order to gather further information on the homogeneity of composition and crystal structure within one single carbide grain, a combination of electron microprobe analysis (EPMA) and microbeam diffraction technique had to be employed. While it would be possible to obtain a concentration profile even of

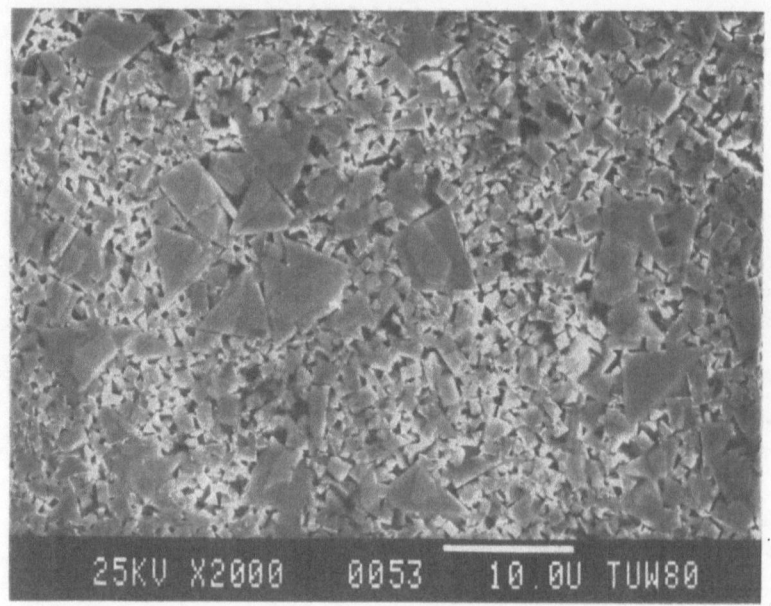

Fig. 1. SEM Micrograph of an etched (Mo,W)C-Co hardmetal

a very small carbide grain only a few microns in size, it is not yet possible
to produce an "in situ" diffraction pattern of satisfactory quality from such
small grains. In order to study the phenomenon of this zoned structure,
larger crystals of (Mo,W)C were grown from a metal bath. The conditions for
the crystal growth were chosen as follows:

Starting materials: WC, Mo_2C, Co and graphite
Mixing ratio: 40 mol% (WC + Mo_2C) 60 mol% Co; nominal composition
of the carbide $(Mo_{0.4}W_{0.6})C$
Temperature: 1800 °C, cooling rate: 2.5 K/min

During the cooling period large triangular crystals of (Mo,W)C are precipitated from the metallic melt. By an acid leach process the auxiliary metal is
removed, while the carbide is not attacked. In Fig. 2 a SEM photograph of
a crystal grown by this technique is shown.

The crystals have a very pronounced crystal growth anisotropy; they are triangular, rather flat plates. The surfaces usually show growth steps and smaller plates often adhere to the triangular face. Evidently these platelets have
grown epitaxially towards the end of the crystallization process.

Fig. 2. Triangular crystals grown from a cobalt bath. Magn. 90x

Electron Probe Microanalysis

The chemical composition of the crystals was determined by wavelength dispersive measurements of the intensities of the characteristic radiations of the the elements in a scanning electron microscope Quantometer SEMQ. The energy dispersive spectrometer (EDS) was from KEVEX, USA. Measurements on the original surface of the crystals gave erratic results, only after grinding and polishing of the triangular face consistent results could be obtained. Using energy dispersive spectroscopy the concentration profiles of the main metallic constituents W and Mo could be measured. Wavelength dispersive spectrometry yields – additionally – the carbon concentrations. Small amounts of iron on the order of 0.1 wt.% could be detected, but cobalt was below the detection limit (0.02 wt.%). Ion microprobe analysis gave the following results: no further impurities besides Fe, the total impurity level of Co and Mn below 10 ppm, Cr and V below 70 ppm.
In Fig. 3 the results of a line scan from point A to point B of the samples are presented. The concentration profiles of the four elements W, Mo, C and Fe were registered simultaneously by four spectrometers. The quantita-

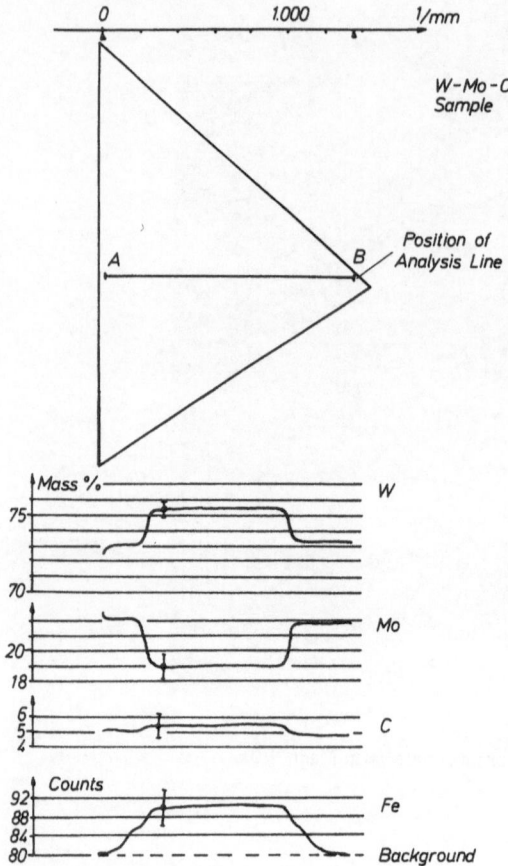

Fig. 3. EPMA of a polished section of the triangular crystal (Mo,W)C

tive analysis was performed by using the program MAGIC IV. Samples of
pure metallic Mo and W were used as standards and carbon was determined
by difference. Apparently the composition of the crystal was constant
over a considerable length of the scan in the centre of the crystal. From a
definite but optically not discernible boundary onwards, the tungsten content, the apparent carbon content and the Fe content decreased. The molybdenum content showed a concurrent increase. A line scan perpendicular to
the first one along the line C–D (Fig. 4) showed the same features, indicating that the crystals have a zoned structure. The centre has a composition
$(Mo_{0.33}W_{0.67})C$, the outer zone is richer in molybdenum – $(Mo_{0.38}W_{0.62})C$
– with the approximate Mo/W ratio of the starting mixture. The origin and
distribution of the iron content is not yet clear. The profile of the carbon $K\alpha$
scan seems to indicate a carbon deficient layer in the outer zone of the crystal, which is at variance with the higher Mo-content, if stoichiometry is

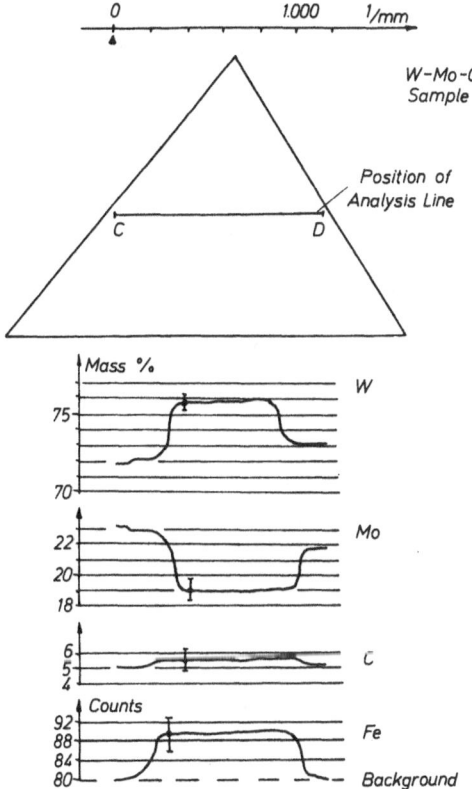

Fig. 4. EPMA line scan perpendicular to the scan in Fig. 3

assumed. It seemed helpful to use microdiffraction methods to obtain further information about the crystal structure of that outer zone.

Kossel Technique

The Kossel technique is a microbeam X-ray diffraction technique, by which the diffracted X-rays are generated in a small region in the sample itself[2]. The most convenient way to generate a point source at a definite site is to use an electron beam. The accelerating voltage should be adjusted to 25 kV. MoC and WC both crystallize in the hexagonal B$_f$-type structure with the lattice parameters

MoC   WC

a = 0.2898 nm   a = 0.2906 nm
c = 0.2809 nm   c = 0.2837 nm

They form a continuous series of solid solutions[1]. As far as we know, no attempts have been made as yet to generate Kossel patterns from these materials. Molybdenum carbide gives rise to an extremely great number of Kossel lines from the Mo $K\alpha$ radiation because of its short wavelength. The intensity of the lines in comparison to the background level is too weak to give Kossel patterns, from which the lattice type and the lattice parameters could be determined. Tungsten carbide on the other hand is a suitable material for obtaining Kossel patterns, the wavelength of the characteristic $WL\alpha$ radiation being in a suitable region in relation to the lattice parameters[2].

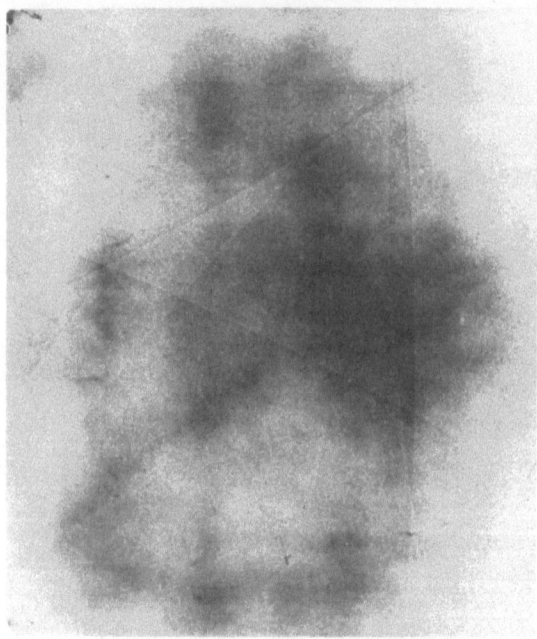

Fig. 5. Kossel pattern of (Mo,W)C (25kV, $WL\alpha/WL\beta$ lines)

There can be another reason for unsatisfactory Kossel patterns; if the concentration of lattice defects, predominantly dislocations, is too high[3], only ill defined and diffuse lines can be obtained. In Fig. 5 a typical Kossel pattern of a Mo substituted (Mo,W)C is shown, in which the lines can easily be distinguished from the background. The fine structure of the lines, especially the bright/dark structure is so well resolved that the dislocation density must be below $10^6/cm^2$. In Fig. 6 the Kossel pattern was obtained with the film plane parallel to the trigonal growth face of the crystal. The hexagonal symmetry of the Kossel pattern indicates that the trigonal growth

Fig. 6. Kossel pattern of (Mo,W)C, film plane parallel to the crystallographic plane (0001) of (Mo,W)C-crystals

face must be the (0001) face. Here again the very good line resolution is a result of a low dislocation density. In order to calculate the lattice parameters from such Kossel patterns, we used the method of direct simulation on an interactive computer-controlled display[4]. In the first step a complete stereographic projection (Fig. 7) of the expected high intensity Kossel lines upon the half dome is generated. The operator compares the actual experimental pattern with the whole reflex system on the display and chooses the region with identical line topology. By proper choise of "a" and "c" input values for the hexagonal lattice the computer-generated Kossel patterns on the display can be adjusted by sequential steps of trial and error, until complete correspondence between display pattern and Kossel pattern is achieved.

Such a computer-simulated pattern is shown in Fig. 8. Mathematically it is a gnomonic projection of the Kossel cones on a projection plane at a distance corresponding to the distance between point source in the specimen and the film plane. In this case the lattice parameters were evaluated as

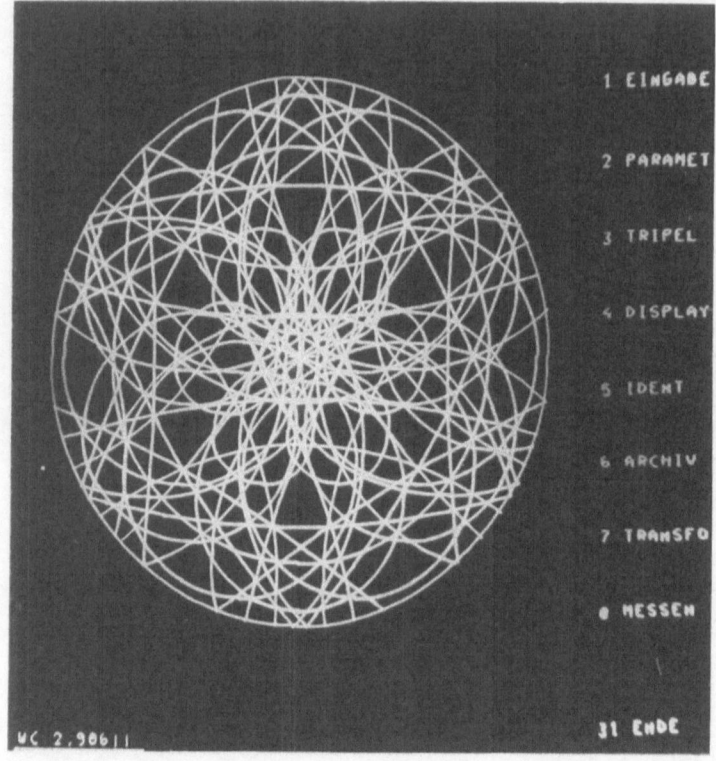

Fig. 7. Stereographic projection of $WL\alpha$ Kossel cones of (Mo,W)C

$a = 0.2904_5$ nm
$c = 0.283_5$ nm

These lattice parameters correspond to a composition of $(Mo_{0.3}W_{0.7})C$ according to Rudy's curve of the lattice parameters vs. composition in the MoC-WC system[1]. Kossel patterns taken from the centre zone and the outer layer revealed the fact that the outer layer has the same crystal structure and and orientation and only slightly different lattice parameters.

From these results and taking into account the results of the EPMA, it can be concluded that the outer zone differs from the centre part only with respect to the Mo/W ratio but not with respect to the phase structure. The slightly lower carbon content in the outer zone must be attributed to other factors. Further investigation is needed to clarify and explain the zoned structure seen in Mo substituted hardmetals sintered under technical conditions. We believe that the surface topography of hardmetals etched by Murakami's solution is to be attributed at least partially to the higher dislocation density caused by the misfit between zones with different compositions and lattice dimensions.

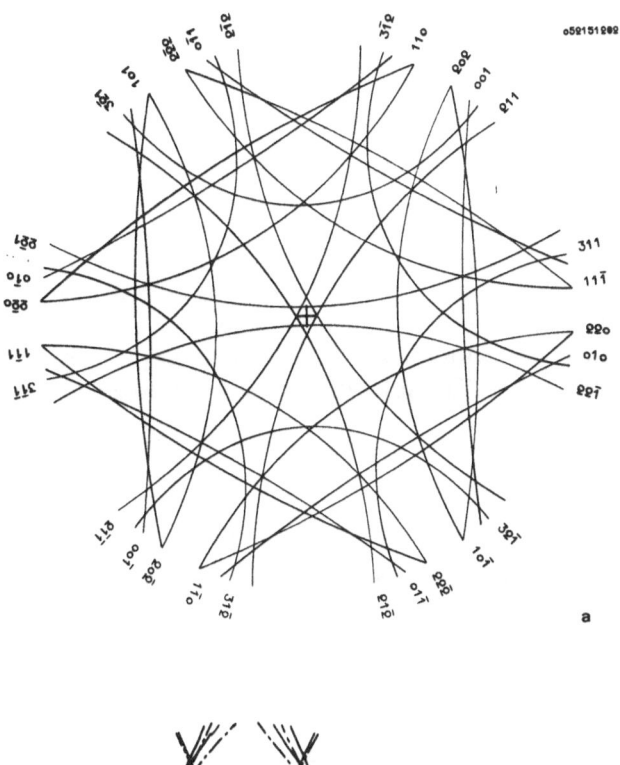

a

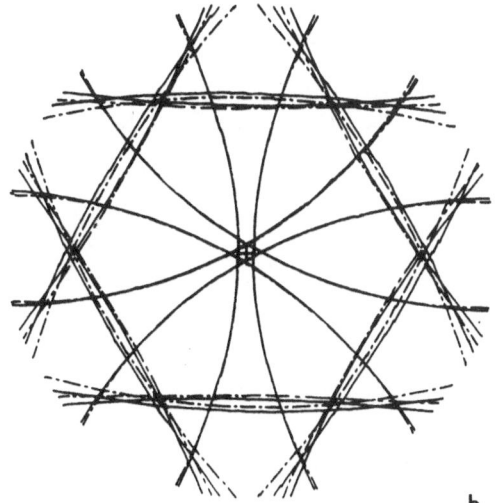

b

Fig. 8. Gnomonic projection of Kossel cones of (Mo,W)C. (a) Digital plotter drawing as a direct simulation of Fig. 6. (b) Superimposed digital plotter drawings with slightly different input data (lattice parameters)

Summary

Investigation of (Mo,W)C Mixed Carbides by Electron Probe Microanalysis and Kossel Technique

Crystals of (Mo,W)C grown from a auxiliary metal melt were investigated by EPMA and Kossel technique. The crystals were shown to have a zoned structure with a nucleus of nearly constant composition and an outer zone with a higher molybdenum concentration. This is due to the fact that molybdenum carbide has a higher solubility in the auxiliary metal (Co) than tungsten carbide does. Kossel microbeam diffraction has revealed the same structure for the whole crystal with just a slight shift to lower parameters from the core to the outer zone. The dislocation density in the core is in the order of $10^6/cm^2$.

References

1. W.H.St. Yih, Sam.A.Worcester Jr., and E.Rudy, Cemented Carbides Containing Hexagonal Molybdenum Carbide, US Pat. 4.040 380 (1977).
2. H. J. Ullrich and G. E. R. Schulze, Kristall und Technik 7, 207 (1972).
3. H.J. Ullrich, A. Herenz, E. Friedrich, W. Schatt, and C. Döring, Mikrochim. Acta [Wien] 1983 I, 175.
4. H.J. Ullrich, S. Däbritz, W. Quellmalz, and H. Schreiber, Abhandlungen des Staatlichen Museums für Mineralogie und Geologie zu Dresden, Bd. 31, p. 7–50 (1982).

Correspondence and reprints: Prof. Dr. P. Ettmayer, Institut für chemische Technologie anorganischer Stoffe, Technische Universität Wien, Getreidemarkt 9, A-1060 Wien, Austria.

Mikrochimica Acta [Wien], Suppl. 10, 261−269 (1983)

Max-Planck-Institut für Metallforschung, Institut für Werkstoffwissenschaften, Stuttgart, Federal Republic of Germany

X-Ray Emission Spectroscopy by Means of Electron-Microprobe for the Determination of the Density of States with Binary Amorphous Alloys

By

Siegfried Falch and **Siegfried Steeb**

With 5 Figures

(Received January 19, 1983)

1. Introduction

Up to now only very few investigations of the electronic structure of amorphous alloys were performed using X-ray emission spectroscopy. By this method the specimen is excited for example within an electron microprobe by electrons and the position and shape of the emitted characteristic X-ray line is measured.

Since the band structure within the metals and alloys is determined mainly by the component atoms and their arrangement, by the method of band spectroscopy one can obtain information about fundamental properties of the solid state. By electron spectroscopy (XPS or UPS) the energy of binding as well of the core electrons as of the shell electrons can be determined. However, it is not possible to determine the partial density of states, i.e. the origin of the analyzed electrons. Furthermore there exist certain problems concerning the extreme sensitivity of XPS or UPS for surface contamination and in the energy determination caused by charging the specimen[1,2]. Contrary to electron spectroscopy, X-ray emission spectroscopy yields information about the local partial densities of states of the valence bands[3], and thus represents an interesting method of investigation especially for binary systems. We have to pay attention to the fact, however, that the access to the density of states is not so direct as with electron spectroscopy.

By means of quantum mechanics one can learn that the shape of X-ray emission spectra represents the density of states modulated by the transition probability. Assuming this transition probability to be constant referred to the valence band, a true picture of the local partial density of states within the valence band spectrum can be obtained[4]. For a precise analysis of the X-ray emission all the factors must be taken into consideration which influence the transition. These are mainly the atomic arrangement and broadening effects caused by Auger transitions. Also distortions by the instrument, the influence of the acceleration voltage and the absorption within the specimen should be mentioned[5, 6]. During the present work the fact is used[7, 8] that the X-ray spectra emitted by the atoms of a certain kind depend in their position as well as in their shape extremely from the chemical and topological surrounding of the corresponding atom. Especially the difference between the crystalline and the amorphous state will be treated. In a preceding paper[9] the results with metallic glasses of the systems Fe-B, Cu-Ti, Cu-Zr, Co-B, Co-P, Fe-P, Ni-P, Mn-Si-P, Fe-Si-B, and Fe-Ge-P were reported. The present paper is concerned with amorphous phases from the Mg-Cu- and the Mg-Zn-systems.

2. Experimental Fundamentals

It was necessary to measure the K- and L-spectra of single elements in metallic glasses and in crystalline phases. For these measurements an electron microprobe is especially convenient since such an instrument normally is capable to perform quantitative analytical work and therefore works sufficiently accurate. With the instrument JSM 50A (Jeol, Tokyo) it was necessary for the elimination of carbon-contamination to use a turbo pump instead of a diffusion pump. Furthermore these measurements only could be done with a special software which controls the experimental runs and the data acquisition by means of a computer (PDP11/05; DEC) via a corresponding electronic control system (Datanim; Canberra). This equipment enabled to run up to ten spectrometer scans successively with different specimens without manual contact.

The adjustment of the specimen surface onto the Rowland circle was done in two steps. The first was to adjust the surface with the installed optical microscope. The second consisted in an automatic adjustment with the aid of the X-ray intensity at the fixed $K\alpha$-position of the spectrometer. During this process the specimen position was shifted up and down within ± 0.02 mm around the "optical" position by steps of 0.001 mm until optimum X-ray intensity was obtained.

The contamination of the specimen surface and the instabilities of the high voltage as well as of the beam current lead together with an eventual drift of the elctronic equipment to long time variations of the intensity. To sup-

press these effects the following normalization procedure was done:
The intensity I_θ measured at each position θ of the first spectrometer was related to the intensity I_{max} obtained within a second spectrometer at the position of the maximum of the emission line under investigation. Thus for each point a relative intensity I_{rel} was obtained:

$$I_{rel} = \frac{I_\theta}{I_{max}}$$

with I_θ = intensity at the angle θ, I_{max} = intensity at the maximum, and I_{rel} = relative intensity.

By this method it was possible to work without further beam stabilizing facilities up to beam currents of 10^{-7} A which allowed the measurement of weak lines with sufficiently low statistical error. The accelerating voltage for the best peak to background ratio amounted to 5 kV.

Concerning the influence of self absorption on the shape of the emission line it should be mentioned, that this effect changes mainly the high-energy side of the emission line and to a smaller extend also the line position (see later). Since, however, during the measurements the accelerating voltage remained unchanged and since the take off angle of the spectrometer was constant for instrumental reasons, a correction of the effect of self absorption could be avoided.

The smoothing of the experimental curves was done according to the method of cubic spline fit. The smoothing was favorable in point of view of the desmearing. Otherwise the physically irrelevant statistical fluctuations would have been intensified too much by the mathematical desmearing process.

Since the experimental spectrum represents the convolution with the instrumental distribution[6,10], the experimental spectrum must be desmeared to obtain the true spectrum. The instrumental distribution results from the entrance slit, the variations in the curvature of the crystal[6], the mosaic spread of the crystal, the deviation from the geometry (Rowland-circle), and the play of gears.

For the determination of the instrumental distribution the higher orders ($n = 2, 4, 5, 6$) of the Ti-$K\alpha$-line were measured exactly and assumed that the measured line-width corresponds to the instrumental distribution. The theoretical basis for the software-development of the desmearing procedure[11] is mainly based on the references[10,12].

3. Performance of the Experiments, Results, and Discussion

The amorphous specimens were prepared as ribbons (breadth 2–4 mm, thickness < 35 μm) within a melt-spin apparatus[13]. With all elements the

$L\alpha_{1,2}$-line was obtained, to which within the atoms corresponds the transition from the 3d-bands to the corresponding L_{III}-level. In the case of Mg, the $K\beta$-line was investigated. For analyzing the Mg-, Cu-, and Zn-emission lines a RAP-crystal was used. The spectra were obtained with a step width of 0.02 mm, corresponding to a wavelength step of about $1.9 \cdot 10^{-3}$ Å. To reach maximum reproducibility, the play of the gears had to be kept low. Therefore all positions were approached with slowing down velocity and only from one direction.

a) Mg-Cu-System

Fig. 1 shows the $L\alpha$-spectra of copper in the pure element copper, in the intermetallic compounds Cu_2Mg and $CuMg_2$, and in amorphous $Mg_{86}Cu_{14}$. The main maximum of the Cu-$L\alpha$-line shows a slight shift from its position with crystalline Cu_2Mg via the crystalline $CuMg_2$ to the amorphous phase in the direction to higher binding energies of about 0.75 eV. All the lines besides that of the element Cu furthermore show a shoulder in the energy region between 930 and 932 eV which grows with increasing Mg-concentration.

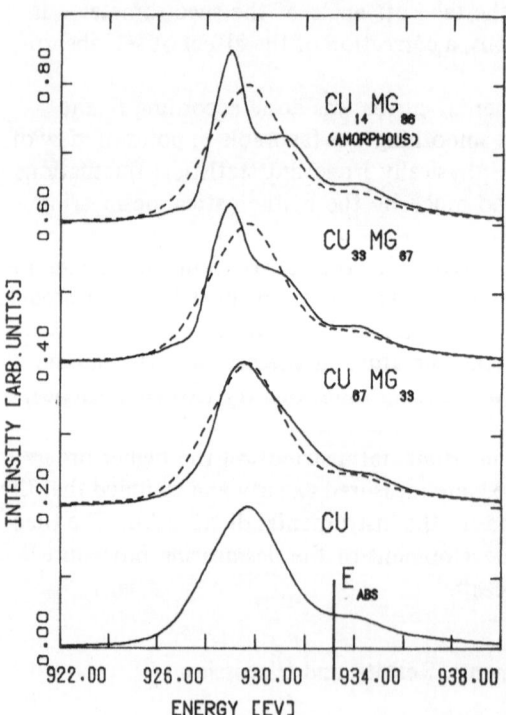

Fig. 1. $L\alpha$-spectra of Cu in crystalline Cu, Cu_2Mg, $CuMg_2$, and in amorphous $Cu_{14}Mg_{86}$

Fig. 2 shows the $K\beta$-spectra of magnesium in the element, in the intermetallic compounds Cu_2Mg as well as $CuMg_2$ and in amorphous $Mg_{86}Cu_{14}$. The position of the main maximum of the Mg-$K\beta$-line remains unchanged, however a rather pronounced shoulder can be observed in the region of 1298 eV which becomes more and more pronounced proceeding from the amorphous phase via crystalline Mg_2Cu to $MgCu_2$. All the spectra are normalized to the same area below the emission line.

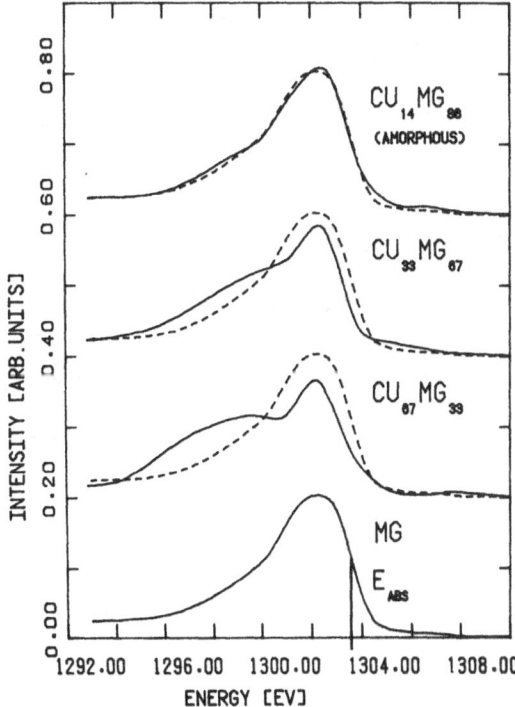

Fig. 2. $K\beta$-spectra of Mg in crystalline Mg, Cu_2Mg, $CuMg_2$, and in amorphous $Cu_{14}Mg_{86}$

As mentioned above, the X-ray-emission spectra can be interpreted in a certain way as a copy of the local partial density of states. Thereby the low energy part of the spectrum is formed by electrons from the lower band edge, the higher energy part in a corresponding way by electrons from the Fermi-edge. Thus follows, that the binding energy of the corresponding electrons becomes smaller with increasing X-ray energy. The position of the Fermi-energy can be determined by XPS-measurements of the next lower (L_{III}) electron states. For crystalline copper the value of the Fermi-energy is taken from [14]

$$2.6 \text{ eV} \cong \text{distance of } (E_F - N_{\max}(E)).$$

The small bump of the spectra in the region of the absorption edge may be caused by the self absorption. This effect also can cause an asymmetry of the spectrum[7].

Self absorption effects could be avoided by using very small acceleration voltages. The instabilities of the electron optical system, however, would then become too large.

The shift of the Cu peak to lower energies in Fig. 1 can be interpreted as a shift of the maximum density of states to higher binding energies within the copper d-band. It should be paid attention to the fact that with increasing Mg- and Cu-concentration, respectively, a shoulder or a peak is formed at larger and smaller binding energies, respectively.

In the Cu-Zr-system this can be understood as a splitting of the density of states, caused by the overlapping of Cu- and Zr-d-states, which is in accordance with the calculated Cu- and Zr-density of states[15].

In the case of Mg-Cu, the shift of the Cu-spectra to higher binding energies is much lower than the shift measured in the Cu-Zr-system, although the electronegativity difference is higher by about 0.2.

The valence band of Mg consists of overlapping $3s$- and $3p$-states. Caused by the transition probability, the $K\beta$-spectrum results mainly from electron transitions out of p-states. With increasing Cu-concentration the $3s$-part of the spectra becomes more dominant. Together with the change in the copper

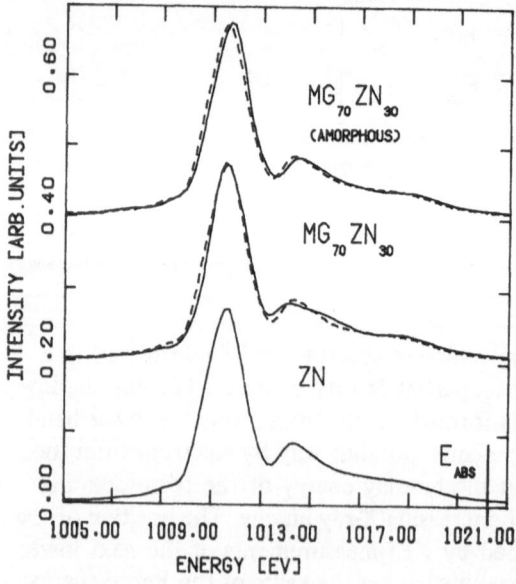

Fig. 3. $L\alpha$-spectra of Zn in crystalline Zn, $Mg_{70}Zn_{30}$, and in amorphous $Mg_{70}Zn_{30}$

$L\alpha$-spectra which show the same effects as in the alloy with a transition metal, there seems to be a hybridization of Mg 3p- and Cu-3d-orbitals in the upper part of the d-band.

b) Mg-Zn-System

Fig. 3 shows the Zn-$L\alpha$-spectra as obtained with the pure element Zn, and the intermetallic compound $Mg_{70}Zn_{30}$ in the crystalline and the amorphous state. In the crystalline as well as in the amorphous phase the spectra remain unchanged on alloying. This is in accordance with the results of Neddermayer[16].

Fig. 4 shows the Mg-$K\beta$-spectra as obtained with the pure element Mg as well as in the crystalline and amorphous Mg-Zn-alloys. In the crystalline and less pronounced in the amorphous phase the peaks of the Mg-spectra are shifted about 1.0 eV towards higher binding energies in comparison to the pure element spectrum. Additionally there is formed a shoulder in the region of higher binding energies in the spectra of the alloys. This can be understood as an hybridization of Mg-3p-states with Zn-states just below the Fermi-boundary.

The present results obtained with the Mg-Cu- and the Mg-Zn-system yield strong evidence for rather pronounced chemical short range ordering in the amorphous phases of these systems. This is in good accordance with X-ray-diffraction results recently reported[18].

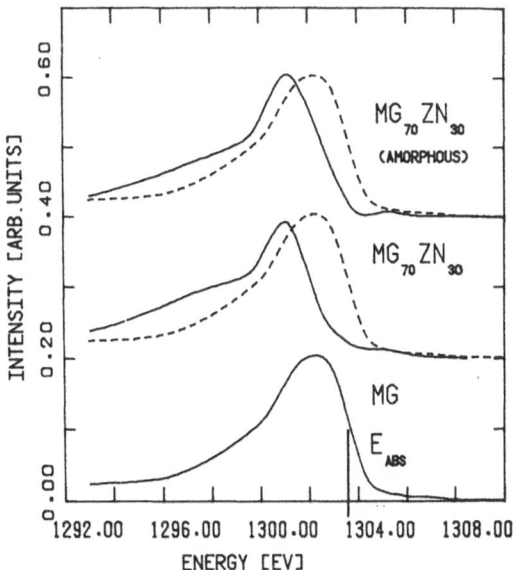

Fig. 4. $K\beta$-spectra of Mg in crystalline Mg, $Mg_{70}Zn_{30}$, and in amorphous $Mg_{70}Zn_{30}$

c) Charge Transfer

To get more quantitative informations on the electronic structure of the alloys under consideration, the method proposed by Wenger et al.[17] was applied to the measured spectra in order to calculate the relative charge transfer in the d-band. Fig. 5 shows the preliminary results.
During the discussion of this Figure it should be kept in mind that uncertainties may come into these data by a possible change of the transition probability for certain parts of the bands on alloying and furthermore from reabsorption effects which may not be suppressed by the normalization procedure.

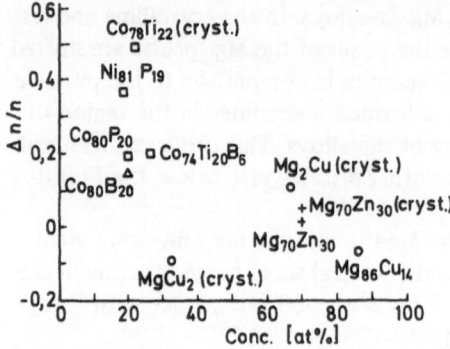

Fig. 5. Charge transfer $\dfrac{\Delta n}{n}$ versus concentration. The differences in the electronegativities are: 0.7   0.4   0.3   0.2
 $\quad\quad\quad\quad\quad\quad\quad\quad\quad\quad\quad\quad$ O $\quad$ + $\quad$ □ $\quad$ △

Summary

X-Ray Emission Spectroscopy by Means of Electron-Microprobe for the Determination of the Density of States with Binary Amorphous Alloys

Using an electron microprobe the position and shape of the characteristic X-ray emission lines from the elements Cu, Zn and Mg in metallic glasses of systems Mg-Cu, Mg-Zn were studied. In each case the lines were compared with those of either the crystalline element or the corresponding crystalline compound.
Compared to electron spectroscopy methods such as XPS or UPS, the method used during this work has the advantage that the spectra obtained are specific for the corresponding element and independent of minor surface impurities. The general trend of the results can be described in such a way that the energy of binding of the electrons to the corresponding atomic

species is larger in the amorphous state than in the crystalline element, whereas the energy of binding of the electrons in the amorphous state is similar to that of the corresponding crystalline compound. This means an electronic stabilization of the amorphous structure by compound formation, i.e. chemical short range order can be concluded from this result which is in accordance to the results of recent diffraction experiments.

References

1. G. Dräger in Handbuch der Festkörperanalyse mit Elektronen, Ionen und Röntgenstrahlen, O. Brümmer, J. Heydenreich, K.H. Krebs, and H.G. Schneider (eds.). Braunschweig: Vieweg. 1980.
2. H. Binder and E. Fluck in Mod. Phys. Chem., E. Fluck and V.I. Goldanskii, Vol. 2. London: Academic Press. 1979.
3. P.J. Durham, D. Ghaleb, B.L. Györffy, C.F. Hague, J.M. Mariot, G.M. Stocks, and W.M. Temmerman, J. Phys. F.: Metal Phys. 9, 1719 (1979).
4. L.M. Watson in Band Structure Spectroscopy of Metals and Alloys, D.J. Fabian and L.M. Watson. London: Academic Press. 1973.
5. C. Bonnelle, Ann. Phys. 1, 439 (1966).
6. H. Drack, Doctor-Thesis, TU Wien. 1977.
7. H. Drack, S. Kosina, and M. Grasserbauer, Z. analyt. Chem. 295, 30 (1979).
8. A. Faessler, Angew. Chemie 83, 51 (1972).
9. S. Falch, G. Rainer-Harbach, F. Schmückle, and S. Steeb, Z. Naturforsch. 36a, 937 (1981).
10. M.A. Blochin, Methoden der Röntgenspektralanalyse. München: Sagner. 1964.
11. S. Falch, Diplom-Thesis, University Stuttgart. 1981.
12. E.O. Brigham, The Fast Fourier Transform. Englewood Cliffs, N. J.: Prentice-Hall. 1974.
13. W. Sperl, Diplom-Thesis, University Stuttgart. 1980.
14. C. Bonnelle, in Soft X-Ray Band Spectra and the Electronic Structure of Metals and Materials, D.J. Fabian. London: Academic Press. 1968.
15. H.J. Güntherodt, P. Oelhafen, R. Lapka, H.U. Künzi, G. Indlekofer, J. Krieg, T. Laubscher, H. Rudin, U. Gubler, F. Rösel, K.P. Ackermann, B. Delley, M. Fischer, F. Greuter, E. Hauser, M. Liard, M. Müller, J. Kübler, K.H. Bennemann, and C.F. Hague, J. physique C8, 41, 381 (1980).
16. H. Neddermayer in Proceedings of the International Symposium on X-Ray Spectra and Electronic Structure of Matter, Vol. II. München: 1973.
17. A. Wenger, S. Steinemann, Helv. Phys. Act. 47, 321 (1974).
18. E. Nassif, P. Lamparter, W. Sperl, and S. Steeb, Z. Naturforsch. 38A, 325 (1983).

Correspondence and reprints: Prof. Dr. S. Steeb, MPI-Metallforschung, Institut für Werkstoffwissenschaften, Seestrasse 92, D-7000 Stuttgart 1, Federal Republic of Germany.

Mikrochimica Acta [Wien], Suppl. 10, 271–279 (1983)
© by Springer-Verlag 1983

Research Centre of the Standard Elektrik Lorenz AG, Stuttgart, and Max-Planck-Institut
für Metallforschung, Stuttgart, Federal Republic of Germany

X-Ray Excited Fluorescence Spectroscopy Within SEM for Trace Analysis

By

R. Eckert and S. Steeb

With 6 Figures

(Received January 19, 1983)

1. Methodical Fundamentals

1.1 Electron Excited Radiation

Within the scanning electron microscope (SEM) the electron beam scans line by line across the sample. The reflected electrons which emerge from the primary beam and the secondary electrons which are knocked out of the surface of the sample are used to produce the electron optical picture. The X-rays generated at the corresponding points of electron impact enable a chemical analysis of the surface of the sample within a depth of information of about 1 μm. The positionable fine electron beam with its diameter of nearly 0.1 μm produces a clear X-ray signal on particles of 1 μm, e.g. 10^{-12} g. More limited is the detectability of uniformly spread material traces, as alloy ingredients. Here the Bremsstrahlung, generated by the stopping of primary electrons hides in its statistical fluctuations the weak signals of trace elements. This is the case especially for energy dispersive spectrometers with their relatively poor energy dispersion, as they are commonly used at the SEM.

1.2 X-Ray Excited Radiation

Illuminating the sample with X-rays will generate characteristic radiation without generation of Bremsstrahlung. Therefore in trace analysis X-ray fluorescence is widely used. Accordingly SEM-workers try to adapt the methods of X-ray fluorescence on their instruments.

2. Experimental Fundamentals

2.1 X-Ray Source with a Foil Anode

The hitherto proposed and used constructions[1-5] adapt the X-ray transmission tube to the SEM, see Fig. 1. The electron beam hits a thin metal foil and

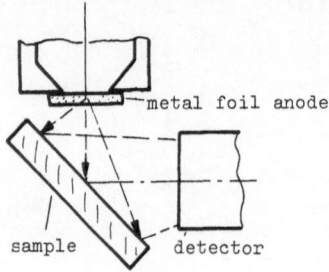

Fig. 1. Foil anode

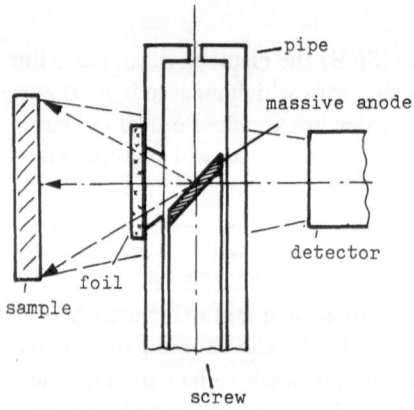

Fig. 2. Massive anode

generates in the upper $1-\mu$m-slice X-radiation. The X-rays penetrate the remaining foil and illuminate the sample. An advantage of the construction is the high intensity of the X-ray beam produced by a moderate electron current. In addition to this the foil acts as an X-ray filter. With suitable choice of the material and of the thickness of the foil this anode yields nearly monoenergetic radiation which irradiates the sample. A certain problem arises by the fact that usually only part of the primary electrons is absorbed and excites X-rays and that another part of the primary electrons transmits the foil. Especially for the detection of elements with low atomic number, i.e. in the low energy range of the spectrum ($E < 4$ keV) problems arise by the transmitted electrons which produce Bremsstrahlung on the sample, which hinders the detection of trace elements. Using a foil thick enough for 100 % absorption of the primary electrons yields only highly energetic X-rays which are inconvenient for the excitation of light elements ($Z \leqslant 14$). Nevertheless the constructions with an anode foil yield finally an X-ray spectrum from the specimen with markedly reduced background compared to an electron-excited spectrum. To the time, at least one construction is used commercially[6].

2.2 X-Ray Source with a Massive Anode

In the following an X-ray source for the SEM is presented which is equivalent to the usual X-ray tube using a massive target (Fig. 2). In this construction the electron beam incidents from the electron optical column onto the anode, which is formed by a small (4×3 mm^2) metal sheet of about 1 mm thickness. The anode is placed within a pipe with a small opening on the top and a larger opening at the side for the emerging of the generated X-rays. The window at the side is closed with a thin foil for the absorption of scattered electrons. Thereby the set up corresponds to the usual equipment for X-ray fluorescence analysis. Most of the facilities needed for performing this kind of analysis are contained within a commercial SEM-equipment, namely an electron source with $V = 0$ to 40 kV, $I = 0$ to 10 μA; a vacuum equipment for pressures down to 10^{-5} mbar; a vacuum chamber with mechanical controls for the sample, and an energy dispersive spectrometer with analysis system.

In the present construction[7,8] the anode sheet metal is sticked onto the top of a screw. The pipe with the foil window (that means the whole filter) is slipped on the screw. It works as Faraday cage. Anode and filter can be changed without problems. The anode is screwed into a plate together with a movable sample holder. To protect the sample from electrons and X-rays scattered from the walls of the vacuum chamber of the SEM the whole arrangement consisting of pipe and sample is surrounded by a metal enclosure which contains in the center of its upper side a small hole for the incoming electron beam. Furthermore this enclosure contains in its side wall directed to the spectrometer (Si(Li)-detector) a larger opening for the X-rays emerging from the sample.

The ground plate is screwed onto an usual sample holder of the SEM. Thus the switching over from the usual examinations in the SEM to a trace analysis in the ppm-range reduces to a changing of a sample holder only. All electrical conducting and vacuum resistent materials may be used as construction material for the anodes, i.e. nearly all metals. Since the electrical power dissipation is limited almost to about 0.1 W, different from usual X-ray tubes even brittle materials such as silicon may be used. Also the filter foils with their area of about 1 cm² may be used in a large variety since there is no atmospheric pressure difference on their surface. Corresponding to this fact the mechanical strength is not the limiting factor. The user of a SEM has for his investigations with the X-ray source the benefits of better use of the existing equipment and of good adaption to the analysis conditions with a wide range of selection of anode- and filter-materials.

Certain limitations result by limited sample size (area 0.1 to 5 cm², thickness up to 1 cm), vacuum resistance of the sample, and poor counting rates for a beam current < 1 μA.

2.2.1 Peak to Background Ratio

The ratio of the X-ray net signal peak to the background (P/B) serves as a

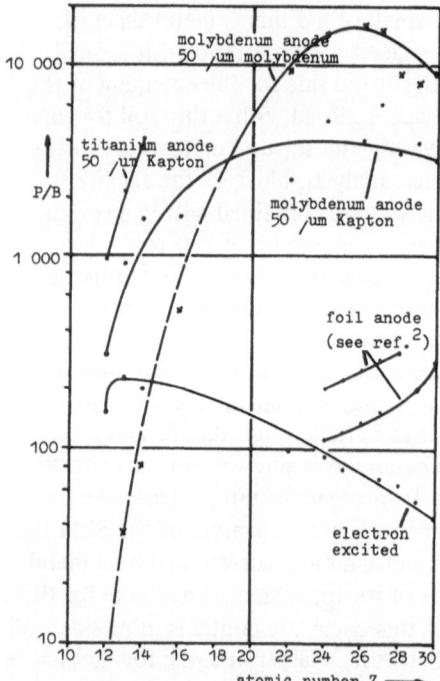

Fig. 3. P/B-ratio versus atomic number Z for the K-emission lines of elements. Primary electron acceleration voltage $V = 20$ kV

scale for the achieved decrease of the X-ray background. For example, with electron excitation pure elements exhibit P/B = 20 to 200, depending on the element, the beam voltage, and the take off angle of the radiation. Constructions with foil anodes reach up to P/B = 100 to 300 for elements with an atomic number $Z > 20$. The presented X-ray source produces for elements with Z = 12 to 16 a P/B = 1000 to 3000 and for elements with $Z > 16$ a P/B = 3000 to 15000, see Fig. 3. Thus the enhancement compared to electron excitation amounts for elements with $Z \leqslant 16$ to about 10 times and for $Z > 16$ to about 20 to 200 times.
The achieved P/B-ratio depends on the element and the impinging X-radiation.

2.2.2 Filtering

Concerning the choice of the filter material, the following facts should be observed.
A thin plastic filter foil absorbs the reflected electrons from the anode but transmits nearly the entire anode X-radiation. This radiation will generate the fluorescence radiation of the sample, but is also scattered by the sample and superimposes on the desired fluorescence spectrum. Thus a complex photon spectrum finally reaches the detector and causes a nearly constant ratio peak to background of about P/B = 3000 over a wide range of elements with satisfactory count rates. However, in the case of crystalline samples, the sample X-ray spectrum in addition contains also strong Bragg-reflexions (see Fig. 4). These reflexions may overlap sometimes with the signals of the trace elements to be detected. Given a certain angle 2 θ between the direction of the primary beam and the connection line between specimen and detector only X-rays with quantum energy $E = \frac{nhc}{2d\sin\theta}$ are reflected by net planes with distance d into the detector. With single crystalline material or with coarse grained material there is a chance to diminish the intensity of the reflected beam by turning the specimen.

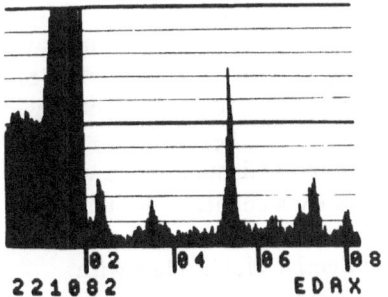

Fig. 4. Silicon spectrum; Ti-anode; 7.5-μm Kaptonfilter

With a higher absorbing metal filter foil the high energy part of the anode spectrum reaches the sample, whereas the low energy part is absorbed in the foil. Therefore the sample spectrum stays free from reflected low energy X-rays. With such a filtered X-ray excitation for elements with $Z = 24$ to 40 a ratio peak to background of about P/B = 10000 up to 15000 is reached. Thus P/B is three to four times larger than with unfiltered excitation. However, the larger P/B values which can be achieved using filtered excitation are only valid for elements with absorption energies not too far from the excitation energy. The curves P/B $= f(Z)$ are steeper than with unfiltered radiation (see Fig. 3).

Furthermore, the low background is obtained only in the high energy part of the peak, which corresponds in Fig. 5 to the right hand side. In this region the signal reaches 50440 and the background 3 thus yielding P/B = 16800. In the low energy range on the left hand side of the peak the background is enlarged as a result of a minor absorption of the sample radiation in the material together with Compton scattering, i.e. inelastic scattering of the X-radiation in the sample and the detector[9]. This is to be considered for doing trace analysis.

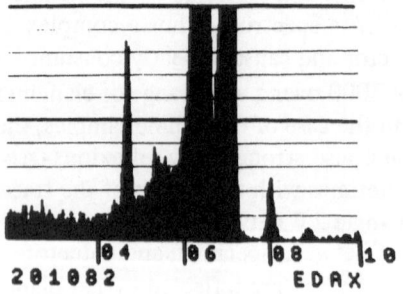

Fig. 5. Iron spectrum; Mo-anode; 50-μm Molybdenumfilter

2.2.3 Counting Rate

The counting rate S in the detector is proportional to the electron beam current I. Furthermore, after passing a limiting voltage V_0, the counting rate is nearly proportional to $(V-V_0)$, V being the accelerating voltage of the electron beam. Thus the relation stands: $S = a \cdot I \cdot (V-V_0)$ with V_0 = const.; a = proportionality factor. As an example, a molybdenum anode with a 50 μm-molybdenum filter produces at $V = 40$ kV a counting rate ten times higher than at $V = 20$ kV. Furthermore, heavy trace elements ($Z > 30$) in a light matrix are effectively excited by the intensive Mo($K\alpha$) line with 17.5 keV. Therefore the user of a SEM will select in such cases the higher beam voltage region. Since a number of SEM's have a high voltage limit of

20 to 25 kV, both examples in chapter 3.2 were measured with a beam voltage of only 20 kV.

3. Application of the X-Ray Source with Massive Anode to Trace Analysis

3.1 Strategy in Trace Analysis

If there is a sample with unknown traces one should use as a first step the unfiltered excitation. As the second step, to compare the traces with the contents in a standard, the filtered excitation should be applied, using such a radiation energy, that only the traces are excited and not the major elements. In this way a larger part of the counting rate (usually about 2000 counts per second = cps) falls into the energy interval of the trace, see Fig. 6. The anode material should be chosen in such a way that the region in the energy scale showing the highest intensity lies only a small amount higher than the absorbing energy E_{abs} of the element to be detected. In addition for the case $E_{abs\ trace} < E_{abs\ matrix}$ the energy of the anode radiation should be chosen lower than $E_{abs\ matrix}$ (compare with lower part of Fig. 6).

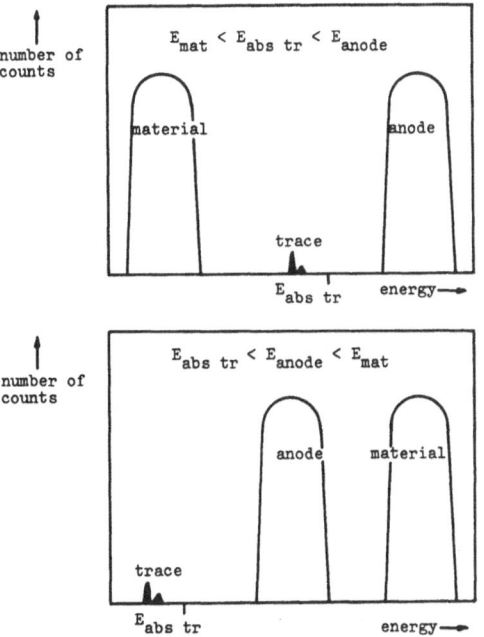

Fig. 6. Schematic diagram for the choice of the primary X-radiation in trace analysis

3.2 Detectable Minimum Concentration

For the present paper the following definition for the detectable minimum concentration c_{min} was used (see, for example[9]):

$$c_{min} = \frac{3\sqrt{N_B} \cdot c_0}{(N-N_B)}$$

with N_B = number of background counts
$\quad N$ = total number of counts of a characteristic line of a reference element
$\quad c_0$ = concentration of the reference element

Hereby is assumed, that c_{min} corresponds to three times the standard deviation of the background in a certain energy interval ΔE. The energy interval is assumed to be 1.2 times the full width at half maximum (FWHM).

Two materials were analysed for evaluating the detection limits: an aluminum standard with a number of trace elements in the 2000 ppm range and the NBS glass 612 with traces in the 50-ppm range. For both standards the same measurement conditions were applied. The results are listed in Table 1 and Table 2.

Table 1. Aluminum Standard, Molybdenum Anode with 50-μm Molybdenum Filter, 20 kV, Counting Time 20 min

Trace	Ti	Cr	Mn	Fe	Cu	Zn
c_0 [ppm]	530	1800	3700	6700	430	1900
c_{min} [ppm]	1310	59	64	37	18	13

Table 2. Glass Standard NBS 612, Molybdenum Anode with 50-μm Molybdenum Filter, 20 kV, Counting Time 20 min.

Trace	Ca	Fe	Ni	Rb	Sr	Ag(L)	Tl(L)	Pb(L)
c_0 [ppm]	86000	51	38.8	31.4	78.4	22.0	15.7	38.6
c_{min} [ppm]	110	3.8	4.1	8.8	14	50	7.0	3.2

Summary

X-Ray Excited Fluorescence Spectroscopy Within SEM for Trace Analysis

Since the excitation of X-ray fluorescence by primary X-rays shows advantages compared to primary electrons, also for the use in a SEM an X-ray excited fluorescence attachment is proposed. This works with massive anode which shows advantages compared to the well known foil anode method. This X-ray source with a massive anode permits the performance of trace analyses in the SEM. Apart from an additional use of the existing equipment a large variety of anodes and filters enables the user to adapt the measurement conditions to the analysis problem. P/B-ratios of mostly 3000 to 15000 in pure elements are usual with this method and therefore detectable concentrations in the ppm range are secured.

References

1. L.M. Middleman and J.D. Geller, Scanning Electron Microscopy 1, 171 (1976).
2. B. Linnemann and L. Reimer, Scanning 1, 109 (1978).
3. R.M. Weiss, Beitr. elektronenmikroskop. Direktabb. Oberfl. 12/1, 209 (1979).
4. I. Pozsgai, Proc. 10th Int. Congress on Electron Microscopy. Hamburg 1982. p. 681.
5. A. van Riessen and K.W. Terry, Jeol News 20E, 19 (1982).
6. Tracor Northern, Prospect "Microtrace", 1982.
7. W. Plannet, AGAR-catalog 4 (1983), Plano GmbH, Friedrichsplatz 9, D-3550 Marburg, Federal Republic of Germany.
8. R. Eckert, Beitr. elektronenmikroskop. Direktabb. Oberfl. 15, 41 (1982).
9. R. Woldseth, X-Ray Energy Spectroscopy, Kevex Corporation. Burlingame, 1973.

Correspondence and reprints: Dipl.-Phys. R. Eckert, SEL-Forschungszentrum, Hellmuth-Hirth-Strasse 42, D-7000 Stuttgart 40, Federal Republic of Germany.

Mikrochimica Acta [Wien], Suppl. 10, 281–295 (1983)
© by Springer-Verlag 1983

Gemeinschaftslabor für Elektronenmikroskopie der RWTH Aachen, and Siemens Forschungslaboratorien, München, Federal Republic of Germany

STEM-EDX Measurements on Grain Boundary Phenomena of Sensitized Chrome-Nickel Steels

By

M. Pohl, H. Oppolzer, and S. Schild

With 13 Figures

(Received February 19, 1983)

1. Stainless Steels

Under the collective heading "Stainless Steels" fall those steels which, in general, contain at least 12% chromium, and are therefore resistant to oxidising attack. Higher chromium contents and other alloying additions improve the anti-corrosion properties, and also have a beneficial effect on the mechanical properties.

Since the first patent by Krupp in 1912 on stainless steel with 18% chromium and 8% nickel a wide variety of alloy compositions has been developed for special purposes. Based on the fundamental steel 304 with 18% chromium and 8% nickel Fig. 1 shows for AISI steel specifications the relation between alloy composition and application. DIN sheet 17440 comprises 29 steel types, and a further 21 types are given in the Iron Material Specification Chart Stahl-Eisen-Werkstoffblatt 400. Internationally, stainless steels are covered by Euronorm 88–71 and ISO 683/T13. Their increasing importance is reflected in the rise in production between 1950 and 1980 to 800.000 t/a, corresponding to more than a twenty-fold increase.

The choice of a stainless steel for a particular purpose is made in the first instance with reference to its behaviour towards corrosion stripping, which in the case of chrome-nickel steel is usually slight, due to its passivity. However, under certain conditions this group of steels can be sensitive towards local corrosion phenomena, such as pitting, inter- and transgranular stress corrosion cracking, and intergranular corrosion (IK).

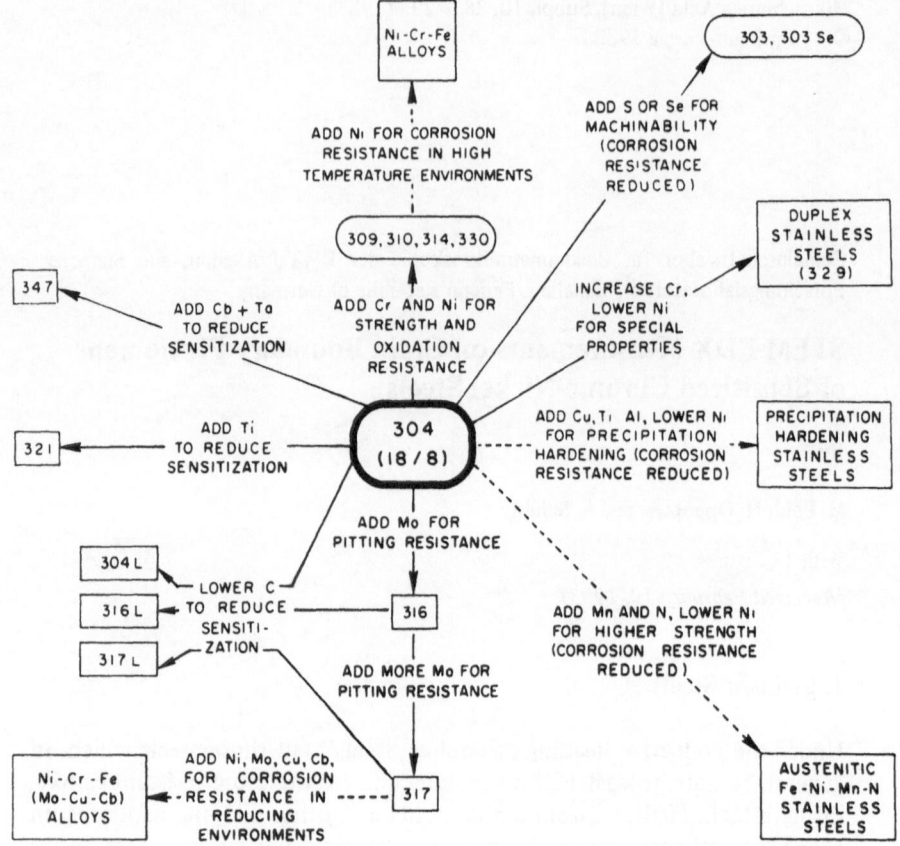

Fig. 1. Compositional modifications of 18/8 austenitic stainless steels to produce special properties[1]

Already in 1930 Strauß, Schottky, and Hinnüber had recognized the effect of carbon on the sensitivity towards IK^2, but at this time there was no metallurgical process for producing economically carbon contents less than 0.08%. For this reason Hoedremont and Schafmeister developed in 1933 the principle of carbide stabilization, using principally additions of niobium and titanium[3].

2. Hot-Strength Chrome-Nickel Steels

Due to their resistance to corrosion and scaling and to their reasonable creep strength over long periods, austenitic chrome-nickel steels are well adapted for high temperature use in the chemical industry, and for steam boilers and nuclear reactors. The high creep strength of austenitic vis-à-vis ferritic stainless steels arises from their markedly lower stacking fault energy[4]. The use of austenitic chrome-nickel steels for operating temperatures above 400 °C can be traced back to 1928, the principal application then being in petro-chemical plant construction[5].

The development of austenitic steels for power plants, which operate today with live steam temperatures of up to 650 °C and pressures of up to 350 bar, has followed two distinct paths.

In USA the titanium stabilized steel AISI 321 was used so successfully that the maximum permitted stresses for components, as laid down in the ASME boiler and pressure vessel code, were up-graded. Any subsequent creep defects in structural parts were countered by using the unstabilized steels 304 H and 316 H.

In the Federal Republic of Germany there was preference in the fifties for the niobium stabilized steels (X 8 Cr Ni Nb 16 13 to X 8 Cr Ni Mo V Nb N 16 13). However, HAZ cracking in the welds of thick-walled structural parts recurred repeatedly, due to the finely dispersed niobium carbides preventing any take-up of welding stresses by plastic deformation of the base material[6]. Following the good experiences in the USA it was logical to develop the fully austenitic unstabilized steels X 6 Cr Ni 18 11 (similar to AISI 304) and X 6 Cr Ni Mo 17 13 (similar to AISI 316) for the new generation of reactors, the so-called fast breeders[7].

Because of the required creep strength extra low carbon qualities could not be used, and this placed new emphasis on the sensitivity of these steels towards *IK*, despite the claim with the water now used in power stations no *IK* should occur[8].

3. Intergranular Corrosion

The sensitivity towards *IK* can usually be detected early by applying the relevant corrosion tests.

3.1 Damage Arising from IK

During pressure testing of fuel-element sheaths for nuclear reactors, some sheaths burst (Fig. 2). Examination showed that the presence of small amounts of hydroxides in the test-fillings had led to *IK*, weakening the sheath walls and causing premature failure (Fig. 3). Such defects can be countered by adopting another heat-treatment for the materials.

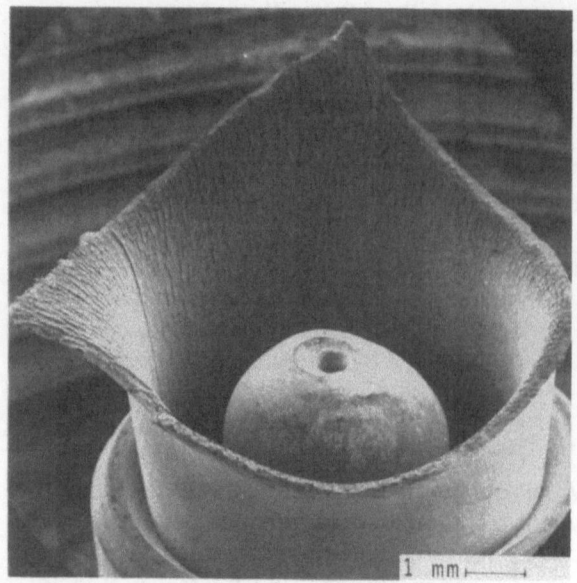

Fig. 2. General view of a burst fuel-element sheath

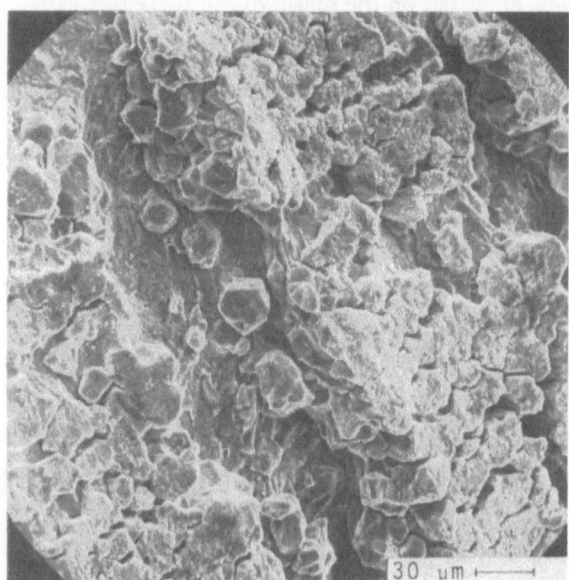

Fig. 3. Part of Fig. 2, left

3.2 The Relationship Between Material Structure and Sensitivity

Stainless steels in the as-delivered state are usually solution heat-treated, this being the state of optimum corrosion resistance. At the solution heat-treatment temperature of $\geqslant 1050\ °C$ the carbon is dissolved homogeneously in the austenite lattice. On water quenching this state is preserved, giving a carbon supersaturated non-equilibrium system. Fig. 4 shows the drastically

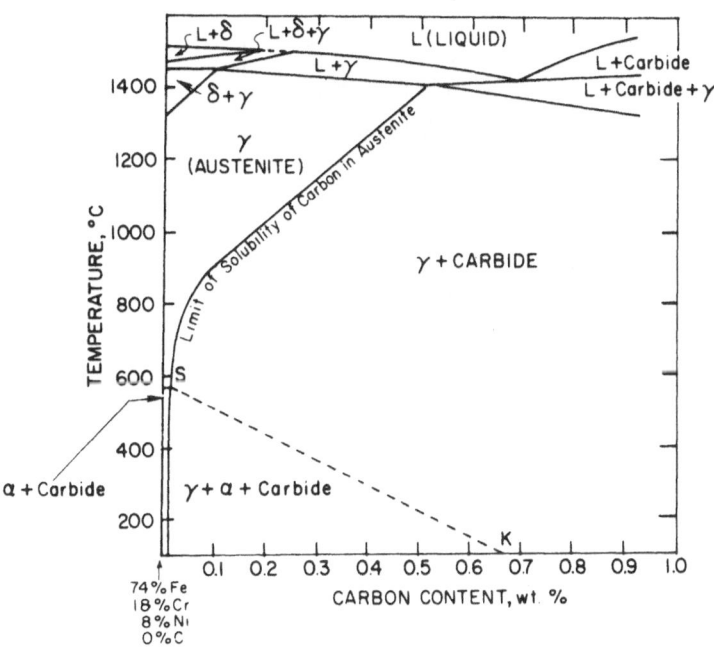

Fig. 4. Pseudobinary phase diagram for 18/8 Cr-Ni-steels with varying carbon content[9]

reduced carbon solution equilibrium down to 600 °C. As the carbon contents of technical steels are generally higher than their equilibrium solubilities at temperatures of application, precipitation of carbides of the type $M_{23}C_6$[10] takes place. The formation of new phases is diffusion controlled, so that with increasing temperature the degree of carbon supersaturation will diminish. The precipitation regions give then the so-called C-curves, the extent and position of which are governed by the carbon content and such structure parameters as grain-size[7]. Fig. 5 shows the tendency towards grain disintegration of 18/8 stainless steels as a function of the carbon content. It should be noted, however, that the criterion for carbide precipitation, and hence for sensitivity towards IK, is not the carbon concentration but the carbon activity. This is raised by nickel and silicon, and lowered by manga-

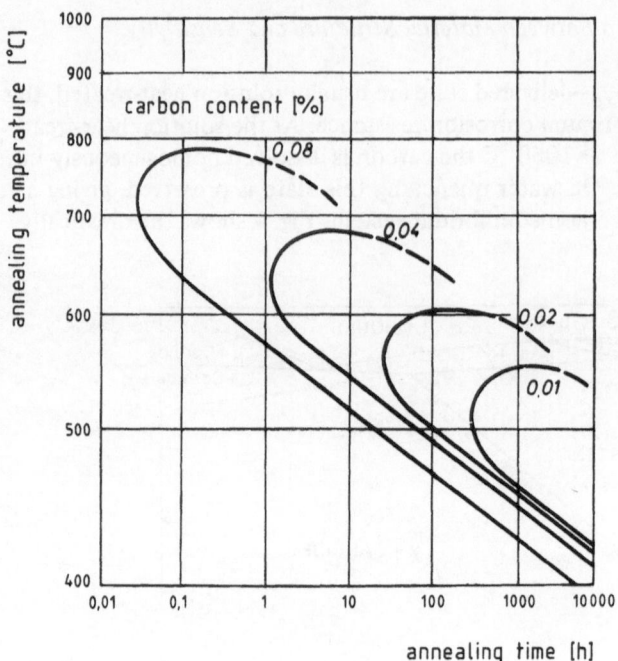

Fig. 5. The relation between sensitization and carbon content[11]

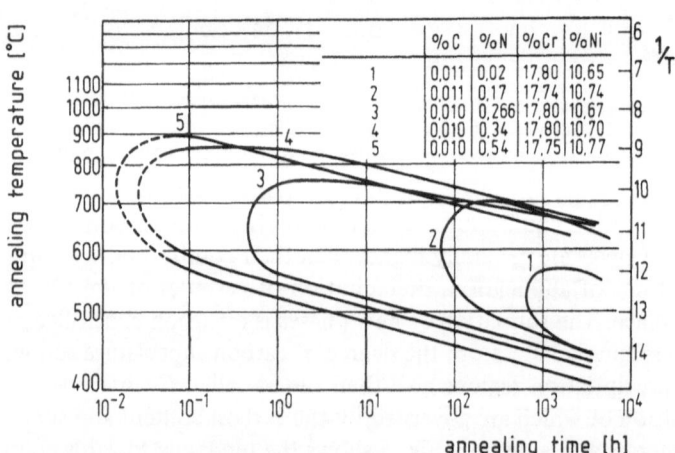

Fig. 6. The relation between sensitization and nitrogen content at constant carbon concentration[13]

nese and nitrogen[12], giving a corresponding increase or decrease in the sensitivity towards *IK*.

Nitrogen has in the past been added as alloying element, as, similar to carbon, it forms an interstitial solid solution, so raising the yield point and tensile strength, also under creep conditions. However, nitrogen tends to give precipitates of the type Cr_2N, which increase the tendency towards *IK* (Fig. 6).

3.3 Detection of Sensitivity Towards IK

Testing for sensitivity towards *IK* can be effected by analogue experiments under conditions of application, or by time-shortened standardised corrosion tests[7].

Huey and Streicher tests (ASTM Standards A 262-70 T) are used preferentially in USA, whereas in FRG the Strauß test (DIN 50914) is applied. In the latter case, where the system copper/copper sulphate/sulphuric acid is used, a potential of 360 mV develops, and this utilizes to the most the maximum potential difference between sensitized grain boundaries and steel matrix. For this reason this testing procedure has been adopted as Euronorm 114-72. In addition to the "go/no go" type of information, a degree of sensitization may be established by SEM-examination[7].

3.4 Discussion of the Causes of Sensitization

It may be taken that sensitization is caused by chrome-rich precipitations, which are located preferentially in the grain boundaries. The following mechanisms have been put forward:

a) Impoverishment of the grain boundary regions in those alloying elements responsible for corrosion resistance.

b) Formation of structural constituents which may be selectively removed in a certain medium.

c) Enrichment in accompanying elements of the grain boundaries, which promote their anodic dissolution.

d) Development of structure stresses in the grain boundary regions, as well as stresses between the differently orientated grains themselves.

e) Production of localized cell effects, where the nobler carbide acts as cathode and the surrounding less noble metal as anode.

The common feature of all these theories is that sensitivity towards *IK* is coupled with the formation of grain boundary precipitations. Special emphasis is placed on the carbides, but nitrides and intermetallic phases also have an influence, as the formation of these phases involves the removal of those elements in the initial structure which are responsible for corrosion resistance i.e. chromium and molybdenum. Special attention is paid therefore to the "chromium impoverishment" theory.

3.5 Detection of Chromium Impoverishment

All available measuring and examination procedures and mathematical methods were tested to assess chromium impoverishment.

3.5.1 Diffusion Calculations

The ratio of diffusion rates of carbon and chromium in austenite is approx. 300 at 900 °C[14], this ratio increasing with decreasing temperature. The constituent with the smallest diffusion coefficient will undergo the greatest concentration drop in front of the phase boundary. The carbon concentration remains practically constant within the metal phase, whereas the chromium concentration shows a decrease towards the site of precipitation. As a result of the differing diffusion rates for volume and grain-boundary diffusion computation studies show lenticular chromium impoverishment zones surrounding the boundary precipitates in the grain-boundary plane (Fig. 7)[15]. Even a low density of boundary precipitations can lead to a continuous chromium impoverishment.

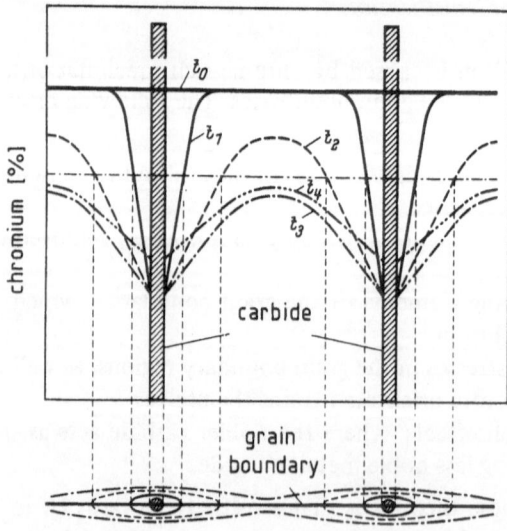

Fig. 7. Chromium content at a grain boundary incorporating two carbides with increasing annealing time $t_0 < t_4$ (schematic). (a) Chromium profile in cross-section; (b) Chromium profile in plan view[15]

3.5.2 Indirect Methods of Detecting Chromium Impoverishment

After tempering, martensitic phases have been observed in the neighbour-hood of the grain boundaries at room temperature[16]. The elevation of the martensite point can be explained by chromium impoverishment up to 12%. Because of the selective attack of the Strauß solution on the grain bound-aries and the contrastingly slight removal of the cut grains, the chromium content of the test solution was determined in relation to the iron content. For 18/8 steels 9–13% chromium[17] and 5–12% chromium[18] were ana-lysed. The layerwise chemical removal of fracture surfaces in sensitized steels gave minimum chromium contents of 7.5%[19].
This indirect detection method does not allow of any direct assignment of the values to a particular site in the structure, giving only a somewhat uncer-tain mean value for the chromium impoverished zones.

3.5.3 Direct Procedures for the Detection of Chromium Impoverishment

Chromium impoverishment has been observed qualitatively by micro-radio-graphic exposures[20] and by magnetothermic methods[21]. Auger spectro-scopy offers an analytical method of the highest sensitivity for the thinnest layers, although measurements are complicated by the fact that the austenite to be examined never shows complete grain boundary fracture, and that on the grain boundaries adjacent to the chromium impoverished zones precipi-tates with high chromium contents occur. Auger analyses have been in-stanced, not only to detect qualitatively chromium impoverishment, but in particular to relate the accumulations of S, Si, and trace elements found on the grain boundaries to tendency towards IK[22].
Electron probe micro-analyses on sensitized steels give different results. In the references[23–25] chromium impoverishment was detected, whereas in references[26–29] it was not.
Our own investigations with SEM showed that the grain boundary regions removed only attained a maximum thickness of 0.15 μm[30,31], and for this reason the EPMA-measurements were carried out with EDX with beam currents of 5 x 10^{-10} A.
A beam diameter of less than 0.1 μm would be expected. Through careful optimisation of all material and experimental parameters it was possible to detect chromium impoverishment, but only for transsensitive structure states (Fig. 8). Here the most pronounced chromium impoverishment is already exceeded with the chromium impoverishment zones tending to spread out. As would be expected the chromium contents detected lay above the resistance limit of approx. 12%. Measurements of sensitized structure on compact sections failed because of the limited lateral resolution of EPMA, in conjunction with the unfavourable nature of the object to be

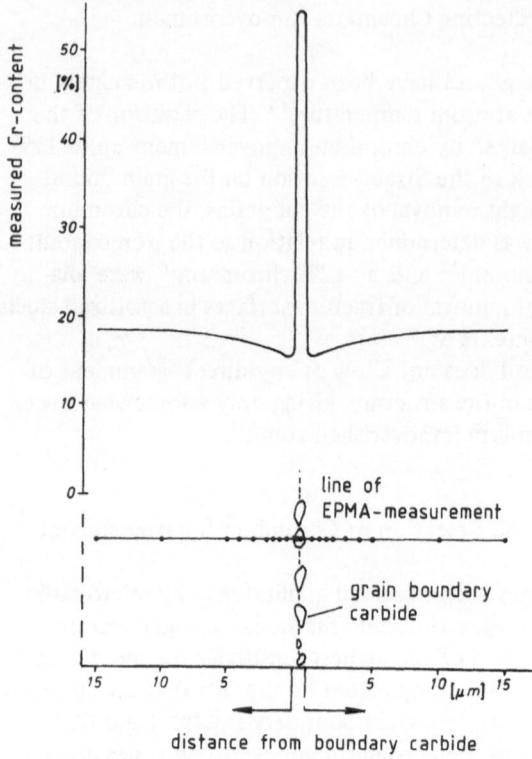

Fig. 8. Chromium impoverisation as measured in compact sample[31]

measured with its extremely high chromium concentrations in the grain boundary precipitations adjacent to the relatively low chromium impoverishment.

These investigations were continued therefore with STEM-EDX measurements, as the thinned metal foils used here allowed of an improvement in the resolution, extending into the order of magnitude of the beam diameter. The feasibility of this measuring procedure has already been demonstrated on one individual sensitized sample[32].

4. STEM-EDX Measurements to Detect Chromium Impoverishment

The knowledge gained from previous investigations on the degree of sensitization as related to the material, temperature, and sensitization time permits an optimization of sample selection[7, 30, 31].

The steel X 6 Cr Ni 18 11 was chosen, as carbides of the type $M_{23}C_6$ are exclusively formed, which precipitate preferentially on the grain boundaries.

Measurements were carried out on a coarse grained quality (Grain-size ASTM 1–3), the region of grain disintegration being correspondingly enlarged. A comprehensive creep-test programme[7] provided a whole series of closely related test samples for all relevant temperatures and for very long operating schedules up to 50.000 h. The actual steel analysis was as follows: C=0.049%; Cr=18.3%; Mo=0.19%; N=0.029%; Ni=11.7%. The measurements were made initially on a STEM of type Hitachi H 700 with EDX System of Kevex Unispec 7000. Subsequently, to improve further the resolution of the STEM, Elmiskop ST 100 F was used, which enables a beam diameter of 5 nm to be attained as a result of the greater directional beam brightness (3 orders of magnitude) of the field emission cathode with beam currents of 10^{-8} A[33] An analysis system of the type Tracor Northern NS 880 with Kevex Si(Li)-detector was connected to this apparatus. This combined equipment then permitted masses of down to 2×10^{-21}g and mass fractions of 0.1% to be detected[34].

The evaluation of the analysis results followed the non-standard method of Cliff and Lorimer[35], whereby the X-ray intensities of the elements are calculated via the K-factors:

$$\frac{C_{Fe}}{C_{Cr}} = K_{(Fe, Cr)} \frac{I_{Fe}}{I_{Cr}}$$

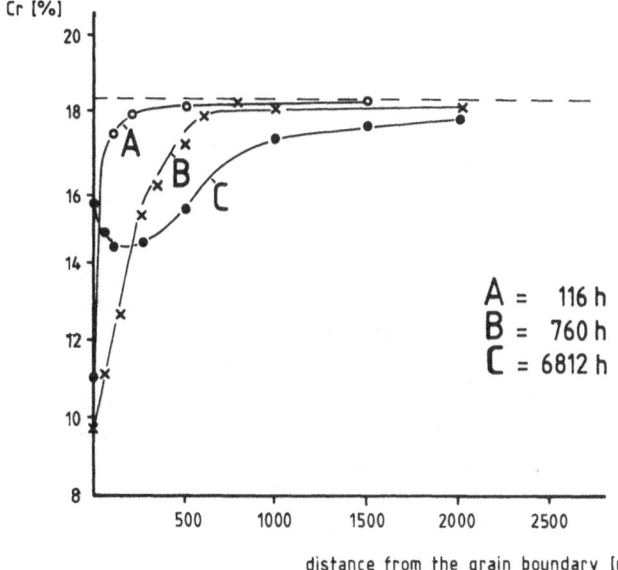

Fig. 9. Chromium profiles as measured in thinned foils

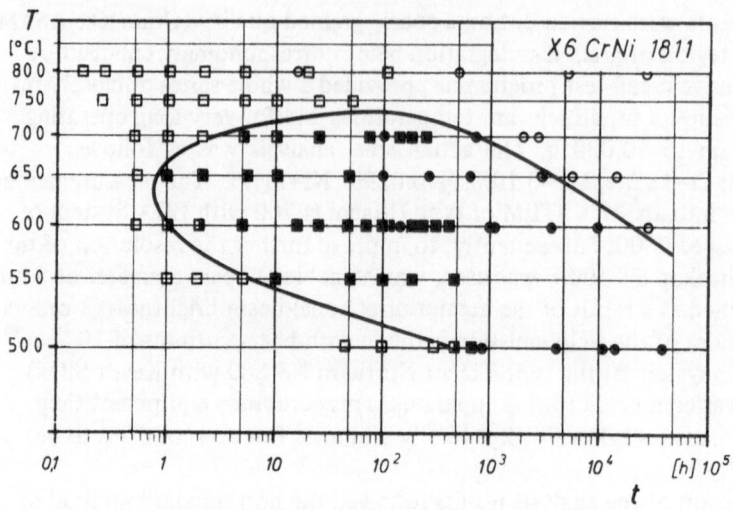

Fig. 10. *IK* region of steel X 6 Cr Ni 18 11. Position of samples as in Fig. 9

The absorption and fluorescence effects are justifiably neglected for thin samples. The *K*-factor is independent of sample thickness, and a *K*-factor of 1.12 was applied for Cr/Fe. The results are shown for an example at 650 °C in Fig. 9, and the siting of the samples can be seen in Fig. 10. As would be expected, chromium impoverishment was found above the resistance limit for sample *C*, as the structure already lies in the transsensitive region. The thickness of this zone was approx. 1 μm. A steeper gradient was measured for structure *B*, and with a resistance limit of 12% chromium this gives a zone width of 0.1 μm for the chromium impoverishment. The measurement of the maximum chromium impoverishment for structure *A* is not absolutely certain, as chromium contents above the chromium content of the steel may be registered with the incorporation of very fine carbides. The problems associated with the measurements are illustrated in Figs. 11 and 12. Fig. 11 shows a grain boundary of structure *C*, where the particle size lies above the foil thickness, permitting clear, unambiguous analysis points. The problem of structure *A* is shown in Fig. 12. Measurements are always carried out on grain boundaries parallel to the electron beam. Although very thin foils are used, randomly orientated fine carbides tend to superpose themselves in the vertical grain boundaries, leading to erroneous measurements. In this structure the most pronounced chromium impoverishment can be registered only on the grain boundaries themselves between the carbides (see Fig. 7). Chromium impoverishment for the individual temperature ranges is shown in Fig. 13 for all temperature ranges. It can be seen that chromium impoverishment increases with decreasing temperature, and this agrees well with quantitative measurements on degree of sensitization[7], where the depth of attack of *IK* increases with falling temperature.

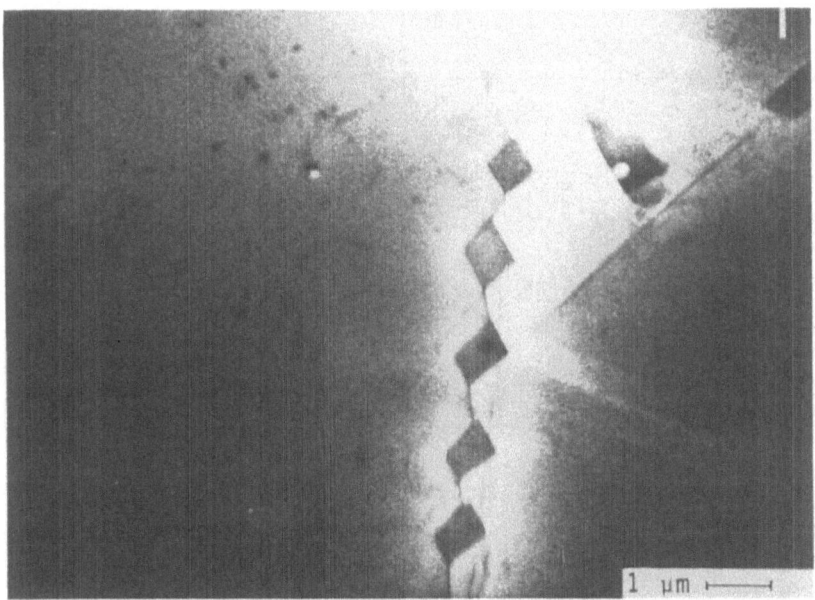

Fig. 11. STEM-image, structure C

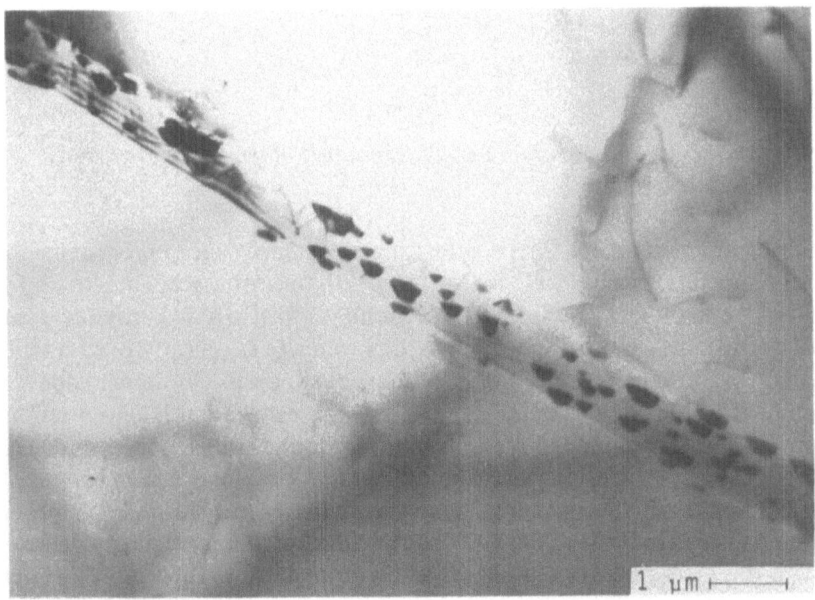

Fig. 12. TEM-image, structure A

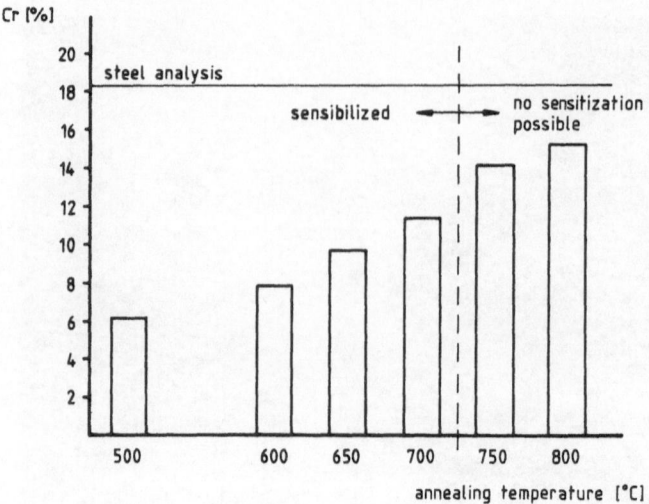

Fig. 13. Measured maximum chromium impoverishment of all structure states from Fig. 10

Acknowledgements

The authors would like to thank Dr. B. Grzemba, VAW-Forschung Bonn and Dr. G. Hinds, Institut für Eisenhüttenkunde, RWTH Aachen for the helpful cooperation.

Summary

STEM-EDX Measurements on Grain Boundary Phenomena of Sensitized Chrom-Nickel Steels

Steel specimens of X 6 Cr Ni 18 11 from an extensive creep programme in the temperature range 500–800 °C, and with operating schedules of up to 50.000 h have been investigated to determine sensitivity towards intergranular corrosion. Based on SEM examination to relate *IK* sensitivity with temperature and time and on EDX-measurements to assess chromium impoverishment STEM-EDX measurements have been carried out. Concentration profiles for chromium impoverishment were found, which corresponded to those based on diffusion calculations. Furthermore a correlation between the maximum sensitivity in various temperature ranges and chromium impoverishment was established. Start of sensitization could not be clearly detected with this measuring procedure, as in this structure state the carbide precipitates and the chromium impoverished zones are much smaller than the foils produced, thereby lying below the analysis resolution attainable in practice.

References

1. J. Sedriks, Corrosion of Stainless Steels. New York: Wiley. 1979.
2. B. Strauß, H. Schottky, and J. Hinnüber, Z. anorg. allgem. Chem. **188**, 309 (1930).
3. E. Houdremont and P. Schafmeister, Arch. Eisenhüttenwes. **7**, 187 (1933).
4. H.J. McQueen and J.J. Jones, Recovery and Recrystallization During High Temperature Deformation, Treatise on Material Science and Technology, Vol. 6. New York: Academic Press. 1975. p. 393.
5. T.M. Krebs and N. Soltys, The Inst. of Mech. Eng. **178**, 621 (1963/4).
6. R. Kautz and H. Gerlach, Arch. Eisenhüttenwes. **39**, 151 (1968).
7. M. Pohl, Elektronenmikroskopische Untersuchungen zum Ausscheidungsverhalten unstabilisierter vollaustenitischer CrNi-Stähle im Temperaturbereich von 500 bis 800 °C, Diss. RWTH Aachen 1977.
8. H. Jesper, H. Kaes, H.R. Kautz, K. Schneemann, and J.J. Schüller, Ber. Bd. VGB-Werkstofftagung **1980**, 241.
9. V.N. Krizobok and A.M. Talbot, Proc. ASTM **50**, 859 (1950).
10. M. Pohl, Prakt. Met. **10**, 361 (1979).
11. H.-J. Rocha, DEW-Technische Berichte **2**, 16 (1962).
12. A. Bäumel, Werkstoffe und Korrosion **26**, 433 (1975).
13. G. Grützner, Stahl u. Eisen **93**, 9 (1973).
14. W. Seith, Diffusion in Metallen. Berlin: 1939. p. 27.
15. H.-J. Schüller, P. Schwaab, and W. Schwenk, Arch. Eisenhüttenwes. **33**, 853 (1962).
16. W. Pepperhoff and H.-E. Bühler, Arch. Eisenhüttenwes. **33**, 711 (1962).
17. P. Schafmeister, Arch. Eisenhüttenwes. **10**, 405 (1937).
18. E. Brauns and G. Pier, Stahl u. Eisen **75**, 579 (1955).
19. J. Voeltzel and J. Plateau, C. r. acad. sci., Paris **252**, 2705 (1961).
20. W. Belteridge and A.W. Franklin, J. Inst. Met. **85**, 473 (1956).
21. T.B. Tokareva, Zaščita Metallov **10**, 245 (1974).
22. E.C. Rollason, J. Iron Steel Inst. **127**, 391 (1933).
23. J. Philibert et al., C. r. acad. sci. Paris **251**, 1289 (1960).
24. J.M. Fleetwood, J. Inst. Met. **90**, 429 (1962).
25. S. Alm and R. Kiessling, J. Inst. Met. **91**, 190 (1962).
26. G. Pomey, Trans. Amer. Inst. Min. Engrs. **218**, 310 (1960).
27. B.E. Hopkinson and K.G. Carroll, Nature (London) **184**, 1479 (1959).
28. C.W. Weaver, J. Inst. Met. **90**, 404 (1961).
29. J. Plateau, G. Henry, and C. Crussard, 3 ième Coll. de Met. Corrosion (1960). p. 185.
30. M. Pohl and W.-G. Burchard, Beitr. elektronenmikroskop. Direktabb. Oberfl. **11**, 119 (1978).
31. M. Pohl, E. Kast, W.-G. Burchard, Mikrochim. Acta [Wien], Suppl. VIII, **1979**, 295.
32. P. Rao and E. Lifshin, Proc. 12th annual Conf. MAS 1977, p. 118.
33. H. Oppolzer and U. Knauer, Mikrochim. Acta [Wien], Suppl. VIII, **1979**, 243.
34. H. Oppolzer and U. Knauer, Scanning Electron Microscopy **1979**, 111.
35. G. Cliff and G.W. Loriner, Proc. Fifth Eur. Congr. Electron. Microscopy, Institute of Physics, London 1972, p. 140.

Correspondence and reprints: Dr.-Ing. Michael Pohl, Gemeinschaftslabor für Elektronenmikroskopie RWTH Aachen, Intzestrasse 5, D-5100 Aachen, Federal Republic of Germany.

Mikrochimica Acta [Wien], Suppl. 10, 297–306 (1983)

Division of Materials Engineering and Research, Kraftwerk Union AG, Mülheim/Ruhr, Federal Republic of Germany

Metallurgical Investigations of Microstructure and Behaviour of High-Alloyed Manganese-Chromium Austenitic Steels for Generator-Rotor Retaining Rings

By

Volker Thien, Jürgen Ewald, and Wolfgang Voss

With 11 Figures

(Received January 19, 1983)

1. Introduction and Problems

The exciter coil of large generator rotors opens up into the so-called coil head at the body ends. The coil, here, must be secured by retaining rings against relative movements resulting from electro-magnetic forces but especially against centrifugal force. Because of their structural function, these retaining rings must have very high strength values – yield point up to max. 1300 N/mm² – and consist of a non-magnetic material. Several extensive reports have been made recently about the demands on generator retaining rings and the properties of materials employed as well as cases of damages[1-3] The austenitic steel X 55 MnCr(N)18K (comp. Table 1) with a nitrogen content of 0.10 %, as examined here, is mainly employed throughout the world. Nitrogen improves the strength and simultaneously stabilizes the austenite.

The retaining rings are forged, solution treated and finally brought to the required strength by cold expanding. The cold deformation, however, causes a reduction of the toughness (Fig. 1) and an increasing anisotropy of the mechanical properties (radial and axial/tangential, resp.) with the cold strengthening.

These properties are influenced to a great extent by heat treatment above 400 °C; Fig. 2 shows the reduction in the values as a function of the annealing temperature. This makes special care essential when handling the retaining rings in manufacture and assembly.

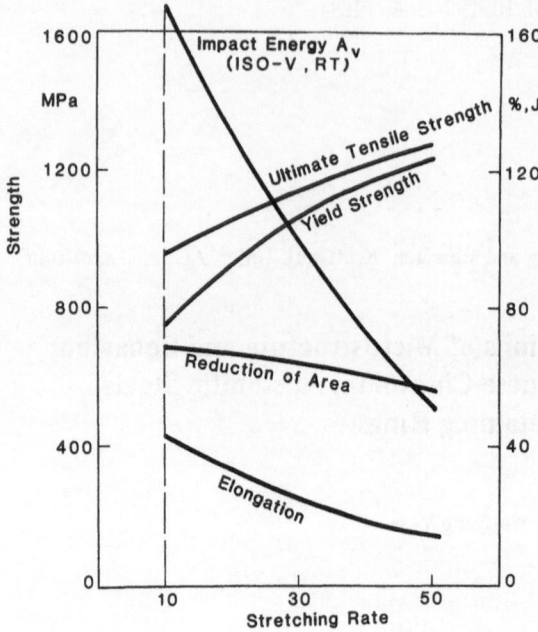

Fig. 1. Mechanical properties of X 55 MnCr(N)18K after cold expanding

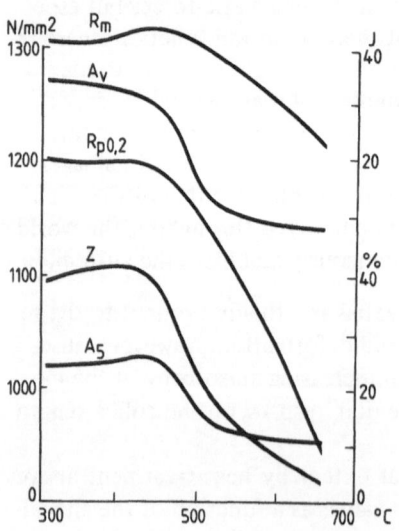

Fig. 2. Influence of the annealing temperature on X 55 MnCr(N)18K

Table 1

	% C	% Si	% Mn	% Cr	% V	% N
X 55 MnCr(N)18K	0.55	0.50	18.0	4.6	0.10	0.10

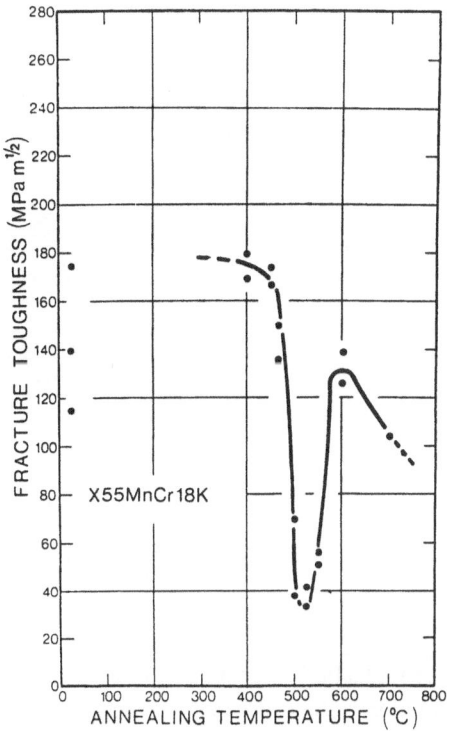

Fig. 3. Fracture toughness as function of the annealing temperature

2. Metallurgical Investigations

As already mentioned, austenitic 18% Mn, 4% Cr steels show a great reduction in toughness at annealing temperatures above 400 °C. Fig. 3 shows in a representation according to Speidel[2] that the steel X 55 MnCr(N)18K has minimum toughness at 500 °C. According to investigations made by Albrecht and Scarlin[4], this toughness minimum is connected with the formation of eutectoid constructions in the grain boundaries; this development

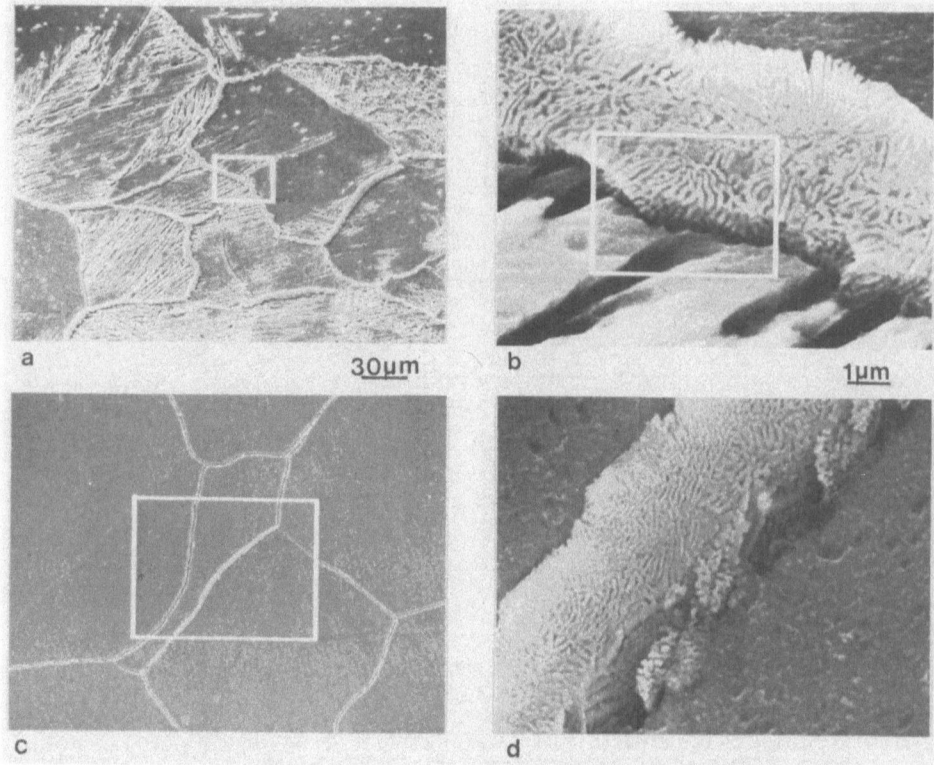

Fig. 4. X 55 MnCr(N)18K after 24 h/525 °C/air; (a+b) cold expanding + 24 h/525 °C;
(c+d) solution treatment 2 h/1120 °C/water + 24 h/525 °C

is interpreted as the breakdown of the metastable γ-matrix into α-Fe and
ϵ-phase. Based on electron diffraction experiments a hexagonal structure is
attributed to this ϵ-phase.

From our own annealing experiments over 24 h at 525 °C, an uninterrupted
grain boundary layer can be seen on a microsection examination in the SEM
(Fig. 4a); with greater magnification, Fig. 4b shows clearly the eutectoid
construction of the grain boundary areas. Eutectoid grain boundaries of this
kind are also obtained in solution treated material — 2 h/1120 °C/water —
after annealing for 24 h at 525 °C; Fig. 4c and 4d show examples of this.

Because of the very fine formation of the eutectoid (lamellar spacing
~ 0.1 μm), a crystallographic and microanalytical examination of compact
samples is not possible. Scanning transmission electron microscopy with
EDX-attachment is a suitable method for this. Fig. 5 shows the STEM
picture of an electrolytically thinned foil. The dark, less attacked, phase of

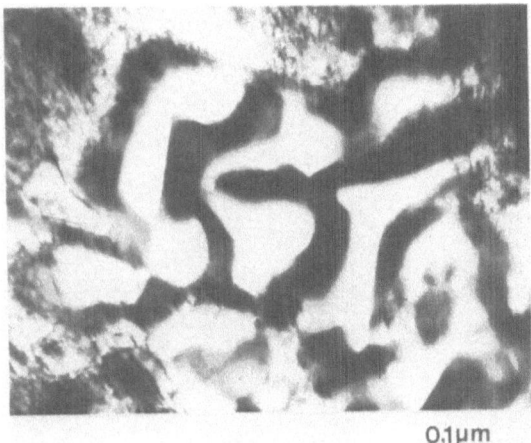

Fig. 5. STEM/EDX micro-analysis after 24 h/525 °C/air

	dark phase	light phase
% Cr	9–10	ca. 1
% Mn	ca. 40	4–5
% Fe	ca. 50	ca. 94
% Ni	0.5–1	ca. 0.5

the eutectoid shows a great increase in Cr and Mn, whereas the light, more heavily attacked, phase consists mainly of Fe. According to micro-small area diffraction, the light phase clearly is α-Fe (measured lattice constant $a =$ 2.87 Å) whereas the dark phase (approx. 10.6 Å measured) very probably is the carbide $M_{23}C_6$. However, an indication of the diffraction patterns of the dark phase in the sense of a hexagonally formed ϵ-structure was not possible. In the investigations mentioned according to[4], the fractographs for annealing temperatures of approx. 500 °C show a purely intercrystalline course, whereas with annealing temperatures < 400 °C – at least with a tangential position of the sample – an exclusively transcrystalline dimple fracture was found. Contrary to this, our own SEM-micrographs of high-strength retaining rings with normal toughness in tangential direction but with more in radial direction yielded a mixed fracture trans-/intercrystalline, as shown in Fig. 6a as an example. With greater magnification, a fine dimple structure is seen on the grain surface (Fig. 6b).

On the other hand, a mainly intercrystalline fracture course was found in various retaining rings cold formed to high strength with low toughness values with the samples in a radial position. Investigations on a retaining ring of a large generator with an output of 1560 MVA broken in the test stand show an example of this; the mechanical properties and the especially high

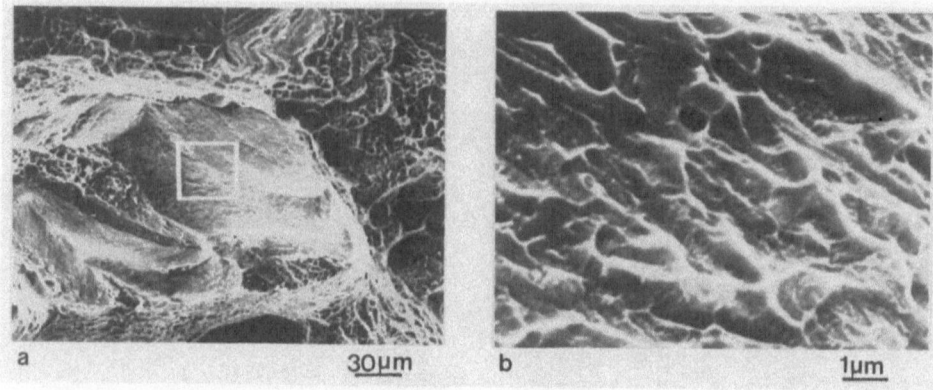

Fig. 6. Retaining ring with normal toughness; notch tensile specimen radial

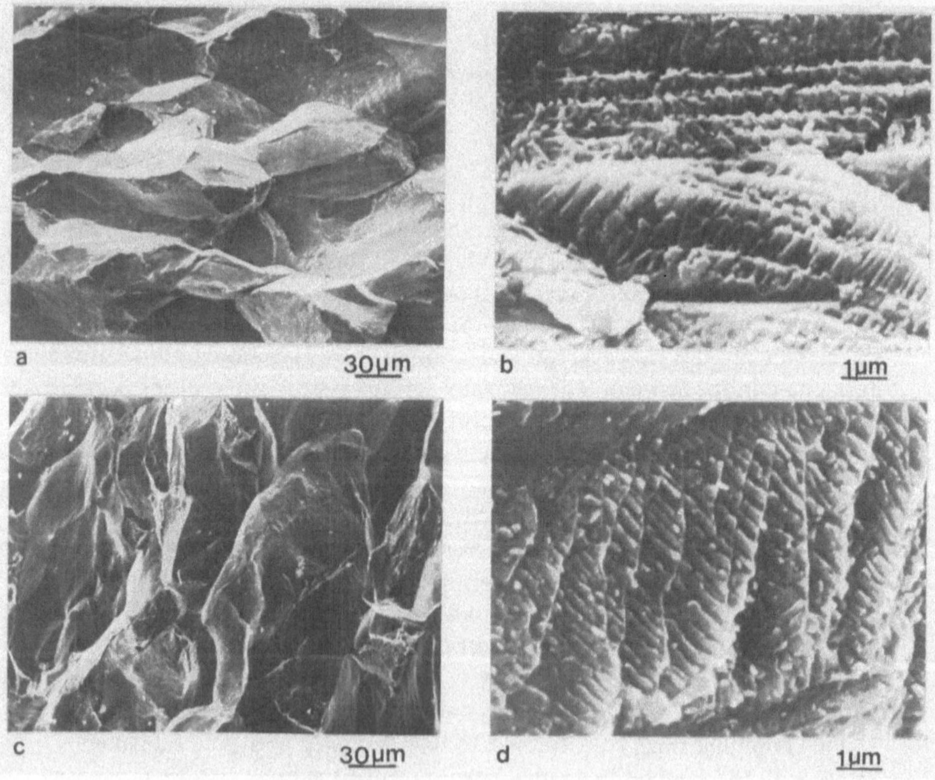

Fig. 7. Retaining ring with low toughness: (a+b) damage fracture, radial; (c+d) notch impact specimen, radial

anisotropy of this component have already been extensively reported by Ewald and coworkers[3].

Fig. 7a from the failure fracture of this retaining ring shows mainly an inter-crystalline fracture course; with greater magnification — Fig. 7b — the grain surfaces show a fine structure which, because of its geometric shape, was first termed rod-shaped.

Corresponding fracture structures show the fracture surfaces of radial notch impact samples; here, too, an intercrystalline fracture pattern (Fig. 7c) and a fine rod-shaped structure of the grain surfaces (Fig. 7d) can be recognized.

Of special importance are rupture pictures from a ring of the same melting which was available at the retaining ring manufacturer but which had not been cold expanded to the final strength (yield strength only 1038 N/mm^2) and pictures from a ring of the same melt which was subsequently cold expanded to the strength of the damaged retaining ring (1230 N/mm^2).

Figs. 8a and 8b are characteristic for both rings; they show an intercrystalline fracture (8a) and a fine, rod-shaped structure of the grain surfaces with high magnification (8b). The toughness values for both retaining rings are far below the scatter band for normal retaining rings.

These observations might be interpreted in terms of extremely fine eutectoid-type deposits forming at grain boundaries affected by cold work; in nucleating the eutectoid, the orientation seems to be determined by intersections of slip-bands with the boundaries. This possibility is supported by the STEM-investigation of thin foils from the failed retaining ring and the rings from the same melting, which is reproduced in Fig. 9. In all cases, finest deposits were found in the grain boundaries and these showed an increase in Cr ($\sim 10\%$) and Mn ($\sim 24\%$). From the lower Mn-concentration in relation to

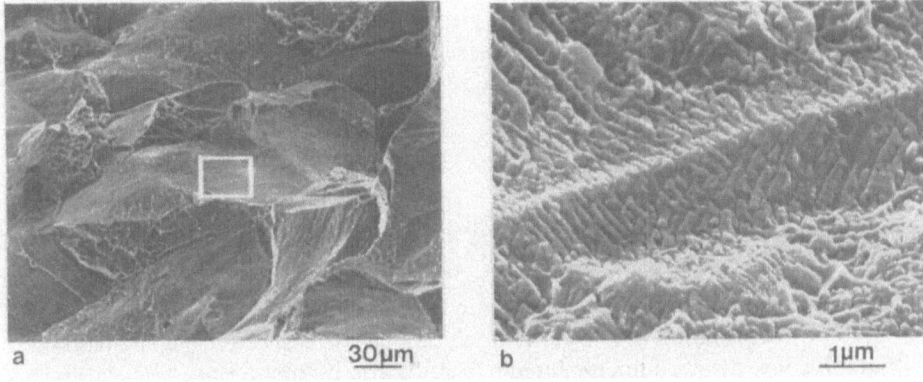

a   30μm   b   1μm

Fig. 8. Ring from same melt as damaged ring, low toughness values

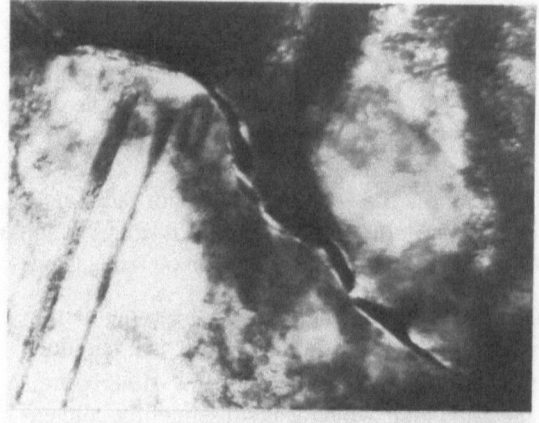

0,5 µm

Fig. 9. STEM/EDX micro-analysis, damaged ring

	matrix	deposits
% Cr	5	9–13
% Mn	18	23–25
% Fe	76	62–66
% Ni	~1	0.5–1

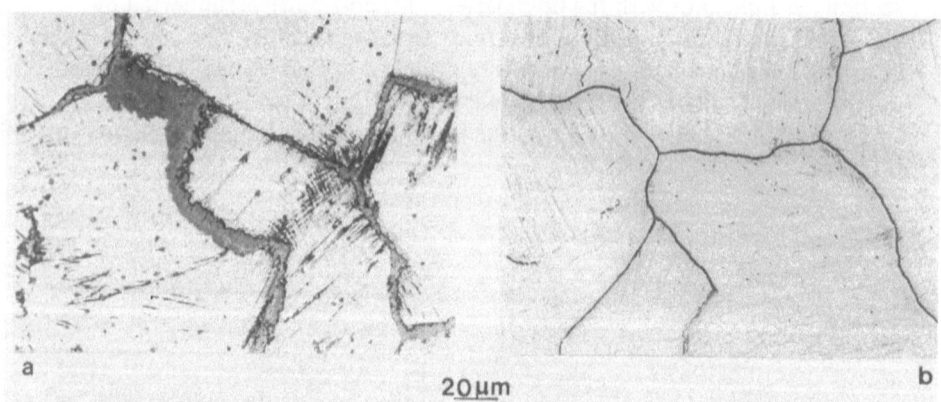

a   20 µm   b

Fig. 10. Metallographic investigation; (a) 24 h/525 °C/air; (b) damaged state

the initially mentioned eutectoid grain boundary layers (annealing approx. 500 °C) a formation at a lower temperature can be suspected. The metallographic investigation documented in Fig. 10 also points to this.

Whereas defined grain boundaries occur (10a) after an annealing at 525 °C, the material from all 3 rings from the damage melt only show attacked grain boundaries (10b) such as expected for finest deposits.

3. Discussion

Fig. 11 shows a schematic isotherm transformation diagram, according to Kroneis and Gattringer[5], by means of which an interpretation of the findings presented here is attempted. Obviously with annealing at about 500 °C, the material enters the eutectoid double-phase area α + carbide which is analogous to the perlite of the low-alloy steels.

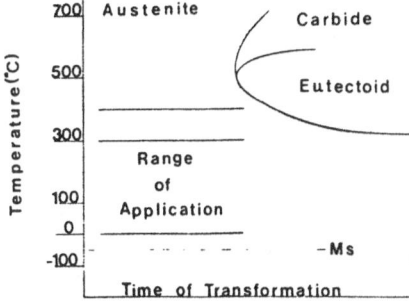

Fig. 11. Transformation diagram, schematic

At annealing temperatures > 600 °C, according to[4], only carbide deposits occur which, however, lead to less reduction in the toughness. At temperatures of 300–400 °C, a commencing eutectoid dissociation of the metastable austenite is conceivable in unfavourable cases. Unfavourable conditions, which can displace the transformation diagram to shorter times or lower temperatures, respectively, are high cold forming and insufficient austenite stability as a result of too low nitrogen content, as was the case with the investigation of the damaged material.

This led on the one hand to a specially strongly defined anisotropy of the mechanical properties – as Ewald and coworkers[3] have extensively shown – and on the other hand to a commencing eutectoid transformation – in the form of finest deposits in the grain boundary (rod-shaped structure) – during the retaining ring manufacture.

Findings from investigations similar to those performed on the damaged retaining rings, which were made on rings from the same melting, which had not been subjected to the assembly stress of the damaged retaining rings have clearly proved that the cause of the component failure is to be searched in a metallurgical peculiarity due to manufacture.

Acknowledgement

The authors thank Mr. Maußner, KWU-Erlangen, for the investigations on the scanning transmission electron microscope and Messrs. J. Albrecht and B. Scarlin, BBC-Zentrallabor, Baden, for the permission to use the diagram in Fig. 3 taken from their work[4].

Summary

Metallurgical Investigations of Microstructure and Behaviour of High-Alloyed Manganese-Chromium Austenitic Steels for Generator-Rotor Retaining Rings

The austenite steel X 55 MnCr(N)18K used for generator retaining rings shows minimum toughness, a purely intercrystalline fracture course and an eutectoid transformation at the grain boundaries after annealing at 525 °C of cold-formed and of solution annealed material.
A retaining ring broken in the test area shows a similar fracture pattern, mainly in the radial direction of stress. This retaining ring which had been cold-formed to high strength and thus was very greatly anisotropic in its mechanical properties accordingly shows specially low toughness values in the radial direction.
From SEM and STEM/EDX investigations of the damage material and of rings from the same melt, it can be assumed that as a result of the coincidence of unfavourable conditions — very high cold-stretching and low nitrogen content — commencing eutectoid dissociation of the metastable austenite occured. The extremely low toughness values of the damage material, especially with radial sample positions, can thus be deduced from the high anisotropy of the mechanical properties and an metallurgical peculiarity due to manufacture.

References

1. E. Heinrich, G. Kröncke, and G. Tacke, Stahl u. Eisen **102**, 1183 (1982).
2. M.O. Speidel, VGB-Kraftwerktechnik **62**, 424 (1982).
3. J. Ewald, H. Hübner, H. Kraemer, and A.W. Schmitz, Sammelband VGB-Konferenz: Werkstoffe u. Schweißtechnik im Kraftwerk 1980, p. 280.
4. J. Albrecht and B. Scarlin, BBC-Forschungsbericht KLR 82—118 C.
5. M. Kroneis and R. Gattringer, Stahl u. Eisen **81**, 431 (1961).

Correspondence and reprints: Dr. Volker Thien, Kraftwerk Union AG, Abt. Technik Werkstoffe/Physikalische Eigenschaften, Postfach 011420, D-4330 Mülheim/Ruhr, Federal Republic of Germany.

Mikrochimica Acta [Wien], Suppl. 10, 307–313 (1983)

Institut für Analytische Chemie der Technischen Universität Wien, Austria

Analytical Electron Microscopy for Interface Characterization – Corrosion of Concrete

By

E. Hoke, G. Eder, and M. Grasserbauer

With 4 Figures

(Received January 19, 1983)

1. Introduction

Anthropogenic acid pollutants, above all SO_2, have been deteriorating large parts of the environment, causing economical damages of approximately 1% of the national budgets[1,2]. Concerning concrete structures, decay is a consequence of local tensional overload. At cement-aggregate interfaces flexural strength is a minimum. These interfaces exhibit also high frost and corrosion sensitivity[3]. Therefore analytical studies of the interfaces are expected to yield significant information on concrete deterioriation caused by acid attack.

2. Samples

In the microanalysis of naturally corroded concrete normally two major shortcomings are encountered: one seldom knows the full history of the sample, and the representativity of any single specimen is usually low.

One way to overcome these problems is a simulation of the corrosion process on specially designed "artificial" concrete: Mineral/cement compounds with a polished interface and defined geometry enable to study the microphase transitions occuring in the corrosion process in detail and with high reproducibility.

Specimen preparation: Aggregate material: Quartzite single crystals, about 1x2 cm, cut in 110-direction to 3 mm thick platelets, polished down to 0.25 μm.

Cement: portland cement 375H, water-cement factor 0.42
curing: 4 weeks at 25 °C, 100% rel. humitidy
Exposure: SO_2-gas, reaction chamber with 82% rel. h., 22 °C
SO_2-concentrations: usually 5%
Time: up to 1 year (normally 1 week)

Microprobe analysis on cross-sections of corroded samples proved enrichments of sulphate at the interface. The penetration depth of SO_2 along the cement/quartzite interface was typically several millimeters after 1 week. The thickness of the sulphate enriched zones adjoining the aggregate ranged from a few μm to some 100 μm.

Penetration of SO_2 into the bulk cement (from the surface of the samples) proved to be less by a factor between 3 and 5, compared to the penetration depth along the interface.

Microprobe analysis furthermore showed that SO_2 did not only cause sulphate formation but also redistribution of potassium. Interfacial maxima of K present before the corrosion were removed from the surface to inner parts of the contact zone, and in the course of further acidification potassium migrated into the cement. Generally this process was accompanied by a loss of tensional strength even until rupture.

Since such changes of mechanical properties are important for concrete structures, their micro-analytical features were investigated further by high-resolution methods.

3. Analytical Electron Microscopy

Bonding between cement and aggregate is determined by the phases in the contact layer. Rupture in concrete usually follows the interfacial areas, revealing the least bonding compounds. Removing them from the underlying material by appropriate techniques makes the application of an analytical TEM or STEM possible.

Fig. 1 depicts this step of sample preparation from a normal concrete and a model sample as described before. This procedure selects the interface phases according to their tensional strength: Only those particles are removed by the replicating plastics (Technovit 3040) which exhibit a low bonding strength. One main advantage of this kind of investigation is the feasibility of high resolution analysis of defined interface positions: replicating a certain spot of the contact area means to select phases corresponding to an easily measureable distance to environmental SO_2, which has migrated to that spot during corrosion. This distance is also characterized by certain S- and K-concentrations, found by microprobe analysis.

The microphases prepared by the replica technique were analyzed by trans-

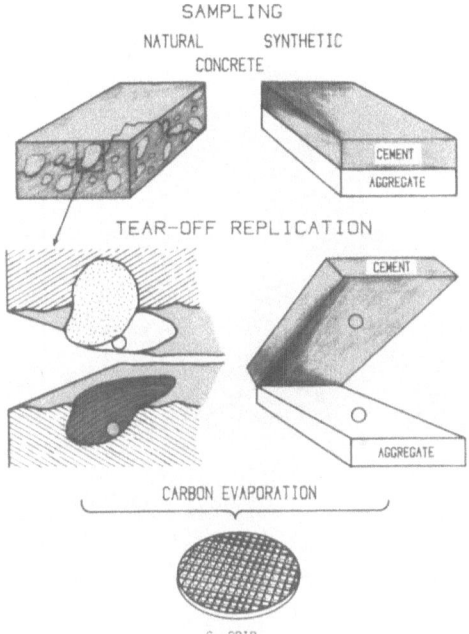

Fig. 1. Scheme of selective sampling of least bonding phases and preparation for TEM

mission electron microscopy (JEOL 100C) using an energy dispersive X-ray spectrometer and electron diffraction (SAED). Particle size was typically in the range between 0.05 to 2 μm.

In order to show the microphase transitions occuring through acid corrosion, the typical phases found in SO_2-exposed concrete were compared with those of low corroded natural samples. The latter ones are shown in Fig. 2, and Fig. 3 contains some of the typical precipitations in heavily corroded samples. This comparison shows that the major components $Ca(OH)_2$ and Ca-silicates (Fig. 2a,b) are gradually replaced by sulphate containing phases (Fig. 3a,b). Potassium rich compounds (Fig. 2d) containing often S (also present in uncorroded cement), Al and Ca exhibit a significant change of morphology accompanied by Ca-reduction. The amount of silica (Fig. 2c, 3c) is heavily increased after SO_2-exposure. In fresh concrete usually some 0.x area percent, in corroded one up to 20% of silicic acid coverage was found. These silica layers appear mainly on smooth contact areas, while $CaSO_4.2H_2O$ is largely deposited in pores. This leads to the conclusion, that the silica phases are largely responsible for the reduction of bonding between aggregate and cement.

Its amount of enhancement by corrosion may serve as an indicator for acid deterioration of concrete.

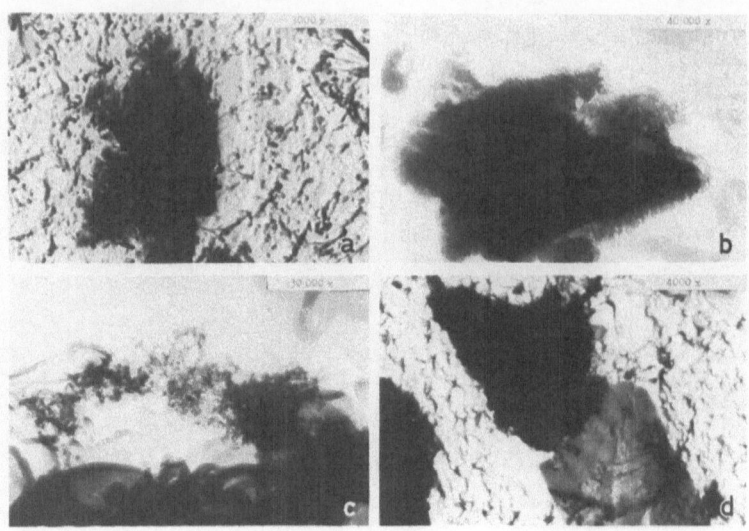

Fig. 2. TEM analysis of torn-off particles from interfaces of typical low corroded concretes

Fig. 2a. $Ca(OH)_2$ -precipitation in three months old concrete

Fig. 2b. Calcium silicate platelet, typical percentages: 60% CaO, 30% SiO_2, 5% Al_2O_3, rest K_2O, Fe_2O_3 and others

Fig. 2c. Formation of practically pure silicic acid in the aggregate-cement interface of a 15 years old, slightly corroded concrete

Fig. 2d. Example of potassium containing phase in slightly corroded natural concrete. Analysis: 54% CaO, 16% SO_3, 11% Al_2O_3, 10% K_2O, rest Fe_2O_3 and others

4. Reference Investigations by SIMS

Depth profiling with SIMS (CAMECA IMS 3f) proved the tendency of silicon to accumulate in the outmost regions of the cement interface (Fig. 4). Additionally potassium, responsible for high pH-values near the aggregate (passivating iron reinforcement), decreases by SO_2 -corrosion by a factor of more than 20 in the interface. With SIMS a relative increase of silicon intensity in the course of corrosion by a factor of about 3 is obtained. The analogous value with electron microscopy was more than 20. The reason for this difference is that the tear-off-replica technique predominatly collects loosely bound microphases (such are silica phases) while SIMS yields an average value without respect to bonding.

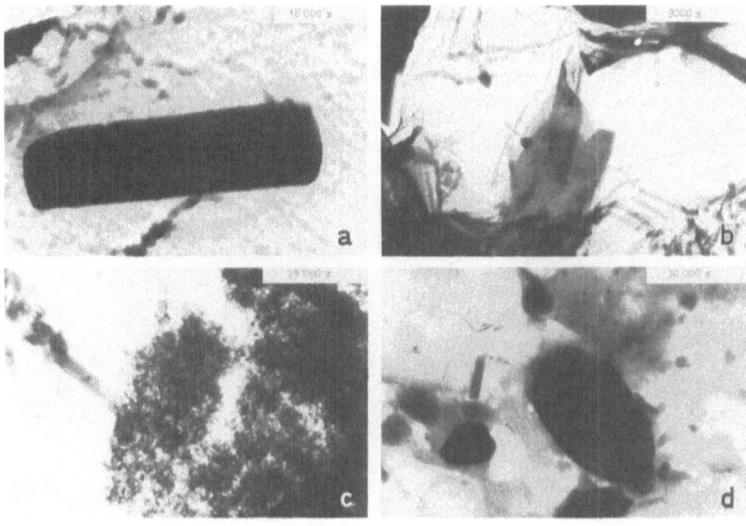

Fig. 3. TEM analysis of torn-off particles from interfaces of SO_2-corroded model concrete specimens

Fig. 3a. Piece of a $CaSO_4 \cdot 2H_2O$-crystal in an interface of a 1 week SO_2-exposed concrete sample

Fig. 3b. Sheets of SO_3-containing Ca-silicates in a 1 week SO_2-exposed specimen (composition about 45% CaO, 25% SiO_2, 20% SO_3, 10% Al_2O_3)

Fig. 3c. Layer of silicic acid in the interface of a 1 month SO_2-exposed sample

Fig. 3d. Inhomogenous phases in 1 month SO_2-exposed specimen: Typical composition: 25% K_2O, 15% SO_3, high but variable amounts of Al-, Si-, Ti- and Ca-oxides

5. Conclusion

Analytical transmission electron microscopy offers a valuable tool for interface characterization. Microphase transitions taking place during the acid corrosion of concrete at the boundary between aggregate material and cement have been studied in conjunction with a replica technique which allows the separation of loosely bound phases from the compound material. The analytical results indicate that the formation of silica layers of submicrometer dimensions at the interface plays an important role in the reduction of flexural strength of concrete under acid attack.

Acknowledgements

The authors thank Prof. Dr. H. Malissa, Dipl.-Ing. P. Wilhartitz (microprobe measurement, design of figures) and Dipl.-Ing. G. Stingeder (ion probe

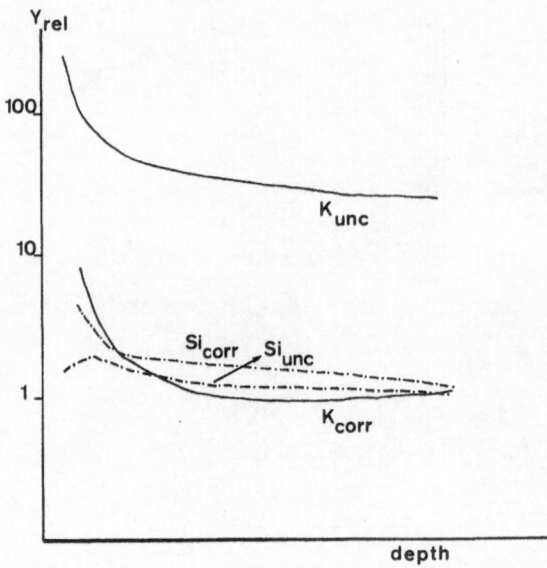

Fig. 4. Relative positive ion yields of ^{28}Si, ^{41}K, compared to the ^{42}Ca-intensity, in a three months old uncorroded (unc) and a one month SO_2-exposed (corr) model concrete sample. Depth profiling by O^--ions, 10 kV, 80 nA

measurement) for their help. Parts of this work have been sponsored by the Bundesministerium für Bauten und Technik, Wien, Wohnbauforschung Projekt F 580.

Summary

Analytical Electron Microscopy for Interface Characterization – Corrosion of Concrete

Tensional strength of concrete is determined by bonding between cement and aggregate. For the characterization of these interface layers a replica-technique was developed which allows a separation of loosely bound micro-phases. The submicrometer particles were characterized by analytical electron microscopy.

Comparing natural, low corroded concrete with samples exposed to wet SO_2, among other findings a significant increase of layers of silicic acid was found, which seems to be responsible for strength reduction. In old concrete the amount of silica might be taken as a measure of previous acid corrosion.

References

1. Kommission der Europäischen Gemeinschaften: Literaturstudie über die ökonomischen Konsequenzen der Schäden und Belästigungen, die durch Luftverschmutzung durch Schwefeldioxid sowohl bei Materialien und der Vegetation als auch bei Mensch und Tier hervorgerufen werden. Luxemburg: 1974.
2. Acidification Today and Tomorrow. Stockholm: Swedish Ministery of Agriculture. 1982.
3. Oldrich Valenta, Durability of Concrete, Proc. of the 5th Symposium on Chemistry of Cement. Tokio, 1968. Washington: NBS. 1969. p. 193.

Correspondence and reprints: Dipl.-Ing. Dr. Ernst Hoke, Institut für Analytische Chemie, Technische Universität Wien, Getreidemarkt 9, A-1060 Wien, Austria.

Mikrochimica Acta [Wien], Suppl. 10, 315–323 (1983)
© by Springer-Verlag 1983

Institut für Angewandte und Technische Physik der Technischen Universität Wien,
Austria

Electron Microscope Characterization
of Highest-Coercivity Magnetic Materials*

By

J. Fidler and P. Skalicky

With 4 Figures

(Received February 1, 1983)

Since the 1960s most of the developments of permanent magnetic materials have been done on the improvement of the magnetic properties. The search for new materials has been shifted from shape anisotropy to crystal anisotropy. New hard magnetic materials based on cobalt rare-earth intermetallic compounds exhibit a considerably higher coercive force and remanence than the traditional Alnico-compounds, hardmagnetic ferrites and Pt-Co magnets. Because of their low content of cobalt the ductile Fe-Cr-Co and Mn-Al-C magnets gain an increasing importance.
The coercive force depends on the magnetization reversal. In magnets consisting of aligned, ideal single-domain particles, the theoretical maximum coercive force is determined by the reverse magnetic field causing a rotation of the magnetic polarization $\vec{J}_s$:

$$_iH_c \leqslant 2 \cdot K / J_s$$

The magnetic anisotropy K is given by the magnetocrystalline anisotropy and the shape anisotropy of the particle, respectively. The coercivity of hard magnetic materials is determined either by the nucleation of reverse magnetic domains at magnetic fields which are lower than the theoretical maximum value, or by the strong pinning of domain walls at crystal lattice defects and precipitates.

---

* This work was supported by the Austrian Forschungsförderungsfonds (project No. 4640).

From all these reasons it is evident that for the preparation of highest quality permanent magnets starting materials with a high saturation polarization, a high magnetic anisotropy and a high Curie temperature are necessary. In the search for new permanent magnets many materials have especially been investigated with regard to their magnetocrystalline anisotropy for more than 15 years.

We have used transmission electron microscopy (TEM) and X-ray microanalysis to characterize crystal defects and precipitates and to study the magnetic domain structure of highest-coercivity magnets, such as cobalt-rare earth, Mn-Al-C and Fe-Cr-Co magnets. The magnetic domains are imaged by defocusing the objective lense (Fresnel method), where the domain walls appear as bright and dark lines, or by displacing the objective aperture (Foucault method), where the domains appear as bright and dark bands. Our investigations have been carried out on a 200 kV microscope, which is equipped with special objective pole pieces for magnetic domain observations.

Microstructure and Magnetic Domains of Highest-Coercivity Magnets

Cobalt-Rare Earth Permanent Magnets (REPM)

It has been found that the compounds of rare earth (RE), preferably samarium, with transition metals (TM), preferably cobalt, combine both characteristic properties, a high saturation polarization and a high magnetocrystalline anisotropy[1]. Based on the intermetallic compounds corresponding to the chemical formulae $TM_5 RE$ and $TM_{17} RE_2$, respectively, hard magnetic materials are prepared with coercive forces (> 1600 kA/m) and energy density products (> 240 kJ/m^3), which are not obtained with any other magnet.

The REPM can be divided into five types depending on whether the magnet has a single-phase or a two-phase microstructure. The ideal microstructure of the "single phase" magnets consists of aligned single-domain particles with a $TM_5 RE$- or $TM_{17} RE_2$-crystal structure. Two types of precipitation hardened magnets can be distinguished: the one type contains 17:2-precipitates in a 5:1-matrix, the other type has 5:1-precipitates in a 17:2-matrix. Besides these magnets there are the bonded magnets, in which the single-domain particles are embedded in a non magnetic phase.

Table 1 summarizes the composition and the magnetic data (saturation polarization, coercive force and energy density product) of the REPM, which have been investigated in our laboratory. Besides single dislocations and inclusions which are dispersed non uniformly in certain regions of the materials only, the investigated single-phase REPM exhibit a low defect density within the grains (Fig. 1). No segregation of precipitates at grain boundaries

Table 1. Chemical Compositions and Magnetic Parameters of the Samples

Composition	J_s T	$_jH_s$ kA/m	$(BH)_{max}$ kJ/m^3	Producer	Micro- structure
Co$_5$Sm	0.98 0.92	1430 1200	200 160	Vacuumschmelze TEW	single phase
Co$_5$(CeMM,Sm)	0.83	960	120	Vacuumschmelze	
Co$_5$(CeMM,Sm,TM)	0.85	1350	130	”	”
(CoFeCuZr)$_{6.9}$Sm	0.48	605	40	Vacuumschmelze	multi
(CoFeCuZr)$_{7.4}$Sm	1.10	517	231	TDK, Shin-Etsu	phase
(CoFeCuZrTi)$_{7.1}$Sm	1.07	400	195	Krupp Widia	”
(CoFeCuHf)$_{6.9}$Sm	1.02	570	205	Krupp Widia	”

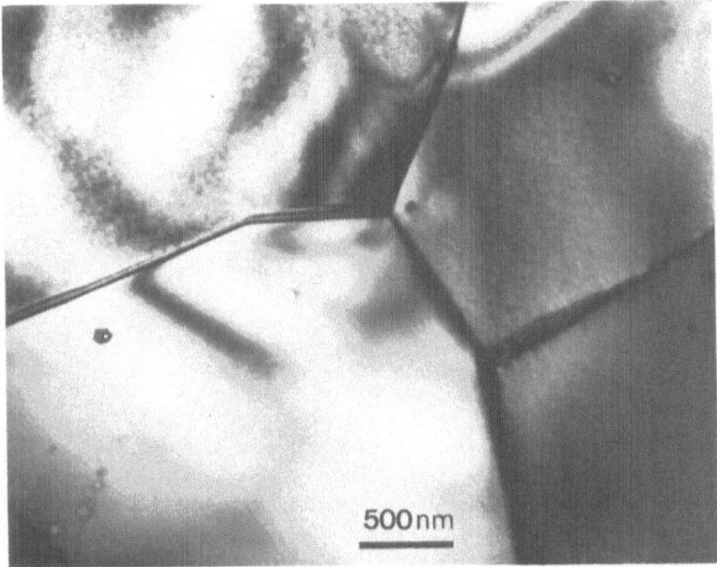

Fig. 1. Transmission electron micrograph showing several grains and precipitates of a "single phase" Co$_5$Sm magnet

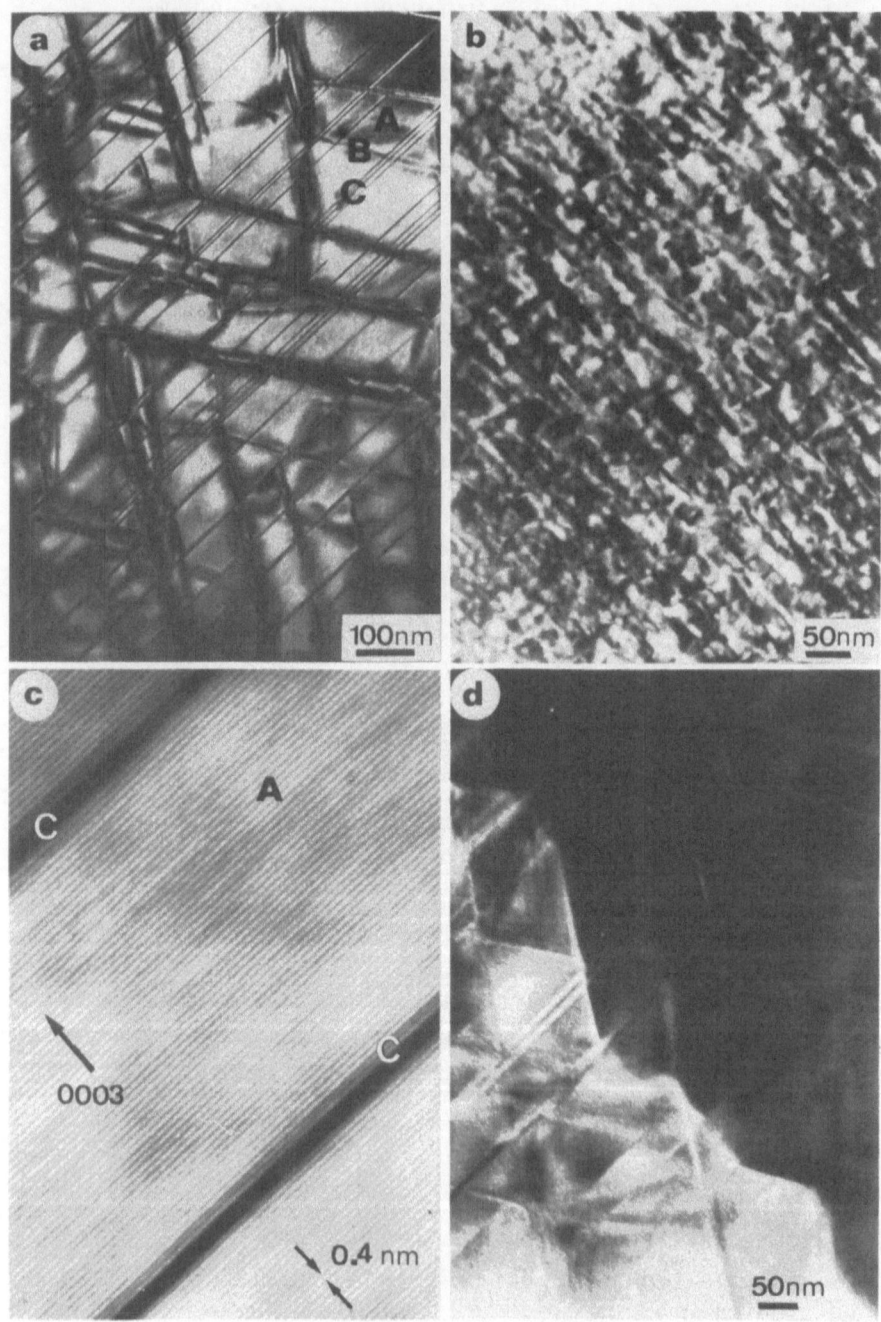

Fig. 2. Microstructure and magnetic domain structure of a $(Co,Fe,Cu,Zr)_{7.5}Sm$ magnet with a heat treatment at (a) 900 °C and (b) 700 °C. (c) is a lattice fringe image and (d) is a Foucault image

is observed. Inclusions were characterized by electron diffraction and dark field analysis as coherent precipitates of the rhombohedral $Co_{17}RE_2$- and of the Co_7RE_2-type. Incoherent precipitates were analyzed as RE_2O_3-inclusions. In some cases the heterogeneous nucleation of precipitates at grain boundaries and dislocations was found[2]. In "single-phase" magnets the magnetization reversal occurs by nucleation and growth of reversed domains. Grain boundaries are found to act as strong pinning centres for domain walls[2].

Copper containing cobalt rare earth magnets with a composition of $(Co,Cu,TM)_{6-8}Sm$ exhibit domain wall pinning controlled magnetization curves[3]. Transmission electron microscope studies of such magnets have shown a fine cell morphology with about 55 nm in diameter at the magnetic optimum state[4,5]. The electron micrograph of Fig. 2a shows the rhombic cells of the type $(Co,Fe)_{17}Sm_2$, (region A), which are separated by a $(Co,Cu)_5Sm$-cell boundary phase, (region B), of a slightly overaged magnet (900 °C). The continuous precipitation structure is determined by the direction of zero deformation strains due to the lattice misfit between the different phases[6]. The microstructure of an underaged magnet (700 °C) with a small cell size is shown in Fig. 2b. The cell interiors (region A) consist of at least two phases. In all of the precipitation hardened magnets thin plates were found perpendicular to the hexagonal $\vec{c}$-axis (region C). Two structures for the 17:2-compounds are distinguished, the hexagonal $Ni_{17}Th_2$-structure (2H) and the rhombohedral $Zn_{17}Th_2$-structure (1R).

From the lattice fringe image of Fig. 2c it is evident that the platelet phase (region C) consists of a (2H)-layer, which is formed from the (1R)-phase by a disorder of the stacking sequence. Combined with a hexagonal $(Co,Fe)_{17}Sm_2$-layer the twinning of the rhombohedral cell interior may occur.

Our electron microscope investigations have shown that in all of the precipitation hardened 17:2-magnets the magnetic domain walls are primarily pinned at the 5:1-cell boundary phase of the rhombic, cellular precipitation structure (Fig. 2d). Occasionally a domain wall pinning at the thin plates within the 17:2-cells perpendicular to the $\vec{c}$-axis is observed[7].

Mn-Al-C Magnets

It has been shown for more than 20 years that a ferromagnetic, metastable phase (named τ) exists in the Mn-Al binary system around the composition of 70 wt.% Mn[8]. The τ-phase has a face centred tetragonal crystal structure and is formed as a transition phase only when the hexagonal, high temperature ϵ-phase is cooled at an appropriate rate. The transformation occurs in a two-stage process; an ordering is followed by a martensitic shear transformation. Additions of carbon have been found to stabilize the τ-phase up to about 700 °C, allowing the production of a partly aligned product by high-

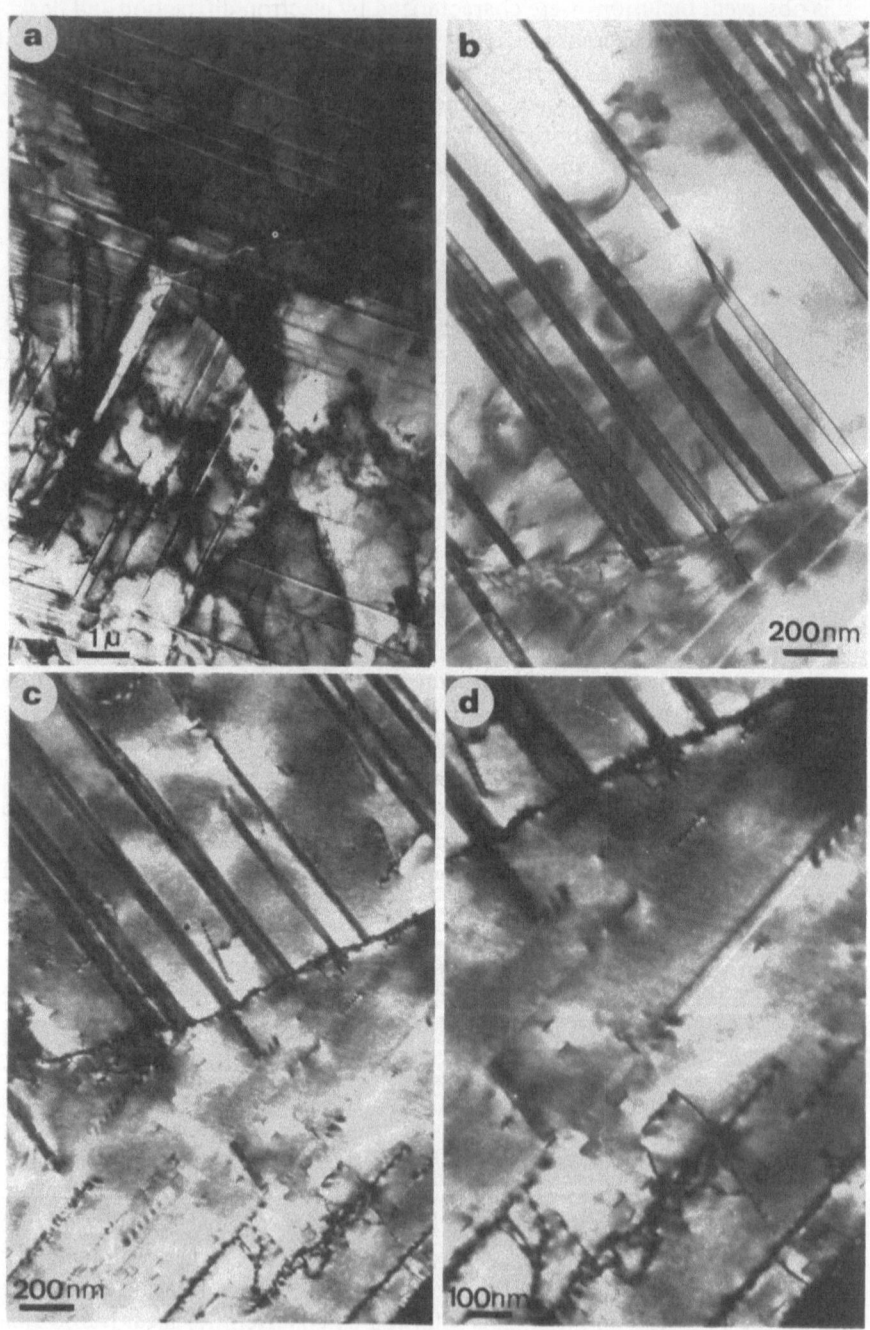

Fig. 3. Electron micrographs showing large grains in a MnAl-magnet containing twins, stacking faults and dislocations. Domain wall pinning at twins and dislocations is shown in the Fresnel images (a) and (d)

temperature extrusion. Mn-Al-C magnets with a grain size of about 1 μm exhibit a higher remanence (0.6 T) and a higher energy density product ($\sim$ 50 kJ/m^3) than the traditional hard magnetic ferrites and Alnico-magnets.

Transmission electron microscopy of Mn-Al-C permanent magnets reveals an essentially single-phase microstructure with no likely wall-pinning features in large quantities other than the grain boundaries[9]. The coercivity of such magnets is controlled by reverse-domain nucleation and is determined by the magnetocrystalline anisotropy. Grain boundaries act as pinning centres for domain walls and isolate the grains from their neighbours. Magnets with a large grain size contain antiphase boundaries (APB) and microtwins[10]. The planar defects act as nucleating and pinning defects for magnetic domains (Fig. 3a and d). Our TEM-investigations show that the microtwins are combined with stacking faults (Fig. 3b). The purpose of our investigations is to determine the influence of microtwins, APB and stacking faults on the coercivity of this type of magnetic material.

Fe-Cr-Co Magnets

Ductile permanent magnet alloys based on the Fe-Cr-Co system have received considerable attention in recent years. The magnet material is quite similar to Alnico alloys in many aspects, because they belong to a kind of so-called "single domain particles" magnet in which the dispersed ferromagnetic fine particles are produced through a certain metallurgical process based on the spinodal decomposition[11]. The magnets consist of 10–30 wt.% Co, 30–50 wt.% Cr and balance Fe. The alloy decomposes spinodally into the Fe-rich α_1-phase and the Cr-rich α_2-phase within the miscibility gap[12]. The microstructures of magnets consisting of 47Fe-30Cr-23Co and 55Fe-28Cr-15Co-1Al-1Nb are shown in the micrographs of Fig. 4. Several different mechanisms may contribute simultaneously to the coercivity of these alloys. It was shown that the particle morphology (shape anisotropy) is the dominant factor in determining the coercivity of the alloy rather than the particle size[13]. Besides this a domain wall-particle interaction (Fig. 4c) also contributes to the coercivity.

The magnet characteristics have been gradually improved by adding various minority elements such as Mo,Si,Nb,Al,V,Ti and Cu. The electron microscope investigation of such magnets reveal a two-phase microstructure with randomly distributed inclusions (Fig. 4b). At the moment an energy density product of 78 kJ/m^3 and a remanence of 1.5 T may be obtained.

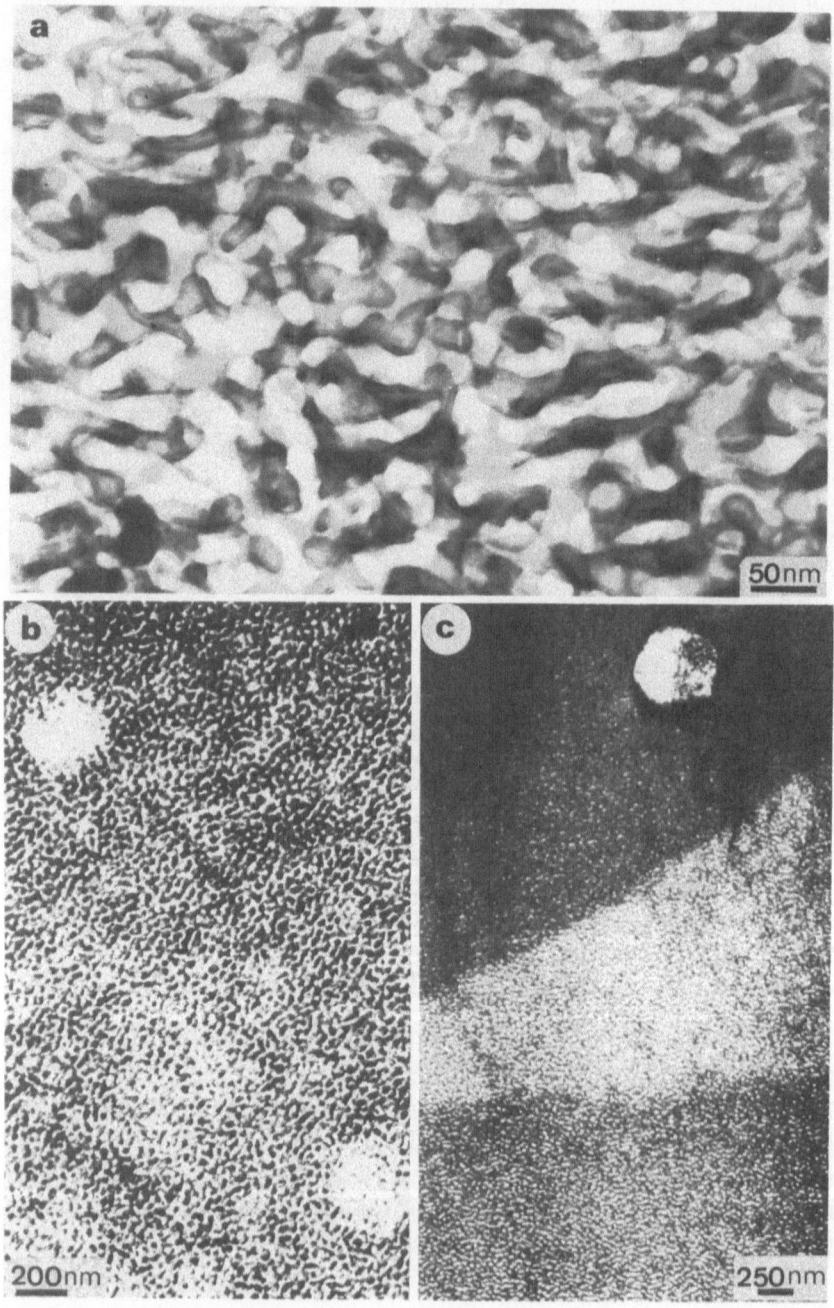

Fig. 4. Multiphase microstructure of 47Fe-30Cr-23Co (a) and 55Fe-28Cr-15Co-1Al-1Nb magnets (b) and (c). Magnetic domains are visible in the Foucault image (c)

Summary

Electron Microscope Characterization of Highest-Coercivity Magnetic Materials

Investigations of the microstructure and the magnetic domain structure are necessary for the understanding of the coercivity of magnetic materials. Transmission electron microscopy has been used to study the new-highest-coercivity magnetic materials based on cobalt-rare earth intermetallic compounds, Fe-Cr-Co and Mn-Al-C alloys. In single-phase cobalt-rare earth and Mn-Al-C magnets the coercivity is controlled by nucleation of reverse domains at low anisotropy defects and by domain wall pinning at grain boundaries. In precipitation-hardened cobalt rare earths the coercivity is controlled by the pinning of domain walls by a continuous, cellular precipitation structure of a second ferromagnetic phase. The shape anisotropy of the FeCo-rich phase is the dominant factor in determining the coercivity of Fe-Cr-Co magnets.

References

1. K. Strnat, G. Hoffer, J. Olson, and W. Ostertag, J. Appl. Phys. 38, 1001 (1967).
2. J. Fidler, Phil. Mag. B 46, 565 (1982).
3. A. Menth, H. Nagel, and R.S. Perkins, Ann. Rev. Mater. Sci. 8, 21 (1978).
4. J. Fidler and P. Skalicky, J. Magn. Magn. Mat. 27, 127 (1982).
5. R.K. Mishra, G. Thomas, T. Yoneyama, A. Fukuno, and T. Ojima, J. Appl. Phys. 52, 2517 (1981).
6. J. Fidler, J. Magn. Magn. Mat. 30, 58 (1982).
7. J. Fidler and P. Skalicky, Proc. 3rd Int. Symp. on Magn. Anisotropy and Coercivity in Rare Earth-Transition Metal Alloys, J. Fidler (ed.), 1982, Baden/Austria, p. 585.
8. H. Kano, J. Phys. Soc. Japan, 13, 1444 (1958).
9. E.L. Houseman and J.P. Jakubovics, Proc. 6th Int. Conf. on High Voltage Electron Microscopy, Antwerp, Belgium, 1980, Vol. 4, p. 380.
10. A.J. Lapworth and J.P. Jakubovics, Proc. 3rd European Conf. on Hard Magn. Mat., Amsterdam, 1974, p. 174.
11. H. Kaneko, M. Homma, and K. Nakamura, AIP Conf. Proc. 5, 1088 (1972).
12. R. Tahara, Y. Nakamura, M. Inagaki, and Y. Iwama, Phys. Stat. Sol. A41, 451 (1977).
13. S. Jin, D. Brasen, and S. Mahajan, J. Appl. Phys. 53, 4300 (1982).

Correspondence and reprints: Univ.-Doz. Dr. J. Fidler, Universitätsassistent, Institut für Angewandte und Technische Physik, Technische Universität Wien, Karlsplatz 13, A-1040 Wien, Austria.

Mikrochimica Acta [Wien], Suppl. 10, 325–335 (1983)

Abteilung für Chemie, Ruhr-Universität Bochum, Federal Republic of Germany

Multi-Element Preconcentration from Technical Alloys

By

Ewald Jackwerth and Horst Mittelstädt

With 5 Figures

(Received January 19, 1983)

In recent years the multi-element analysis of technical alloys has become an important field of interest in routine analysis. This is valid not only for some of the new developed materials with very special qualities, but also for many well proved alloys with a wide spread application. As main reason for this development in analysis one may consider the increasing demand in the purity of materials as a consequence of the often found correlation between mechanical, electrical etc. properties of metals and their content of trace elements. Additionally the increasing use of recycled metals leads to an increase of the number of elements present in considerable amounts in the alloying materials. So, even the search for elements generally considered exotic suddenly may become an important task in routine analytical work.

Though simultaneously working optical emission and X-ray fluorescence spectroscopy are the most favoured methods in routine metal laboratories, efficient "chemical" techniques of trace preconcentration and determination are also indispensable in this field: Such methods are of great use as reference procedures for the periodical accuracy control of the instrumental analyses. As those procedures are easily calibrated, they are helpful for the certification of standard reference materials. Finally, easy to handle procedures of that kind may be very necessary for smaller laboratories which do not have a costly instrumental equipment at their disposal.

Zinc Casting Alloys and Their Specification

Fine zinc casting alloys have a wide spread application in manufacturing very different things as well of a mass production (i.e. toys, souvenirs, household articles)[1] as of construction parts difficult to produce with highest accuracy requirements[2]. Mostly used for these purposes are zinc base alloys with aluminium, magnesium and (partially) copper as alloying components in different composition, wellknown as "ZAMAK"-alloys[3,4] For these types of alloys it is known for a long time that even very small amounts of certain foreign elements may cause a severe tendency to inter-crystalline corrosion which effects the mechanical properties of the material and, finally, leads to a total destruction of the castings[5-7]. Particularly traces of Bi, Cd, (Cu), Fe, In, Ni, Pb, Sn, and Tl are said to react in this way; their negative influence is intensified under extreme climatic condi-tions such as high humidity and temperature[5]. So, most of these elements with their highest permissible content are listed in the standard specifi-cations: DIN 1743, BS 1004, NF A55-102, UNI 3717, ASTM B240, ISO R301.

To prevent disturbing impurities, in the processing of the alloys, fine zinc metal of 99.99 per cent purity and appropriate pure alloying components are applied. The trace content of the semifinished goods and of the finished products, in general, is controlled carefully.

Trace Preconcentration from ZAMAK-Alloys

In view of trace analysis, alloys are complex analytical systems with multi-component matrices. For a chemical trace preconcentration as a step in the purity control, this implies the search for separation conditions which allow the simultaneous enrichment of numerous elements without grave disturb-ances by the main components of the sample. For most of the unalloyed metals this problem often can be solved by rather simple separation systems if the traces to be determined are not in extremely low concentration ranges. Several alloying components with different chemical behaviour, however, render the separation process considerably difficult. In this paper typical difficulties of this kind will be demonstrated on the zinc alloy ZnAl 4 (Z 400) which contains 0.02—0.06 per cent Mg and about 4 per cent Al.

Complexation of the Matrix

For the preconcentration of some elements from pure zinc it is possible, after having dissolved the sample, to use suitable chelate forming group reagents which favourably react with the desired traces, in combination with extraction or collector precipitation methods[8-11]. Reaction medium mostly is a nearly neutral or a strong ammoniacal sample solution in which the matrix zinc remains unchelated as the metal cation or in form of ammine complex species. In presence of a few per cent of aluminium, as it is found in the alloy ZnAl 4, under such pH conditions, a voluminous precipitate of aluminium hydroxide is formed which disturbs the analysis in many aspects. Experiments with the aim to keep the Al and Mg in solution by use of complexing agents such as EDTE or tartaric acid failed because of the simultaneous complexation and loss of some of the interesting trace elements. Furthermore, the necessary high amount of masking reagents added caused a strong increase of the blank values for some elements. Therefore we have tried to preconcentrate the traces in relative strong acid solution and, without the addition of any foreign masking agent, keeping all the alloying components in solution.

Selection of the Group Reagent

Among the large number of chelate forming group reagents which are normally used in multi-element trace analysis, the different substituted dithiocarbamidates

$$R \diagdown N-\overset{\displaystyle S}{\overset{\|}{C}}-S^- \quad Z^+$$
$$R \diagup$$

probably have the widest application[12, 13]. Essential for the selectivity of complex formation are the two sulfur atoms containing functional group within the reagent molecule which react with more than 30 metal cations to stable 4-ring chelates[14, 15]. Among the alloying components of ZnAl4 zinc is one of these reacting elements, but not aluminium and magnesium[16, 17]. Within the group of chelate forming cations we also find all the trace elements which are normally of interest for the purity control of ZnAl4. To a first approximation the reacting palette of elements is identical to those which form slightly soluble metal sulfides; like those the metal dithiocarbamidates are practically unsoluble in water. Additionally they also can be extracted by different organic solvents, they are volatile without decomposition, they are adsorbable onto activated carbon, they can be flotated at suitable foams etc. These different useful analytical properties of the chelates comprising a large number of reacting elements are the reason for the frequent preference of dithiocarbamidates as reagents in multi-element preconcentration techniques.

Perhaps the most important disadvantage of the dithiocarbamidates as analytical reagents is their low molecular stability in acidic solutions. Preferably the mainly used sodium diethyldithiocarbamidate (Na-DDTC) is known to decompose even at pH 3–4 in a few minutes. The stability of the reagents and their chelates, however, can be increased by suitable substitution of the molecule at the nitrogen atom, and by use of "voluminous" organic cations Z^+ as the counterion of the anionic chelating molecule. So, the tetramethyleneammoniumsalt of tetramethylenedithiocarbamidate (TMA-TMDTC) is rather stable at pH 3–4[18]; using the hexamethyleneammoniumsalt of hexamethylenedithiocarbamidate (HMA-HMDTC), it even is possible to work in solutions with pH < 0[19]. Probably kinetic influences of the substituent are responsible for the increasing molecular stability of the compounds. The selectivity of different dithiocarbamidates is not or, at best, scarcely influenced by these substituents which are bound at the N-atom[16]. Altogether, a sufficient molecular stability of HMA-HMDTC and its chelates, as well as the selectivity of the compound which encompasses all the desired trace elements, are the essential arguments for our selection of this preconcentration reagent.

Selection of the Separation Method

Considering the aspect that, together with the desired traces, the matrix zinc reacts with HMA-HMDTC, two separation techniques seemed to be suitable for us as an efficient preconcentration procedure: Limitation of the added reagent to a few milligrams followed by a) extraction of the traces[20] together with small amounts of the zinc matrix, or b) co-precipitation of the traces with zinc dithiocarbamidate as a collector.

Because of its special advantages, we have preferred to apply the collector precipitation. Compared with the extraction of the traces, using similar complex forming conditions, the recovery of trace elements tends to be larger in collector reactions. This often is true also for trace concentrations far below the solubility products of their co-precipitated compounds. Separating traces from relativ large sample volumes, the collector precipitation, in general, is simpler and faster to handle and implies in many cases lower blank values than it is found in comparable extraction procedures. Finally, as the result of a collector precipitation, a solid trace concentrate in form of a thin layer is obtained which directly can be used as a suitable target in neutron activation analysis, X-ray fluorescence spectroscopy and other methods. Otherwise the precipitate can easily be dissolved and subsequently be analysed by one of the different methods of atomic spectroscopy applicable to trace solutions, such as AAS, ICP-spectroscopy etc.

Selection of the Determination Method

In this paper only the application of flame atomic absorption spectroscopy is described. For the analysis of the preconcentrated traces the precipitate, together with the membrane filter, was dissolved in acid and filled up to a small sample volume. The traces were determined sequentially by the injection technique of flame AAS[21, 22] using aliquots of 50 μl for each element.

Optimization of the Working Parameters

Because of the different behaviour of the matrix elements in the investigated sample material extensive experimental work was required to find the analytical conditions in which the experimental variables such as pH value, the necessary amount of reagent, the reaction time and temperature etc. simultaneously are optimized. Accordingly methods for a multi-element preconcentration from samples with multi-element matrices, in general, require a careful observation of the procedural details.

Permissible pH Range

As mentioned before, among the reacting trace elements, also the matrix zinc precipitates with HMA-HMDTC. The complex stability of the zinc chelate is much lower than that of the interesting traces; zinc, however, compared with the traces, is present in the sample with an excess of about $10^5 - 10^6$. So, according to the law of mass action, it consumes the very

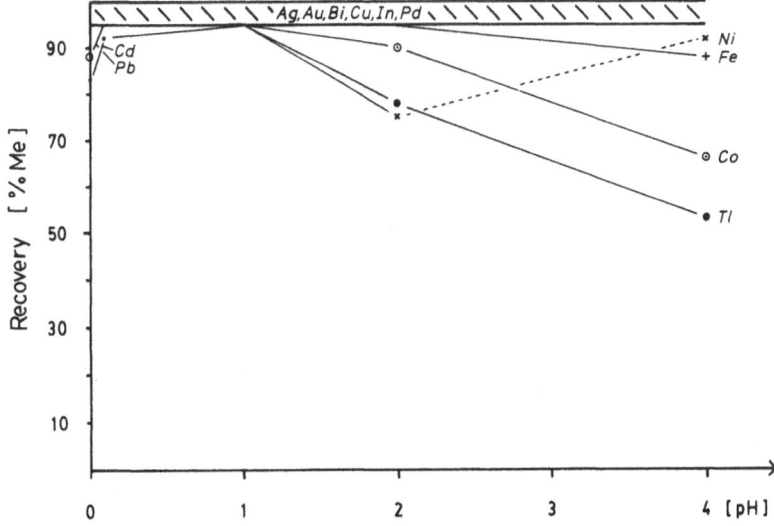

Fig. 1. Dependence of the trace recoveries on the pH value of the sample solution

predominant part of the added reagent. Therefore the question, whether the traces will be preconcentrated or not is mostly dependent on this excess ratio and on the different complex stabilities of all the chelates competing for the reagent. The complex stability, for its part, depends on the pH value of the solution: The more acid the solution the lower are the stabilities of the metal chelates. So, one has to adjust the pH to a range at which the solution is acid enough to allow the zinc chelate in a necessary amount to precipitate as the collector while, on the other side, the complex formation of the traces is not yet prevented by the large excess of zinc. But, because of the molecular decomposition of the reagent, the acidity of the solution should not be too high.

Fig. 1 shows that the optimal pH range is between pH 0.1 and 1. At lower pH values recoveries decrease because of the increasing instability of the chelates and because of the molecular decomposition of the reagent. At higher pH values the excess of the zinc matrix supersedes some traces in their complex forming equilibrium.

Necessary Amount of the Precipitation Reagent

A further optimization process concerns the amount of zinc dithiocarbamidate to be precipitated as the collector.

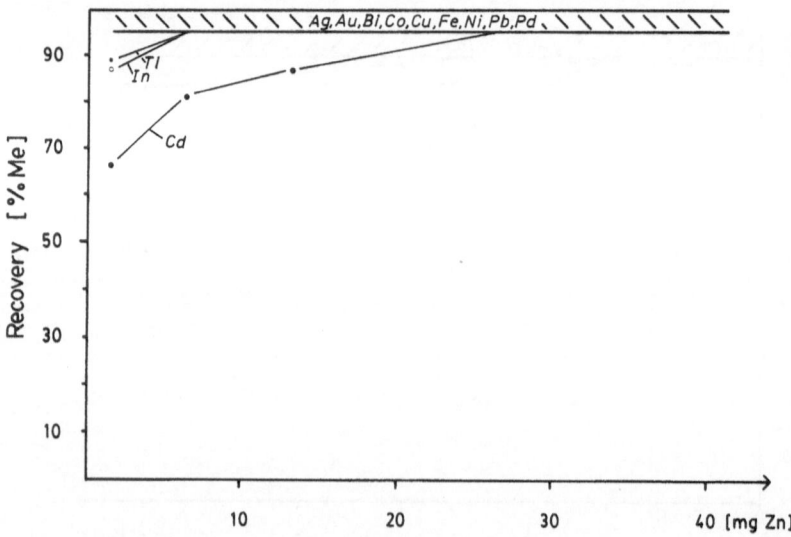

Fig. 2. Dependence of the trace recoveries on the amount of the precipitated collector (reaction temperature: 20 °C; coagulation time: 30 min)

Fig. 2 demonstrates that a minimum quantity of about 20mg of zinc must be precipitated to obtain quantitative recoveries for the desired traces. The actually precipitated collector amount should be kept as close as possible to this quantity in order to obtain a trace concentrate of a small volume with only a small rest of the zinc matrix. This is important for the undisturbed and sensitive determination of the preconcentrated trace elements.

Coagulation of the Collector Precipitate

Some important properties of the collector precipitate such as co-precipitation rate and the filtering behaviour are strongly influenced by the temperature during the precipitation process and by the time which is conceded for the coagulation of the precipitate.

On addition of HMA-HMDTC the collector first precipitates very dispersed. So, during the filtration period, the pores of the membrane filter clog and a further filtration is prevented.

Fig. 3 shows that at 30 °C the precipitate needs about 10 minutes to coagulate and to become filterable; at 20 °C one has to wait at least 30 minutes before starting the filtration process etc. During the time between precipitation and filtration of the collector, the initially quantitative recoveries of the traces decrease in the sequence of the stability of their respective com-

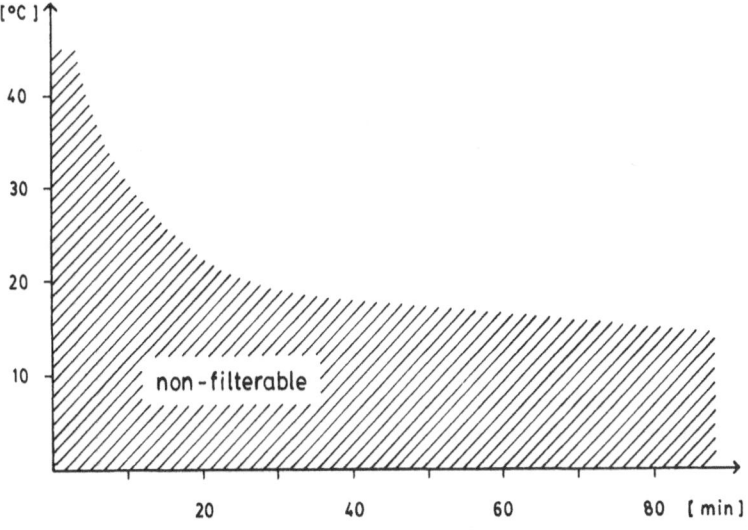

Fig. 3. Dependence of the filtration capability of the collector on the reaction temperature and the coagulation time

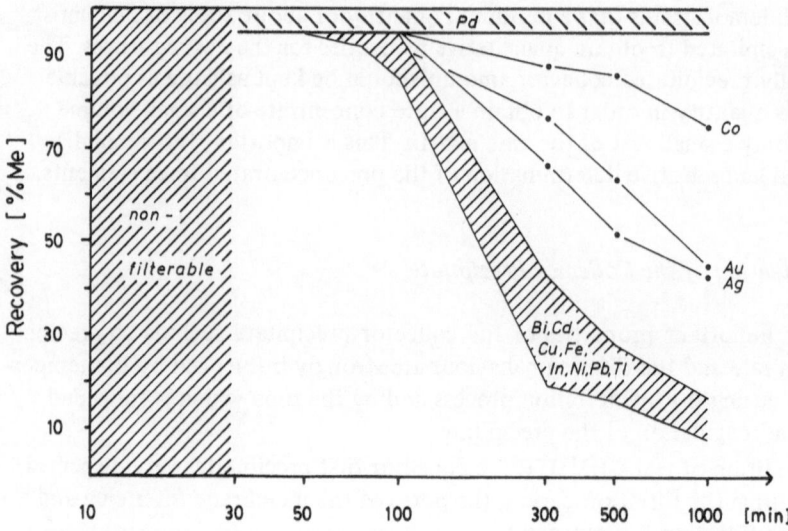

Fig. 4. Dependence of the trace recoveries on the coagulation time of the collector (reaction temperature: 20 °C)

plexes, because of the considerable decomposition rate of reagent and metal chelates in solutions of about pH 0.1. So, at 20 °C, in order to obtain quantitative trace recoveries, after the addition of the reagent and waiting of about 30 minutes, the filtration process must be finished in further 30 minutes.

The investigations have shown that, in spite of the large number of effecting parameters, for all the important process variables, there is a sufficient

The investigations have shown that, in spite of the large number of effecting parameters, for all the important process variables, there is a sufficient large band width of adjustment to obtain a relatively simple to handle preconcentration procedure.

Fig. 5 demonstrates, as one could expect, the strong decrease of the recoveries at increasing temperatures of the suspension.

Procedure

2 g of metal cuttings are dissolved in a mixture of 15 ml of 23 per cent ($7N$) HCl and 7.5 ml of 65 per cent HNO_3 (250-ml beaker). By addition of 150 ml H_2O the pH value is adjusted to 0.1 and 0.2. By use of a thermostat the temperature is set to about 25 °C. Without stirring, the collector is precipitated by addition of 50 ml of reagent solution (5 mg/ml of the hexamethyleneammoniumsalt of hexamethylenedithiocarbamidate dissolved in

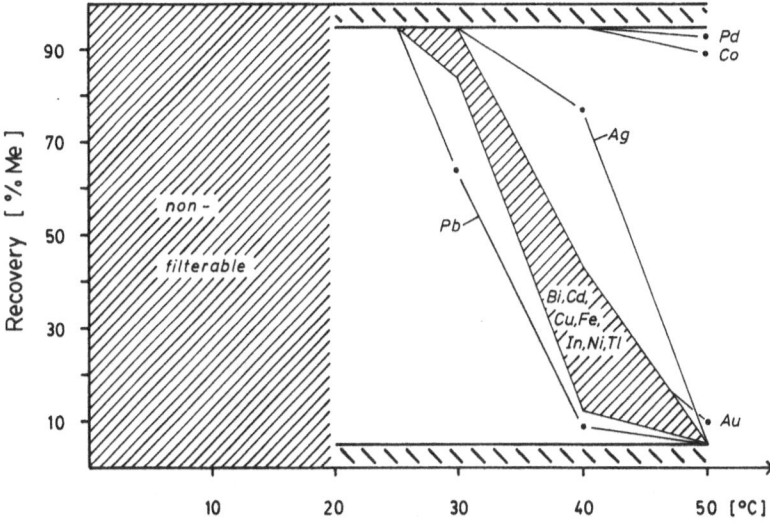

Fig. 5. Dependence of the trace recoveries on the precipitation temperature (coagulation time: 30 min)

water). The suspension is kept at 25 °C for 25−30 minutes and is then filtered through a cellulose nitrate membrane filter (Schleicher & Schüll, Dassel; 3 μm, ϕ 25 mm; Art.-No AE97/1) by use of a demountable suction filter unit. A subsequently appearing turbidity in the filtrate is without importance. The precipitate is washed with 3−5 ml of cold water. The collector, together with the filter, is dissolved in 3 ml of 65 per cent HNO_3 (25-ml beaker). The solution is evaporated until a white precipitate appears. Then 2 ml of 23 per cent HCl are added cautiously, and the solution is evaporated nearly to dryness with only a small white residue remaining. The residue is dissolved in a few drops of 23 per cent HCl which is transferred into a 2-ml volumetric flask. The beaker is rinsed with 23 per cent HCl into the flask to fill the flask to the mark. This solution is used for the trace determination. All the investigated elements are determined by flame AAS using the "injection method"[21,22]. For each element double measurements with 50 μl aliquots of the trace solution were carried out. A Varian spectrometer model AA6 and the air-acetylene flame (Sn: nitrous oxide-acetylene flame) were used.

Results

Table 1 lists the results for the preconcentration and determination of 13 investigated trace elements. Within the selected calibration ranges which have been adjusted to the practical requirements, the recoveries of the ele-

334   E. Jackwerth and H. Mittelstädt:

Table 1

Trace Element	Calibration Range [μg/g]	Recovery [%]	Trace Contents [μg/g]	RSD $s/\bar{x}$ (N=20)	Detection Limit ($3s$) [μg/g]
Ag	3– 11	⩾95	3	0.015	0.01
Au	1– 4	⩾95	1	0.038	0.03
Bi	2– 20	⩾95	5	0.028	0.2
Cd	1– 6	95	1	0.026	0.02
Co	1– 8	⩾95	2	0.019	0.1
Cu	20– 60	⩾95	20	0.018	0.06
Fe	10– 30	⩾95	10	0.036	0.3
In	2– 20	⩾95	5	0.023	0.2
Ni	1– 3	⩾95	1	0.037	0.09
Pb	40–120	⩾95	40	0.027	0.2
Pd	1– 5	⩾95	1	0.035	0.08
Sn	2– 20	⩾95	10	0.040	1.1
Tl	5– 15	⩾95	5	0.018	0.2

ments are found to be quantitative. For the compilation of the relative standard deviation (RSD; $s/\bar{x}$; N=20), spectrochemically analysed sample material was used to which the traces Au, Bi, Co, In, Pd, and Sn were added before dissolving the samples. The limits of detection ($3s$; N=20) were calculated from the fluctuations of the background signals caused by the blanks of the reagents.

Acknowledgement

The authors are thankful to Dr. F. Pottkamp, Preussag-Boliden, Nordenham, for the sample materials and for his stimulating discussions. We also want to thank the Deutsche Forschungsgemeinschaft for the support of this work.

Summary

Multi-Element Preconcentration of Technical Alloys

A method is described for the preconcentration and determination of traces of Ag, Au, Bi, Cd, Co, Cu, Fe, In, Ni, Pb, Pd, Sn, and Tl in ZnAl4 (Z 400). The trace elements are co-precipitated by a collector which is obtained with 20 mg of the zinc matrix (1 per cent of the sample weighing) by precipitation with the hexamethyleneammoniumsalt of hexamethylenedithiocarbamidat (HMA-HMDTC) at pH 0.2. After dissolving the collector in acid the traces are determined by flame AAS ("injection method"). The recoveries for the investigated elements are $\geqslant 95$ per cent; the relative standard deviation ($s/\bar{x}$; $N=20$), in general, being below 0.05. The detection limits ($3s$; $N=20$) are found to be between 0.01 and 1.1 μg/g.

References

1. Mitteilungsblatt der Zinkberatung eV. Düsseldorf, Heft 12, 1977.
2. Zink-Druckguß, Zinkberatung eV. Düsseldorf, 1973.
3. E. Brunhuber, Legierungshandbuch der Nichteisenmetalle. Berlin: Schiele & Schön. 1960.
4. ZAMAK-Feinzinklegierungen, Metallgesellschaft AG, Frankfurt a.M.
5. H.P. Kehrer, Metall 28, 883 (1974).
6. G. Frommey, Z. Metallkunde 67, 361 (1976).
7. H. Johnen, Metall 30, 855 (1976).
8. N. Strafford, P.F. Wyatt, and F.G. Kershaw, Analyst 78, 624 (1953).
9. I.N. Bykova, T.G. Manova, V.G. Silakova, and G.P. Boznyakova, Zh. analit. Khim. (USSR) 28, 1481 (1973); Analyt. Abstr. 28, 1B42 (1975).
10. M. Kimura, Talanta 24, 194 (1977).
11. E. Jackwerth and H. Linke, Erzmetall 31, 275 (1978).
12. A. Hulanicki, Talanta 14, 1371 (1967).
13. O. G. Koch and G. A. Koch-Dedić, Handbuch der Spurenanalyse. Berlin-Heidelberg-New York: Springer-Verlag. 1974.
14. H. Bode, Z. analyt. Chem. 143, 182 (1954).
15. H. Bode and F. Neumann, Z. analyt. Chem. 172, 1 (1960).
16. K. Gleu and R. Schwab, Angew. Chem. 62, 320 (1950).
17. H. Malissa and E. Schöffmann, Mikrochim. Acta [Wien] 1955, 187.
18. H. Bode, Z. analyt. Chem. 142, 414 (1954).
19. A.I. Busev, V.M. Byrko, A.P. Tereschtschenko, N.N. Novikova, V.P. Naidina, and P.B. Terentev, Zh. analit. Khim. (USSR) 25, 665 (1970); Analyt. Abstr. 21, 1705 (1971).
20. E. Kovács and H. Guyer, Z. analyt. Chem. 186, 267 (1962).
21. E. Sebastiani, K. Ohls, and G. Riemer, Z. analyt. Chem. 264, 105 (1973).
22. H. Berndt and E. Jackwerth, Spectrochim. Acta 30B, 169 (1975).

Correspondence and reprints: Prof. Dr. E. Jackwerth, Abteilung für Chemie, Ruhr-Universität, D-4630 Bochum, Federal Republic of Germany.

Mikrochimica Acta [Wien], Suppl. 10, 337–349 (1983)

Institute of Metals Research, Polish Academy of Sciences, Krakow, Poland

Application of the Levitation Melting Technique in the Investigations of Iron and Steel Making

By

J. Foryst and W. Przybyło

With 4 Figures

(Received January 19, 1983)

The principle of levitation melting was first formulated by Muck[1] in 1923 and applied in practice about 30 years later in the fundamental work by Okress et al.[2]. Generally speaking it is a kind of induction melting in which the dynamic interaction of a high frequency electromagnetic field and the currents flowing in the metal permits to counterbalance the weight of the metal and to maintain it in a central position inside the coil of an appropriate shape.

At the time of melting there takes place intensive mixing and thereby the homogenization of the composition and temperature of the liquid metal remaining in contact only with the surrounding gas, atmosphere or vacuum.

The main advantage of this method is the absence of any type of crucible and owing to this, in physico-chemical investigations we can avoid the effect of side reactions resulting from the influence of the active metal on the walls of the ceramic crucible. In many cases this fact makes the correct interpretation of the experimental results difficult and sometimes impossible.

The other advantages are:

— increased rate of chemical reactions between the metal and gas phase due to advantageous ratio of the reacting surface and the mass,
— the possibility of applying an arbitrary atmosphere or vacuum,
— the possibility to obtain extreme cooling rates of the metal (up to ca 10^8 K sec^{-1}).

The factors restricting the applicability of the levitation melting method are as follows: inaccurate determination of the temperature, increased evaporation rate of the liquid metal due to a sharp temperature gradient and uncertainty of the thermal conditions and surface composition.

For this reason the design of the levitation unit as well as the experimentation sequence, i.e. the choice of the parameters of the power input to the coil and of the measurement factors, are of great importance with respect to the applicability of levitation melting for the investigation of equilibrium in the systems: metal-gas phase, or for the purpose of investigating the kinetics of the reactions occurring between a metal droplet and the flowing gas.

The present paper brings a description of the application of the levitation melting technique for the determination of the equilibrium solubility of oxygen in liquid iron, the reduction kinetics of oxygen dissolved in iron in H_2/H_2O atmosphere of low oxygen potential, and desulphurization of Fe-S alloys of low carbon content (in the same gas atmosphere).

Design of the Levitation Unit and Development of the Experimental Technique

With levitation melting in a state of thermal equilibrium the heat radiation power of the metal is equal to the power supplied to the metal by the electromagnetic field. Fogel et al.[3] have defined the relation between the power transmitted to the metal and the melting parameters by the formula:

$$r_s = 0.55 \, K\gamma \, [\pi \rho f/\mu_0] \, r_k/A$$

where:

K — coefficient of the shape of the metal in a liquid state (for a sphere $K = 1$)

r_k — radius of a metal sphere of a given mass,

A — coefficient characterizing the field configuration of a given coil at a given power input.

From the above formula it follows, among others, that the temperature control which is a very important factor in the experiments can be obtained through the proper choice of the mass of the metal, current frequency and the shape of the coil, characterized by the values of the coefficient A. In practice, for a given design of the levitation unit and the type and mass of the metal the shape of the coil may be selected only by the experiment.

For the apparatus built at the Institute of the Metals Research of the Polish Academy of Sciences[4] we used a generator with the power output equal to 10 kW and the frequency 407 kHz; the apparatus was designed in such a way that after its evacuation it allowed free flow of gases of a given composition, the control of the temperature by means of extra devices, melting and casting of three successive metal samples without impairing the tightness of the system. Under the above given conditions it is necessary to apply an outside multi-turn coil what involves the reduction of the volume of the molten metal to ca 0.5 cm^3.

Fig. 1 gives a scheme of an apparatus used in the investigations described in the present paper. The apparatus consists of three main parts: upper head of the coil which is wound around a quartz tube and the bottom part supported on a base. The upper head has a peep hole, an optical device for measuring and controlling the temperature and a gas offtake through a stop cock. In the bottom part there is a turn-table with the metal sample and copper ingot moulds and a mechanism for rotating the table and manipulating the sample. The table is rotated by means of a transmission gear with a rotating rod which is also used to insert the sample into the coil chamber. To regulate the temper-

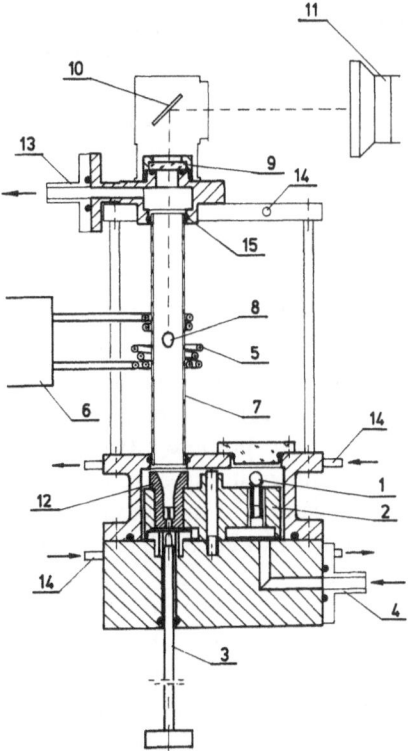

Fig. 1. Diagram of the levitation melting apparatus. 1— metal sample, 2— turn table, 3— manipulating rod, 4— gas inlet, 5— coil, 6— generator of h.f. current, 7— 15 mm o.d. quartz tube, 8— metal droplet, 9— looking window, 10— mirror, 11— pyrometer, 12— copper mould, 13— gas outlet, 14— water cooling, 15— rubber O-ring seals

ature of the apparatus there are water channels inside the head. The bottom part is connected through an elastic vacuum conduit with the plate of the base (not marked in Fig. 1) which connects the whole appliance with the vacuum set.

The principal elements of the apparatus have been made from aluminium alloys, the others — from brass and steel. To seal the joints rubber 0-ring seals have been used. The multi-turn coil enclosing a quartz tube, 15 mm in diameter, constitutes the most essential element of the apparatus responsible for the stable levitation of the metal and the possibility of controlling its temperature. The shape of the coils made of copper pipes is similar to those used by Comenetz and Salatek[4].

The experimental technique of melting the metal was similar in all the experiments described further on. After the air had been removed from the levitation unit the given gas phase (H_2/H_2O, Ar, He or H_2) was let into the apparatus, the rate of the flow being regulated. By means of the sample lift 3 a metal sample 2 was inserted in the coil chamber. After setting the proper power output of the generator the power input to the coil with high-frequency current was switched on. Then the metal was lifted from the carrier and while oscillating in a chaotic way it became heated until it melted, which took from a few to some dozens of seconds, depending on the type of the metal. By lowering and turning the rotating rod by 180° a casting mould was placed under the coil. After melting there followed a considerable reduction of the oscillation of the metal droplet or its position became stable. The droplet may assume the shape of a sphere, an egg or a pear turned upside down. After the required time had passed the flow of the current to the generator was suddenly stopped and then the metal was poured into the casting mould. It is also possible that the metal becomes solidified in a suspended form.

As it has been mentioned in the introduction the main difficulty in the investigations when using a levitation unit, is the measurement of temperature. In the investigations described below we used for this purpose a two-colour Ratio-Scope III pyrometer, calibrated beforehand with respect to molten iron temperature — the difference between the indications of the pyrometer and the temperature measured by the thermocouple PtRh 30, PtRh 6 — did not exceed the nominal measurement error ± 14 K.

By applying coils of different shapes for the particular metals (Fe, Ni, Cu, Mo, Al), different gas atmospheres and different parameters of power input in the levitation unit — temperatures of 1500—2200 K have been obtained. A characteristic feature of levitation melting has been found incidentally: the decrease of the temperature of the metal with increasing coil current. Within a considerable power range this dependence is related to the values of field divergence in the coil: with decreasing power the metal is being displaced to the sphere of a higher though less diversified field strength, and due to this fact the metal is heated more intensely while the supporting power remains the same.

For temperature adjustment of the metal a three-position control was applied utilizing the regulation contacts in an automatic temperature register. With correct adjustment of power and good stability of metal the obtained regulation was better than ± 10 deg.

Equilibrium Solubility of Oxygen in Liquid Iron

In the investigations of the solubility of oxygen in iron we used mixtures of hydrogen and steam which at a given temperature contain also oxygen of a defined equilibrium pressure. Then the reaction between the dissolved oxygen and the gas phase is expressed by the equation:

$$H_{2g} + \underline{O}_{Fe} = H_2O_g \tag{1}$$

and the equilibrium constant:

$$K = \frac{p_{H_2O}}{p_{H_2}} \frac{1}{f_0 \text{ wt.\%} \underline{O}} = \frac{p_{H_2O}}{p_{H_2}} \frac{1}{\% \underline{O}} \quad \% \underline{O} \to 0 \tag{2}$$

For a concentration range of oxygen in which the solution is subject to Henry's Law the equilibrium constant is determined directly from the value of the ratio of the pressure p_{H_2O}/p_{H_2} and the obtained equilibrium concentrations of oxygen.

The occuring of considerable temperature gradients in the gas mixture $H_2 + H_2O$ causes simultaneously a marked segregation of its components and this is connected to thermal diffusion: the gas layer near the metal surface becomes enriched with hydrogen. Consequently, according to Eq. (2), the oxygen content in the metal is lower than that obtained in the isothermal system for a given bulk gas composition. Then, after substituting the experimental values in Eq. (2) we obtain the values of the apparent equilibrium constants K referring to the given conditions of transferring the mass and heat in the gas mixture $H_2 + H_2O$ surrounding the droplet of a liquid metal.

In the present investigations the mixture of hydrogen and steam of a definite composition was obtained in an apparatus, which is schematically shown in Fig. 2.

The hydrogen from a cylinder was directed through a flowmeter for a preliminary purification from oxygen in a furnace filled with copper turnings and next to a steam saturation system. In a glass bulb with distilled water the gas was supersaturated with steam due to intensive mixing with a magnetic mixer and heating the water up to a temperature $10-20$ K higher than the given saturation temperature.

The excess of steam was next condensated in a cooler supplied with water of

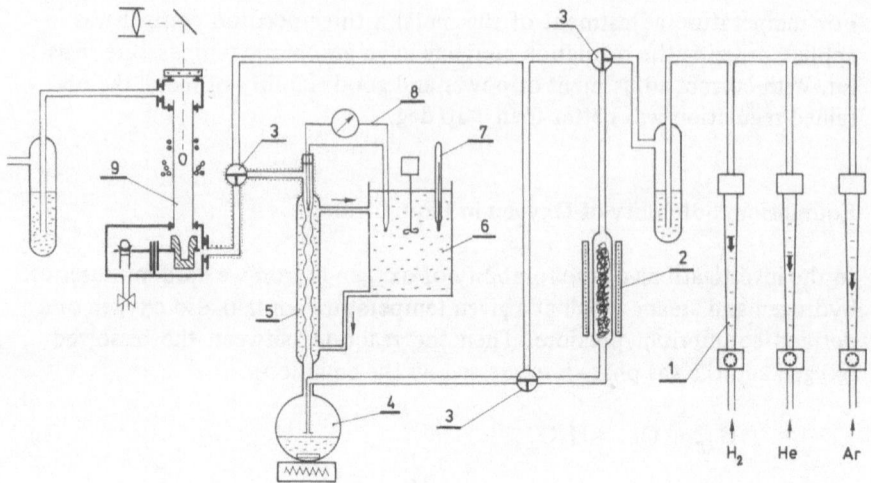

Fig. 2. Diagram of the gaseous H_2 + H_2O mixtures generator. 1— flowmeters, 2— purification furnace, 3— stop-cocks, 4— saturator, 5— cooler, 6— thermostat, 7— mercury thermometer, 8— battery of the copper-constantan thermocouples, 9— levitation melting apparatus

a temperature controlled with the accuracy ± 0.1 K. The saturation temperature T_n was defined as the temperature of the gas at the outlet of the cooler. The obtained hydrogen mixture was directed to the levitation unit. The levitation unit and the gas conduits in the cooler were heated to prevent the condensation of steam from the gas mixture before it reached the metal surface. After the reaction with metal the gas was let into the atmosphere.

The composition of the mixture H_2 + H_2O was controlled by electrochemical method using oxygen cell with solid electrolyte (ZrO_2, CaO). Preliminary measurements revealed that this method allowed to obtain and to control in a continuous way the composition of the mixtures H_2 + H_2O with an accuracy ± 4% with respect to the theoretical value derived on the basis of tabular pressures of saturated steam for a given saturation temperature T_n.

To investigate the solubility of oxygen in iron we used armco iron with the average impurity content in percentage by weight:

C	− 0.015	Cu	− 0.045
Mn	− 0.08	Cr	− 0.01
Si	− 0.005	Ni	− 0.01
P	− 0.009	Al	− 0.003
S	− 0.025	O	− 0.115

From a rod of a 6-mm diameter samples were cut out with a mass of $1 \pm 0.02\,g$.

The experiments were carried out at the following temperatures of the metal: 1873, 1948, and 2023 K and with the composition of the saturated gas phase at the temperatures T_n ca 296, 307, and 315 K which corresponds to 2.8 − 8.3% of the volume of steam content. The melting time was 10−21 minutes.

The total oxygen content in the cast samples was analyzed by the method of vacuum fusion in a LECO apparatus.

From the obtained experimental data, using the least square method, there has been derived an equation for a straight line describing the dependence of the apparent equilibrium constant of reaction (1) on the inverse temperature of liquid iron.

$$\lg k = \frac{5950 \pm 570}{T} - 2.78 \pm 0.29 \tag{3}$$

For each temperatures of the metal − the values of standard deviation were equal to 0.043, 0.028 and 0.023, respectively.

It follows from the obtained data that with increasing temperature of the metal and of the steam content (and thus increased pressure of the oxygen p_{O_2}) in the gas phase, the oxygen content in the metal also increases. At the same time no explicit correlation between $\lg k$ and the oxygen content in metal has been found.

The authors' own results show good agreement with the values of the equilibrium constants obtained from investigations of melting metals in a crucible, carried out by other authors[6-13], in particular by Awierin, Poliakow, and Samarin[7], whereas they are distinctly lower than the values of log of apparent equilibrium constants obtained under conditions of levitation melting by Kershaw, Mc Lean, and Ward[12] (by ca 0.3) or by Larché and Mc Lean[13] (by ca 0.25).

On the basis of metallographic investigations we can say that with a given oxygen potential in the gas phase, part of the oxygen contained in the metal may appear in the form of stable oxide phase; its amount depending on the kind and amount of admixtures in the iron. This causes a simultaneous compensation of the effect of thermal diffusion of the components of gas phase on the oxygen content in the metal and thus a convergence of the apparent values of the equilibrium constants k with the values of K obtained from investigations carried out under isothermal conditions.

This points to the fact that high purity of the initial material to be used in the investigations is important for the reliability of the results.

Deoxidation of Iron in $H_2 + H_2O$ Mixture

The redox potential in $H_2 + H_2O$ mixture depends above all upon the ratio p_{H_2}/p_{H_2O}, and practically on the temperature of hydrogen saturation with steam. For reasons not given in the present paper, we used in the investiga-

tions of the deoxidation of liquid iron with hydrogen a mixture of this gas with steam at a saturation temperature of 273 ± 0.1 K. According to the results described in the preceding chapter the equilibrium concentration of oxygen in iron at a temperature above 1870 K is then equal to ca 50 ppm. As the initial material we used samples of armco iron of a composition typical for this grade, with the initial concentration of total oxygen from 1000 to 1500 ppm. The method of procedure was analogous to that described earlier, and the details can be found in[14].
The results are shown in the diagram in Fig. 3.

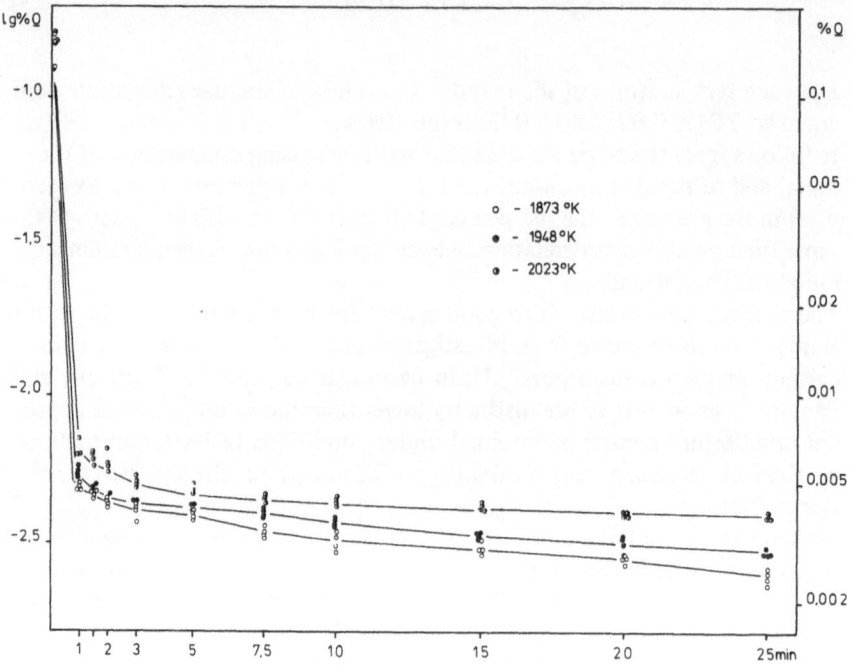

Fig. 3. Kinetics of deoxidation of armco-iron in $H_2 + H_2O$ mixture

As can be seen, the setting time of the final concentration of oxygen is about 1 minute – within the range 1–25 minutes the decrease of the total oxygen content in the iron is very small. This is an indication that under the experimental conditions, i.e. little volume of the iron and small surface area of the droplet the deoxidation proceeds very fast. According to K. Mamro and M. El-Zeky[14] under such conditions, within the range of 1873–2023 K, in the first minute of the reaction, about 90% of oxygen present initially in the metal undergoes reduction, the hydrogen concentration in the sample after the reaction being equal to ca 4 ppm. As it has been anticipated, increased temperature causes a decrease in the final concentration of oxygen.

The above investigations confirm the great usability of the levitation technique to determine the kinetics of the deoxidation process; in the conventional investigation methods, in which a ceramic crucible is used, the progress of the kinetics curves is much slower and allowing for the difficulties of the experiment (the necessity to keep the metal inside the crucible for a long time, the influence of side reactions, the need to maintain constant temperature for a long time) – the results are in considerable error.

Desulphurization of Iron in H_2 + H_2O Mixture

To investigate the kinetics of desulphurizing liquid iron we used a method analogous to that applied when investigating the deoxidation reaction, the only difference being that after melting the initial armco iron in a ceramic crucible in a Tamann furnace sulphur was added so that its total content was 143–10000 ppm[15]. For the investigations we used pipetted samples with the initial sulphur content of ca 160, 2000, 5000 and 10000 ppm S; the gas phase composition was H_2 + H_2O; its flow rate through the levitation unit was analogous to that described earlier. The investigation results are shown in the diagrams in Figs. 4a–c and Table 1 gives the dependence of the velocity constant K_s on the initial sulphur content in metal and on temperature.

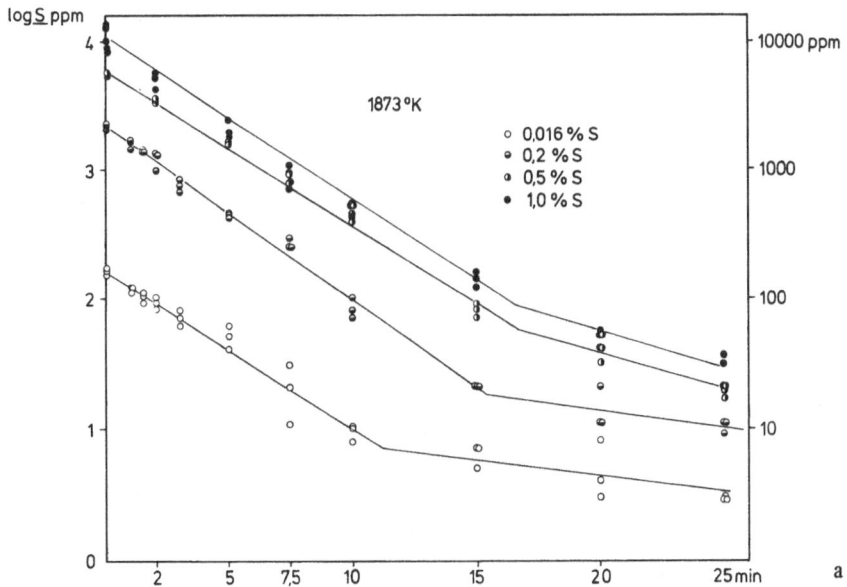

Fig. 4. Kinetics of desulphurization of Fe-S melts in H_2 + H_2O mixtures at temperatures: (a) 1873 K, (b) 1948 K, (c) 2023 K

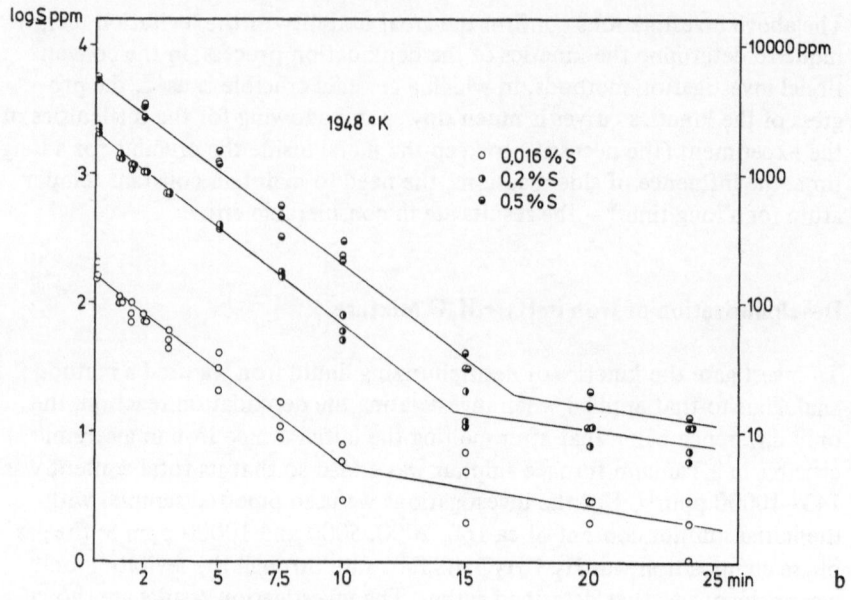

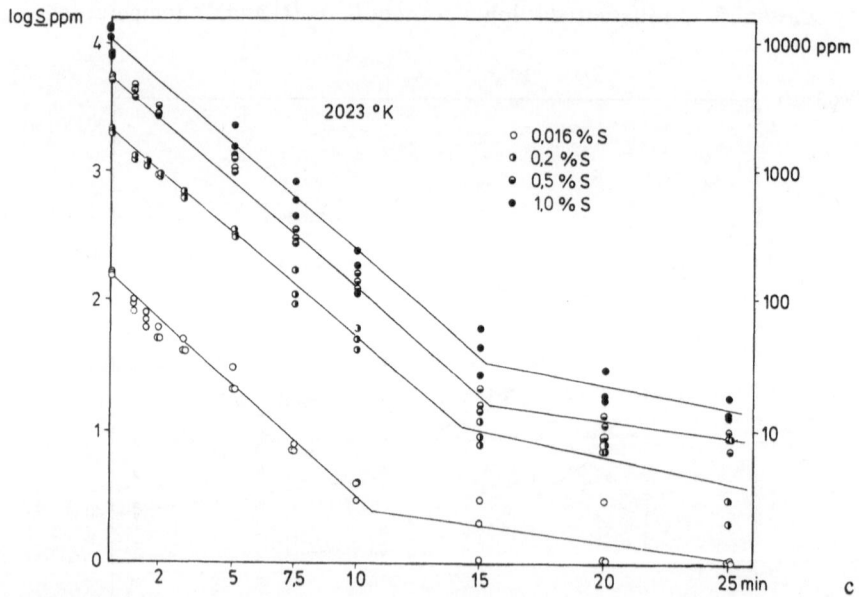

Fig. 4b, c

The behaviour of the desulphurization curves as depending on the reaction time and temperature and on the initial concentration of sulphur is an indication that:

a) ing the dependence of the reaction velocity constant; in the initial period
the velocity constant is independent of the sulphur content (Table 1),
whereas it depends evidently on the temperature of the metal;
whereas it depends evidently on the temperature of the metal,

b) after a definite time of the reaction, depending on the initial concentration of sulphur, we can observe a distinct decrease of the desulphurization
rate; up to 10 minutes this rate is constant for a given temperature, whereas after that time the desulphurization process proceeds more slowly;

c) unlike deoxidation, an increase in the reaction temperature distinctly
favours desulphurization, i.e. its final concentration in the alloy decreases
with increased temperature;

d) the observed différence in the velocity constants may be an indication of
a different mechanism of desulphurization reaction; these differences cannot be explained by a possible difference in the diffusion velocity of
sulphur from the centre of the droplet towards the surface on account of
very intensive mixing. We may assume that a simultaneous influence of
oxygen and sulphur dissolved in iron as well as the influence of other admixtures affecting the duration of sulphur-oxide compounds – may play
a role in this case;

e) the application of hydrogen as a reducer under conditions of levitation
melting makes possible in practice a total removal of sulphur from iron.

It should be noted here that the mechanism of the reaction between hydrogen and sulphur dissolved in metal is a complex one; the investigations of
other authors[16] point to the formation, besides hydrogen sulphide, of
gaseous sulphur and gaseous HS, the above indication is also connected to
the stability of H_2S at high temperatures (thermal dissociation).

Table 1. Desulphurization Rate Constants K_s(min^{-1}) for the Fe-S Melts in
$H_2 + H_2O$ Mixture

Temper- ature [K]	Initial sulphur content in metal [wt.%]				Average value of K_s
	0.016	0.2	0.5	1.0	
1873	– 0.124	– 0.136	– 0.120	– 0.126	– 0.126
1948	– 0.154	– 0.146	– 0.146	–	– 0.149
2023	– 0.173	– 0.163	– 0.169	– 0.172	– 0.169

Perspectives of the Application of Levitation Melting Technique in Metallurgical Investigations

The advantages of the method of investigating such processes as deoxidation and desulphurization, common in iron and steel making, without using a crucible — in the light of the above presented results cannot be questioned. To be accurate it should be added that this refers both to the equilibrium investigations (i.e. the equilibrium between the gas phase and liquid iron) and — above all — the investigations of the kinetics of the processes taking place. Presently, investigations are under way concerning denitrification of alloyed steels with the application of hydrogen and argon as gases reducing the nitrogen content in liquid metal.

On the basis of the results of the above investigations we can conclude that this method is equally useful for the determination of actual solubility of gases (H_2 and N_2) in metal alloys. At the Metals Research Institute of the Polish Academy of Sciences research work has been started on the determination of the solubility of hydrogen in cobalt and next in liquid nickel, aluminium and titanium.

Considering the solidification conditions of liquid metal it will be possible to obtain metals in an amorphous state — then it will be necessary to make some adjustments in the existing apparatus. It is also known that vapour pressures, density and other physical properties of liquid metals examined by the method of levitation melting can be determined in this way.

Summary

Application of the Levitation Melting Technique in the Investigations of Iron and Steel Making

The design of the levitation unit and the adopted experimental technique was described. The results of the estimation of the solubility of oxygen in liquid iron in H_2/H_2O mixtures was presented. The kinetics of deoxidation and desulphurization of liquid iron alloys by streaming hydrogen was measured. It was found that the deoxidation process is very rapid; the rates of deoxidation decreases with temperature, when desulphurization increases. It was observed, that the desulphurization process by hydrogen consists of two stages — up to 10 minutes the rate of the reaction is constant for a given temperature (1870–2020 K) and is probably connected with the reduction of pure sulphide phase in iron. Another application of the levitation method for the kinetic studies in metallurgical process was proposed.

References

1. O. Muck, DR Patent Nr. 422004, 30.10.1923.
2. E.C. Okress, J. Appl. Phys. 23, 545 (1952).
3. A.A. Fogel et al., Izw. AN SSSR Met. 3, 111 (1972).
4. W. Przybyło, Arch. Hntn (pol.) 24, 501 (1979).
5. G. Comenetz and J.W. Salatka, J. Electrochem. Soc. 105, 673 (1958).
6. J. Chipman and M.N. Dastur, Trans. Met. Soc. AIME 185, 441 (1949).
7. W.W. Awierin, A.J. Poliakow, and A.M. Samarin, Izw. AN SSSR OTN 3, 90 (1955).
8. H. Sakao and K. Sano, Trans. J.I.M. 1, 38 (1960).
9. T.P. Floridis and J. Chipman, Trans. Met. Soc. AIME 212, 549 (1958).
10. E.S. Tankins, N.A. Gocken, and G.R. Belton, Trans. Met. Soc. AIME 230, 820 (1964).
11. S. Matoba and T. Kuwana, Tetsu to Hagane 5, 187 (1965).
12. P. Kershaw, A. McLean, and R.G. Ward, Can. Met. Quart. 11, 327 (1972).
13. F.S. Larché and A. McLean, Trans. J.I.S.I. 13, 71 (1973).
14. K. Mamro and M. El-Zeky, Stahl und Eisen 100, 16, Suppl. 54–60 (1980).
15. M. El-Zeky, Praca doktorska AGH Kraków, 1981.
16. Ban-Ta Shiro and J. Chipman, Trans. Met. Soc. AIME 242, 940 (1968).

Correspondence and reprints: J. Foryst, Institute of Metals Research, Polish Academy of Sciences, Kraków, Poland.